AF352231

Progress in Mathematical Physics

Volume 53

Quantum Spaces
Poincaré Seminar 2007

Bertrand Duplantier
Vincent Rivasseau
Editors

Birkhäuser Verlag
Basel · Boston · Berlin

Editors:

Bertrand Duplantier
Service de Physique Théorique
Orme des Merisiers
CEA - Saclay
91191 Gif-sur-Yvette Cedex
France
e-mail: bertrand.duplantier@cea.fr

Vincent Rivasseau
Laboratoire de Physique Théorique
Université Paris XI
91405 Orsay Cedex
France
e-mail: Vincent.Rivasseau@th.u-psud.fr

2000 Mathematics Subject Classification: 81V70, 81R50, 81R60, 81T75, 81T17, 82B23

Library of Congress Control Number: 2007933916

Bibliographic information published by Die Deutsche Bibliothek
Die Deutsche Bibliothek lists this publication in the Deutsche Nationalbibliografie; detailed bibliographic data is available in the Internet at <http://dnb.ddb.de>.

ISBN 978-3-7643-8521-7 Birkhäuser Verlag, Basel – Boston – Berlin

© 2007 Birkhäuser Verlag, P.O. Box 133, CH-4010 Basel, Switzerland
Part of Springer Science+Business Media
Printed on acid-free paper produced of chlorine-free pulp. TCF ∞
Printed in Germany

ISBN 987-3-7643-8521-7

e-ISBN 978-3-7643-8522-4

9 8 7 6 5 4 3 2 1

www.birkhauser.ch

Contents

A.P. POLYCHRONAKOS
Non-commutative Fluids

Foreword

This book is the seventh in a series of lectures of the *Séminaire Poincaré*, which is directed towards a large audience of physicists and of mathematicians.

The goal of this seminar is to provide up-to-date information about general topics of great interest in physics. Both the theoretical and experimental aspects are covered, with some historical background. Inspired by the Bourbaki seminar in mathematics in its organization, hence nicknamed "Bourbaphi", the Poincaré Seminar is held twice a year at the Institut Henri Poincaré in Paris, with contributions prepared in advance. Particular care is devoted to the pedagogical nature of the presentations so as to fulfill the goal of being readable by a large audience of scientists.

This volume contains the tenth such seminar, held on April 30, 2007. It is devoted to the application of non-commutative geometry and quantum groups to physics.

The book starts with a pedagogical introduction to Moyal geometry by Vincent Pasquier, with special emphasis on the quantum Hall effect. It is followed by a detailed review of Vincent Rivasseau on non-commutative field theory and the recent advances which lead to its renormalizability and asymptotic safety. The description of the quantum Hall effect as a non-commutative fluid is then treated in detail by Alexios Polychronakos. Integrable spin chains can be studied through quantum groups; their striking agreement with neutron scattering experiments is reviewed by Jean-Michel Maillet. The book ends up with a detailed description by the world famous expert Alain Connes of the standard model of particle physics as a spectral model on a very simple non-commutative geometry, including the recent progress on the Higgs sector and neutrino masses.

We hope that the continued publication of this series will serve the community of physicists and mathematicians at professional or graduate student level.

We thank the Commissariat à l'Énergie Atomique (Division des Sciences de la Matière) and the Daniel Iagolnitzer Foundation for sponsoring the Seminar. Special thanks are due to Chantal Delongeas for the preparation of the manuscript.

Bertrand Duplantier
Vincent Rivasseau

Séminaire Poincaré X
Espaces Quantiques

Samedi
28 avril 2007

V. PASQUIER : Introduction • *10h00*
V. RIVASSEAU : Renormalisation non commutative • *10h30*
A. POLYCHRONAKOS : Noncommutative Fluids • *11h30*
J.-M. MAILLET : Des groupes quantiques aux expériences
de diffusion de neutrons • *14h30*
A. CONNES : La géometrie non commutative
et le modèle spectral de l'espace-temps • *15h30*

INSTITUT HENRI POINCARÉ • Amphi Hermite
11, rue Pierre et Marie Curie • 75005 Paris

www.lpthe.jussieu.fr/poincare

ÉCOLE
POLYTECHNIQUE

CNRS
CENTRE NATIONAL
DE LA RECHERCHE
SCIENTIFIQUE

cea

FONDATION
IAGOLNITZER

Quantum Spaces, 1–17
© 2007 Birkhäuser Verlag Basel

Quantum Hall Effect and Non-commutative Geometry

Vincent Pasquier

1. Introduction

Our aim is to introduce the ideas of non-commutative geometry through the example of the Quantum Hall Effect (QHE). We present a few concrete situations where the concepts of non-commutative geometry find physical applications.

The Quantum Hall Effect [1–4] is a remarkable example of a purely experimental discovery which "could" have been predicted because the tools required are not extremely sophisticated and were known at the time of the discovery. What was missing was a good understanding of topological rigidity produced by quantum mechanics whose consequences can be tested at a macroscopic level: The quantized integers of the conductivity are completely analogous to the topological numbers one encounters in the study of fiber-bundles.

One can give a schematic description of the Quantum Hall Effect as follows. It deals with electrons constrained to move in a two-dimensional semiconductor sample in a presence of an applied magnetic field perpendicular to the sample. Due to the magnetic field, the Hilbert space of an electron is stratified into Landau levels separated by an energy gap (called the cyclotron frequency and proportional to the applied field). Each Landau level has a macroscopic degeneracy given by the area of the sample divided by a quantum of area (inversely proportional to the field) equal to $2\pi l^2$ where the length l is the so-called magnetic length. It is useful to think of the magnetic length as a Plank constant $l^2 \sim \hbar$. The limit of a strong magnetic field is very analogous to a classical limit.

The electrons behave much like incompressible objects occupying a quantum of area. Thus, when their number times $2\pi l^2$ is exactly equal a multiple of the area, it costs the energy gap to add one more electron. This discontinuity in the energy needed to add one more electron is at the origin of the incompressibility of the electron fluid. The number of electrons occupying each unit cell is called the filling factor, and the transverse conductivity is quantized each time the filling factor is exactly an integer.

We shall stick to this simple explanation, although this cannot be the end of the story. Indeed, if it was correct, the filling factor being linear in the magnetic field, the quantization of the conductance should be observed only at specific values of the magnetic field. In fact, it is observed on regions of finite width called plateaux, and it is necessary to invoke the impurities and localized states to account for these plateaus. Roughly speaking, some of the states are localized and do not participate to the conductance. These states are populated when the magnetic field is in a plateau. We refer the reader to a previous Poincaré seminar for an introduction to these effects [4]. What is important for us here is that (although counter-intuitive) it is possible to realize experimentally situations where the filling factor is *exactly* an integer (or a fraction as we see next).

It came as a surprise (rewarded by the Nobel prize[1]) when the Quantum Hall Effect was observed at non-integer filling factors which turn out to always be simple fractions. To explain these fractions, it was necessary to introduce some very specific wave functions and to take into account the interactions between the electrons. The proposed wave functions are in some sense variational, although they carry no adjustable free parameters.

This approach of a universal phenomenon through the introduction of "rigid" trial wave functions goes in opposite direction to the renormalization group ideas. Nevertheless, the trial wave functions are undoubtedly the most powerful tools available at present and it remains a challenge to reconcile them with the field theoretical point of view. The recent progress in understanding the renormalizability of non-commutative field theories [6] may be a crucial step in this direction.

Another point of contact between field theory and the QHE must be mentioned. Remarkably, many good trial wave functions for the fractional QHE are also correlation functions of conformal field theory (CFT) and integrable models [7], although this relation is not fully understood. The possibility to identify the certain fields such as the currents with the electron arises from the fact that their operator product expansions are polynomial, a property obeyed by the wave functions when electrons approach each other. In some sense, the short distance properties combined with minimal degree constraints arising from incompressibility control the structure of the theory. Another mysterious aspect is that the quasiparticles of the Hall effect carry a fractional charge, as if an electron breaks up into pieces. A very similar phenomenon occurs in CFT where the current can be obtained as the short distance expansion of other fields. In the study of the XXZ spin chain, the magnon is known to break up into two spinons. This leads to a rich variety of phenomena which can be studied by exact methods. J.M. Maillet explains these aspects in his contribution to this book.

The non-commutative geometrical aspects that occupy us here are preeminent in the fractional QHE when the lowest Landau level is partially filled. A great simplification and a source of richness comes from the Lowest Landau Level (LLL) projection. When the energy scales involved are small compared to the cyclotron

---

[1]Horst L. Störmer, Daniel C. Tsui. Robert B. Laughlin, in 1998.

gap, one can study the dynamics by restricting it to the LLL. As a consequence of this drastic reduction of the degrees of freedom, the two coordinates of the plane obey the same commutation relations as the position and the momentum in quantum mechanics. The electron is therefore not a point-like particle anymore and can at best be localized at the scale of the magnetic length.

In a series of experiments [20, 21], it was realized that the electron behaves as a neutral particle when the filling factor is exactly 1/2 [13]. We shall present an image of the $\nu = 1/2$ state where electrons are dressed by a companion charge of the opposite sign. When the filling factor is 1/2, the two charges conspire to make a neutral bound state with the structure of a dipole [17]. It is this very specific experimental situation which we advocate to be a paradigm of non-commutative geometry [23]. We hope to convince the reader that it is deeply connected to the non-commutative field theory aspects developed by V. Rivasseau in this book. We present our understanding of the theory, being conscious that it cannot be the complete story.

We also review the example of the Skyrmion [31] which consists of the non-commutative analogue of the nonlinear sigma model solitons. This gives an exactly solvable model where the classical concept of the winding number has a quantum counterpart which is simply the electric charge of the soliton. This gives a physical application for the non-commutative geometry developed by A. Connes [37]. In particular, the topological invariant which measures the winding number has a non-commutative analogue which evaluates the electric charge of the Skyrmion [33].

A remarkable aspect of the QHE physics is that it becomes a matrix theory. The study of this theory has suscitated many very interesting works, in particular in relation with the Chern-Simon theory [8]. These aspects have been extensively studied by A. Polychronakos [9] who develops them in this book.

2. Lowest Landau level physics

2.1. Single particle in a magnetic field

Let us first recall some basic facts about the motion of a particle in a magnetic field. We consider a charge q particle in the plane subject to a magnetic field transverse to the plane.

It is convenient to define the magnetic length l by

$$l = q^{-\frac{1}{2}}. \tag{1}$$

To simplify the notation, we take units where the magnetic field is equal to q.

We introduce a vector potential A for the magnetic field:

$$q = \partial_x A_y - \partial_y A_x. \tag{2}$$

The vector potential A is defined up to a gauge transformation $A \to A + \nabla \chi$. The action from which the equations of motion of a mass m and charge 1 particle (confined to the plane) in presence of the magnetic field $B\hat{z}$ derive, is given by:

$$S = \int \left(\frac{m}{2}\dot{r}^2 - A\dot{r} \right) \, dt. \tag{3}$$

Using the canonical rules, we obtain a Hamiltonian:

$$H_0 = \frac{1}{2m} \left(\mathbf{p} + \mathbf{A} \right)^2 = \frac{\pi^2}{2m}, \tag{4}$$

where $\pi_i = m\partial_{x_i}/i - A_i$ is the momentum conjugated to x_i; the so-called dynamical momenta

$$\pi_x = p_x + A_x, \quad \pi_y = p_y + A_y \tag{5}$$

obey the commutation relations

$$[\pi_i, \pi_j] = iq\epsilon_{ij}, \quad [r_i, r_j] = 0, \quad [\pi_i, r_j] = -i\delta_{ij}, \tag{6}$$

where ϵ_{ij} is the antisymmetric tensor $\epsilon_{xy} = -\epsilon_{yx} = 1$.

We consider the case where $q > 0$. If we define creation and annihilation operators as linear combinations of the two dynamical momenta,

$$a = \sqrt{\frac{1}{2}}(\pi_x + i\pi_y), \quad a^+ = \sqrt{\frac{1}{2}}(\pi_x - i\pi_y), \tag{7}$$

obeying the Heisenberg relations

$$[a, a^+] = q, \tag{8}$$

the Hamiltonian is

$$H_0 = \frac{1}{2m} \left(a^+ a + \frac{q}{2} \right). \tag{9}$$

Its spectrum is that of an oscillator:

$$E_n = \frac{q}{2m}(n + \frac{1}{2}), \tag{10}$$

with $n \geq 0$. Each energy branch is called a Landau level.

A useful gauge is the so-called symmetric gauge defined by

$$A_x = -\frac{qy}{2}, \quad A_y = \frac{qx}{2}. \tag{11}$$

In this gauge

$$a = \frac{i}{\sqrt{2}}(\partial_{\bar{z}} + qz), \quad a^+ = \frac{i}{\sqrt{2}}(\partial_z - q\bar{z}), \tag{12}$$

where we set $z = (x + iy)/2$, $\bar{z} = (x - iy)/2$.

We can define new coordinates R_x, R_y which commute with the dynamical momenta

$$R_x = \frac{qx}{2} - p_y, \quad R_y = \frac{qy}{2} + p_x, \tag{13}$$

with the commutator given by

$$[R_i, R_j] = iq\epsilon_{ij}, \quad [\pi_i, R_j] = 0. \tag{14}$$

The so-defined coordinates are called guiding centers. The guiding center coordinates can be combined into two oscillators $b^+ = \frac{1}{2}(R_x + iR_y)$, $b = R_x - iR_y$:

$$\begin{aligned}
2b^+ &= qz - \partial_{\bar{z}}, \\
b &= q\bar{z} + \partial_z,
\end{aligned} \tag{15}$$

obeying

$$[b, b^+] = q. \tag{16}$$

The lowest Landau level wave functions are obtained upon acting onto the ground state of (9) with the l power of b^+. In this gauge, the angular momentum $L = b^+b/q = z\partial_z$ is a good quantum number and they carry an angular momentum $L = l$. Their expression is

$$\langle z|l \rangle = (q^{1/2}z)^l/(2\pi l!)^{1/2}e^{-qz\bar{z}}. \tag{17}$$

The wave functions are normalized so that $\langle l|l' \rangle = \delta_{ll'}$. The parameter q has the dimension of a length^{-2}. By taking its modulus: $|\langle z|l \rangle| \sim r^l e^{-qr^2}$, we see that a wave function is peaked around circular shells of radius $\sqrt{2l/q}$ around the origin. Thus if we quantize the system in a disk of finite radius R, we recover the expected degeneracy (18) by keeping only the wave functions confined into the disk $l \leq l_0 = 2R^2/q$.

If $q < 0$, the role of the creation and annihilation operators is reversed, and the lowest Landau level wave functions carry a negative angular momentum $L = -bb^+/q = -l$ and are given by the complex conjugated expression of (17).

The fact that the guiding center coordinates commute with H_0 implies that its spectrum is extremely degenerate. The two coordinates $q^{-1}R_x, q^{-1}R_y$ do not commute with each other and cannot be fixed simultaneously. There is a quantum uncertainty $\Delta R_x \Delta R_y = q$ to determine the position of the guiding center. Due to the uncertainty principle, the physical plane can be thought of as divided into disjointed cells of area $2\pi/q$ where the guiding center can be localized. The degeneracy per energy level and per unit area is $\frac{q}{2\pi}$ so that in an area Ω, the number of degenerate states is

$$N_\Omega = \frac{q\Omega}{2\pi}, \tag{18}$$

so that electrons behave "as if" they acquire some size under a magnetic field, the area being inversely proportional to the charge q.

In the strong magnetic field limit, one projects the dynamics onto the lowest Landau level LLL $n = 0$. In other words, we impose the constraint $a|\text{states}\rangle = 0$ and the dynamics is fully controlled by the guiding center coordinates.

2.2. Non-commutative product

Let us show how the non-commutative product on functions arises from the projection in the LLL. For future convenience, we keep the charge of the particle equal to q in this section.

The idea is to transform a function $f(x)$ into a one body operator: $\int d^2x |x\rangle f(x)\langle x|$, and to project this operator in the LLL:

$$\hat{f} = \sum_{n,m} |n\rangle \int \langle n|x\rangle f(x)\langle x|m\rangle d^2x \ \langle m|, \tag{19}$$

where n, m are the indices of the LLL orbitals. In this way, we transform a function into a matrix.

Conversely, using coherent states, a one body operator acting in the LLL can be transformed into a function. The coherent states $|z\rangle$ are the most localized states in the LLL. They are the adjoint of the state $\langle z|$ defined by $\langle z|n\rangle = \psi_n(z)$ where $\psi_n(z)$ are the LLL wave functions (17). They form an overcomplete basis, and transform a matrix into a function by

$$f(z, \bar{z}) = \langle z|\hat{f}|z\rangle. \tag{20}$$

To see this work in practice, it is convenient to use the symmetric gauge. The guiding center coordinates (13) are combined into the oscillators (16) acting within the LLL. It is convenient to absorb the factor $e^{-qz\bar{z}}$ in the measure. Thus, we recover the Bargman–Fock representation of the operators on analytical functions:

$$b = \partial_z, \ b^+ = qz. \tag{21}$$

The coherent states $|\xi\rangle$ are defined as the eigenstates of b: $b|\xi\rangle = q\bar{\xi}|\xi\rangle$. Thus, their wave function is proportional to

$$\langle z|\xi\rangle = e^{q\bar{\xi}z}. \tag{22}$$

The Q-symbol [36] of a one body operator $\hat{A}$ acting within the LLL consists in bracketing it between coherent states:

$$a(z, \bar{z}) \quad = \quad \langle z|\hat{A}|z\rangle/\rho. \tag{23}$$

In particular one has

$$e^{iPX} = e^{i(\bar{p}z + p\bar{z})} = \langle z|e^{i\frac{\bar{p}b^+}{q}} e^{i\frac{pb}{q}}|z\rangle. \tag{24}$$

Therefore, the Q-symbol induces a non-commutative product on functions which we denote by $*$:

$$a * b = \langle z|\hat{A}\hat{B}|z\rangle. \tag{25}$$

If we apply this to plane waves, we obtain the product

$$e^{iPX}e^{iRX} = e^{-\frac{P.R}{q}} e^{i\frac{P\wedge R}{q}} e^{i(P+R)X}. \tag{26}$$

This algebra is known as the magnetic translation algebra [18] and plays an important role in the theory of the QHE. We shall see next that it originates from the fact that the coordinate X has a dipolar structure.

In the limit $q \to \infty$, the $*$-product coincides with the ordinary product. Using (24) one can evaluate the first-order correction to the ordinary product given by

$$a * b = ab + \frac{1}{q}\partial_{\bar{z}}a\partial_z b + O(\frac{1}{q^2}). \tag{27}$$

It is straightforward to establish the following dictionary between the commutative space and non-commutative LLL projected space:

$$\frac{1}{\pi} \int \; . \; d^2x \to \frac{1}{q} \mathrm{Tr} \; ., \quad \partial_z. \to q[b,.], \quad \partial_{\bar{z}}. \to -q[b^+,.] \tag{28}$$

where Tr now stands for the trace of the matrix in the LLL Hilbert space.

3. Interactions

3.1. Spring in a magnetic field

Let us first consider a simple model for particles in interaction. A pair of particles of opposite charge $\pm q$ is coupled by a spring. In the Landau gauge:

$$A_x = 0, \quad A_y = x, \tag{29}$$

their dynamics follows the Lagrangian

$$L = (x_1 \dot{y}_1 - x_2 \dot{y}_2) - k/2((x_1 - x_2)^2 + (y_1 - y_2)^2), \tag{30}$$

where we have have taken the strong B field limit which enables to neglect the masses of the particles. The Hamiltonian is therefore:

$$H = \frac{k}{2}((x_1 - x_2)^2 + (y_1 - y_2)^2) = P^2/2k. \tag{31}$$

Its eigenstates are simply plane waves:

$$\Psi_P(X) = e^{iP.X}, \tag{32}$$

with X the center of mass coordinates. The magnetic translation commutations (26) arise because the plane waves are extended bound states and not point particles as we have seen on this simple example.

The momentum $\vec{P}$ and the relative coordinate are then related by $-P_X = y_1 - y_2$, $P_Y = x_1 - x_2$ so that the bound state behaves as a neutral dipole with a dipole vector perpendicular and proportional to its momentum. Note that since the strength k of the spring enters the Hamiltonian (31) as a normalization factor the wave function (32) which describes the two charges is independent of k.

It is a general fact, that if the spring is replaced by a rotation invariant potential $V(r)$, one can refine the preceding approach to show that the wave function of the bound state is *independent* of the potential. To see it we need to consider the problem of the particle and the hole interacting in their respective LLL.

We mention that the recent developments in non-commutative string theory have the same origin. Indeed, one of the fundamental fields of string theory called the B field is the exact analogue of the magnetic field. When this field acquires an expectation value, the open strings behave very much like a spring in an external magnetic field [38].

3.2. Structure of the bound state

It is instructive to consider the dynamics of two particles within the lowest Landau level. The two particles interact through a potential $V(\mathbf{x_1} - \mathbf{x_2})$ which is supposed to be both translation and rotation invariant. In a physical situation the potential is the Coulomb interaction between the electrons, but it can in principle be any potential. For reasons that will become clear in the text we consider particles with a charge respectively equal to q_1 and q_2. Our aim is to show that the properties of the dynamics are independent of the detailed shape of the potential. More precisely, the potential interaction is a two-body operator which can be projected into the lowest Landau level. The projection consists in replacing the coordinates, $\mathbf{x_1}$, $\mathbf{x_2}$, with the guiding center coordinates, $\mathbf{R_1}$, $\mathbf{R_2}$. After the projection is taken, the potential becomes an operator which is the effective Hamiltonian for the lowest Landau level dynamics. Using a simple invariance argument, we can see that the eigenstates of the potential do not depend on it, as long as it is invariant under the isometries of the plane. In other words, the two-body wave functions of the Hall effect are *independent* of the interactions. The case of the spring studied in the preceding section corresponds to $q_1 = -q_2$.

The guiding center coordinates for a particle of charge $q > 0$ and the angular momentum, $L = \frac{b^+ b}{q}$, generate a central extension of the algebra of the isometries of the plane:

$$[b, b^+] = q, \quad [L, b^+] = b^+, \quad [L, b] = -b. \tag{33}$$

This algebra commutes with the Hamiltonian H, and therefore acts within the lowest Landau level. It plays a role similar to the angular momentum in quantum mechanics, and the operators b, b^+, L are the analoga of the angular momentum operators J^-, J^+, J^z. (Indeed, the wave functions can be obtained by taking the limit $J \to \infty$, J^z finite, called contraction, from the wave functions on the sphere in presence of a monopole field.) The Landau level index n plays the same role as the representation index j in the rotation group, and it can be recovered as the eigenvalue of a Casimir operator: $C = 2b^+ b / q + L$. The states within each Landau level can be labeled by their angular momentum $m \leq n$.

When two particles of positive charge q_1 and q_2 are restricted to their respective lowest Landau level and interact through a translation and rotation invariant potential, we can form the operators $b^+ = b_1^+ + b_2^+$, $b = b_1 + b_2$ and the total angular momentum $L = L_1 + L_2$. Since the normalization (13) is such that these sums are proportional to the total momentum, the interaction potential commutes with these operators. They obey the commutation relations of the algebra (33) with the charge $q = q_1 + q_2$. Thus, as for the angular momentum, a product of two representations decomposes into representations of the isometry of the plane (33). As a result, the bound state structure is given by Clebsch-Gordan coefficients and does not depend on the potential.

A first physically interesting case is when the two charges are equal to the electron charge ($q_1 = q_2 = 1$). It is easy to verify that each representation is constructed from a generating state annihilated by b: $(b_1^+ - b_2^+)^n|0\rangle$, and the value of the Casimir operator is $C = n$. The corresponding wave functions are:

$$\Psi_n(z_1, z_2) = (z_1 - z_2)^n, \tag{34}$$

an expression that plays an important role in the theory of the fractional Hall effect. The potential being invariant under the displacements, it is a number V_n in each representation. Conversely, the V_n are *all* the information about the potential that is retained by the lowest Landau level physics. The numbers V_n are called pseudopotentials, and turn out to be fundamental to characterize the different phases of the fractional Hall effect (see the contribution of D. Haldane in [1]).

The other case is when the two particles have charges with opposite signs, $q_1 > 0$ and $q_2 < 0$, $|q_2| < q_1$. Because of the sign of the second charge, b_2^+ and b_2 become respectively annihilation and creation operators and the lowest Landau level wave functions are polynomials in $\bar{z}_2$ instead of z_2. For a negative charge, after we factorize the exponential $e^{q\bar{z}z}$, the expressions of the oscillators (21) are

$$b = 2q\bar{z}, \ b^+ = \frac{1}{2}\partial_{\bar{z}}. \tag{35}$$

The same analysis can be repeated, but now the Casimir operator has a negative value n exactly as for the Landau levels. The physical interpretation is that a couple of charges with opposite sign behave exactly like a bound state of charge $q^* = q_1 - |q_2|$. The wave functions are independent of the potential. They organize into *higher Landau levels within the lowest Landau level.*

In particular, the states annihilated by $b = b_1 + b_2$ have the wave function

$$\Psi_n(z_1, \bar{z}_2) = \bar{z}_2^n \exp(-2q_2 z_1 \bar{z}_2). \tag{36}$$

They are the nth Landau level's wave functions with the smallest possible angular momentum $L = -n$. By taking the modulus of these wave functions, we see that in presence of these bound states the space is divided into circular sectors of area $2\pi/q^*$ each.

In the case where the two charges have the same magnitude $q_1 = -q_2 = q$, we recover the plane waves of the last section:

$$\Psi_{\mathbf{p}}(z_1, \bar{z}_2) = \exp\left(2q z_1 \bar{z}_2 + i\bar{p}(z_1 + z_2) + ip(\bar{z}_1 + \bar{z}_2)\right). \tag{37}$$

3.3. Bosons at $\nu = 1$

Let us indicate how a scenario involving these composite particles enables to apprehend the QHE plateaux in the region of a magnetic field around $\nu_0 = 1/2$. To further simplify, consider bosonic particles interacting repulsively in a magnetic field at a filling factor $\nu = 1$. Although both problems first look different, they are essentially the same.

The LLL particles interact with a repulsive potential $V(\vec{x} - \vec{y})$. After projection the Hamiltonian takes the form

$$H = 1/2 \int \rho(\vec{x}) \, V(\vec{x} - \vec{y}) \, \rho(\vec{y}) \, d^2x \, d^2y \tag{38}$$

where $\rho(\vec{x})$ is the projected density operator. The projection relates a field $\rho(\vec{x})$ to a matrix $\hat{\rho}$ as we saw earlier, therefore this is a problem of *matrix* quantum mechanics. Let us be as schematic as possible and consider a local repulsive potential:

$$H = 1/2 \int \rho(\vec{x})^2 d^2x. \tag{39}$$

Since the density is an operator, the Hamiltonian is not trivially diagonal as it would be in the absence of a magnetic field.

If we use the matrix formulation, we introduce N by N matrices M_{ij} for the density operator ρ. The commutation relations of the matrix elements are given by

$$[M_{ij}, M_{kl}] = \delta_{jk} M_{il} - \delta_{il} M_{kj}, \tag{40}$$

and one recognizes the $GL(N)$ commutation relations; this algebra is trivially realized by the matrix units $M_{ij} = |i\rangle\langle j|$.

We can as well decompose the field $\rho(x)$ into plane waves: $\rho(x) = \sum_{\vec{k}} e^{ikx} \rho_k$. The matrix-plane waves $\hat{e}^{ikx}$ obey the magnetic translation relations (26). To give their matrix expression, it is convenient to introduce the matrices U, V defined by $U|i\rangle = e^{\frac{2i\pi}{N}}|i\rangle$ and $V|i\rangle = |i+1\rangle$. By choosing a normal ordering convention, one has $\hat{e}^{ikx} = U^{k_1} V^{k_2}$. The decomposition of the density operator M on the density Fourier modes is given by $M = \sum_k \rho_k \hat{e}^{ikx}$. Substituting this expression into (40), we see again that the modes ρ_k obey the magnetic translation algebra (and do not commute as in the case where there is no field).

The constraint that the filling factor is equal to one is realized by

$$\hat{N}|\text{state}\rangle = N, \tag{41}$$

where $\hat{N} = TrM$ is the number operator.

We can find a representation of the algebra (40) given by: $M_{ij} = a_i^+ a_j$, where the operators a_i^+ creates a LLL particle in the orbital i, with $1 \leq i \leq N$. In the bosonic case we want to diagonalize the Hamiltonian (39) in the space containing exactly N bosons. The dimension of the Hilbert space is very large, and there is no good approximation scheme available.

If the commutation relations of a_i, a_i^+ are fermionic, however, there is a unique state obeying the constraint (41) given by $a_1^+ \ldots a_n^+|0\rangle$, and the ground state is trivially the full Slater determinant of the LLL orbitals. Its wave function is the Vandermonde determinant of the variables z_i.

If we proceed naively and replace the interaction $\rho(\vec{x})^2$ by the star product: $\rho(\vec{x}) * \rho(\vec{x})$, the Hamiltonian (39) becomes

$$H = TrM^2, \tag{42}$$

which is a Casimir operator of the algebra (40). Thus, in the bosonic case, it is constant and equal to $\hat{N}(\hat{N} + N - 1)$ which is equal to $(2N - 1)N$ in the subspace obeying the constraint (41).

To find an approximation scheme for the ground state of (39), the approach proposed in [23], developed in [24–27] is to build the theory of these particles on top of the ground state having the opposite statistics. The "composite" particle is made of the original particle and a "hole" in the ground state.

The ground state is filled with bound-states introduced above made of an electron of charge 1, and a hole in a ground state filled with charges $-q$. The charge of the bound state is thus q^*, with

$$q^* = 1 - q. \tag{43}$$

To obtain the values of the filling factor that give rise to a Hall effect, we use the fact that for a fixed magnetic field and a fixed density, the proportionality relation between the charge and the filling factor is given by

$$\text{charge} \propto \frac{1}{\text{filling factor}} . \tag{44}$$

When the magnetic field is varied, we take as a postulate that the charge q adjusts itself so that the filling factor of the ground state of the q charge is always equal to 1. Thus, $q \propto 1$. An integer quantum Hall effect will develop when the filling factor of the bound state is an integer p. So, when $q* \propto 1/p$. We can recover the normalization coefficient through the relation between the filling factor of the electrons and their charge: $1 \propto 1/\nu$. Substituting these relations in (43) we obtain the following expression for the filling factors giving rise to a Hall effect:

$$\frac{1}{\nu} = 1 + \frac{1}{p}. \tag{45}$$

We can represent the composite particles by fermionic rectangular matrices $\Psi_{\alpha i}$, $\Psi_{i\alpha}^+$ where i labels the lowest Landau level orbitals of a charge 1 particle and α labels the lowest Landau level orbitals of a charge q particle. Thus, $\Psi_{i\alpha}^+$ creates particles in the orbitals i and a hole in the orbital α. They obey the commutation relations

$$\{\Psi_{i\alpha}^+, \Psi_{\beta j}\} = \delta_{ij}\delta_{\alpha\beta}. \tag{46}$$

The density operators M'_{ij} for the charge 1 particle and $M''_{\alpha\beta}$ for the charge q particle can be represented as

$$M'_{ij} = \Psi_{i\alpha}^+ \Psi_{\alpha j}, \quad M''_{\alpha\beta} = \Psi_{\alpha i}\Psi_{i\beta}^+, \tag{47}$$

and it is straightforward to see that they obey the $GL(N') \times GL(N)$ commutation relations.

We can then convert these matrices into functions $\rho'(x)$ and $\rho''(x)$, which represent the particle and the hole density respectively. The total density is the sum of both: $\rho = \rho' + \rho''$. The dynamics is obtained by substituting these representations in the Hamiltonian (39).

At $\nu = 1$ exactly, Ψ is a neutral fermionic bound state. In this limit, the density is approximated by

$$\rho(\vec{x}) =: \{\Psi^+\star, \Psi\}(x) :\approx \vec{\nabla} \times \Psi^+ i\vec{\nabla}\Psi(x) \tag{48}$$

The anticommutator originates from the fact that we add the two contributions ρ^b and ρ^f treating the pairs as composite particles. As a result, the dominant term in a gradient expansion is the right-hand side of this equality.

The Hamiltonian (39) can be expressed in terms of these operators to obtain an effective description in terms of neutral dipolar particles.

The theory can be adapted to the $\nu = 1/2$ fermionic case relevant for the QHE. The background charge adjusts so that its filling factor is $1/2$, and the filling factors are given by $\frac{1}{\nu} = 2 + \frac{1}{p}$. These filling factors are those predicted by Jain [10], and fit well with the Hall effect observed at $\nu = 3/7, 4/9, 5/11, 6/13$. In the region, close to $\nu = 1/2$, the quasiparticles have practically zero charge, and therefore see a weak magnetic field. They behave very much like a neutral Fermi liquid in agreement with several experiments [20]. One of them measures directly the charge q^* of the quasiparticles through the cyclotron radius of their trajectory [29]. At $\nu = 1/2$ exactly, the quasiparticles are neutral dipoles with a dipole size of the order of the magnetic length. The main difficulty for the theory is that the separation between two dipoles is of the same order as their size, and this is therefore a strong interaction problem. It would be extremely interesting if the new developments in NCFT [6] can help to make progress in this theory.

4. Skyrmion and nonlinear σ-model

As another application, we review the Skyrmion of the Hall effect [31] which relates the topological charge of the classical soliton to the electric charge of the quantum state. In a given topological sector, the solitons which minimize the action are in one-to-one correspondence with the degenerate eigenstates of a quantum Hamiltonian in the same charge sector. Moreover in both cases, the energy is equal to the modulus of the charge.

Belavin and Polyakov [32] have considered the classical solutions of the two-dimensional nonlinear σ-model on the sphere S_2. The field configurations $\vec{n}(x, y)$ can be characterized by their stereographic projection on the complex plane $w(x,y)$. One requires that the spin points in the x direction at infinity, or equivalently $w(\infty) = 1$. The minima of the action are rational fractions of $z = x + iy$: $w(z) = f(z)/g(z)$. The soliton's winding number and its classical action are both given by the degree k of the polynomials f, g. The soliton is thus determined by the positions where the spin points to the south and the north pole given by the zeros of f and g.

In quantum mechanics, the wave function for a spin $1/2$ particle constrained to the Lowest Landau Level (LLL) is fully determined by the positions where the spin is up or down with probability one. Up to an exponential term, the two components of this wave function are polynomials in z vanishing at the positions where the spin is respectively up and down. The wave functions:

$$\langle z_1, \ldots, z_{N_e}|\Phi\rangle = \prod_{i=1}^{N_e}(f(z_i)\uparrow + g(z_i)\downarrow)\prod_{i<j}(z_i - z_j)^m \tag{49}$$

have been considered (for $m = 1$) by MacDonald, Fertig and Brey [34] in the Hall effect context. They represent the ground states of spin $1/2$ particles at a filling fraction close to $1/m$. We show that the correspondence with the nonlinear σ-model can be made precise in the case $m = 1$. One has the following correspondence table:

$\int(\vec{\nabla}\vec{n})^2 d^2x$	$\sum_{i<j}\delta^{(2)}(\vec{x}_i - \vec{x}_j)$
$\vec{n}(\vec{x})$	$|\Phi\rangle$ a Slater determinant
winding number	electric charge

4.1. Non-commutative Belavin–Polyakov soliton

Let us show that the Skyrmion is the exact non-commutative analogue of the Belavin–Polyakov soliton [33].

A point on the sphere S^2 is a unit vector $\vec{n}(\vec{x})$ with which we construct the projector $p(\vec{x}) = (1 + \vec{n}\vec{\sigma})/2$. p is a two by two rank one hermitian projector $p^2 = p$, $p^+ = p$, and the action for the nonlinear σ-model takes the form:

$$S = \frac{1}{\pi}\int \mathrm{tr}\ \partial_z p\,\partial_{\bar{z}}p\ d^2x. \tag{50}$$

To obtain the solitons which minimize the action let us substitute $\partial_z p^2$ for $\partial_z p$ to rewrite the integrand as $\mathrm{tr}\ p(\partial_z p\,\partial_{\bar{z}}p + \partial_{\bar{z}}p\,\partial_z p)$ and add to (50) the topological term

$$K = \frac{1}{\pi}\int \mathrm{tr}\ p(\partial_z p\,\partial_{\bar{z}}p - \partial_{\bar{z}}p\,\partial_z p)\ d^2x \tag{51}$$

so that the sum takes the form

$$S' = S + K = \frac{2}{\pi}\int \mathrm{tr}\ (p\,\partial_{\bar{z}}p)^+(p\,\partial_{\bar{z}}p)\ d^2x. \tag{52}$$

(52) is positive and the solutions with $S' = 0$ must obey $p\partial_{\bar{z}}p = 0$. If we parameterize p by a unitary vector v, $v^+v = 1$, $p = vv^+$, it is solved for $v = N^{-1}(f(z), g(z))$ where f, g are holomorphic functions and $N = \sqrt{|f|^2 + |g|^2}$. If one requires that $p(\infty) = (1 + \sigma_x)/2$, f and g are polynomials with the same highest coefficient z^k. The integrand of K is the field strength of the gauge potential $\omega = -v^+dv/2i$ which goes to a pure gauge far from soliton. The topological term is therefore given by the contour integral of ω at infinity equal to $-k$ and thus $S = k$.

The quantum analogous problem we consider here consists of finding the degenerate ground states of electrons interacting by a δ repulsive potential in the lowest Landau level (LLL). The electrons are confined in a finite disc thread by N_ϕ magnetic fluxes. When the number of electrons N_e differs from N_ϕ by an integer equal to the winding number k the quantum eigenstates coincide with the classical solitonic field configurations if the scale of variation of the soliton is large compared to the magnetic length.

The second quantized field that annihilates (creates) an electron with a spin σ at position $\vec{x}$ in the LLL can be constructed in terms of the fermionic operators $c_{l\sigma}$ ($c_{l\sigma}^+$) which annihilate (create) an electron in the lth orbital:

$$\Psi_\sigma(\vec{x}) = \sum_l \langle z|l\rangle c_{l\sigma}. \tag{53}$$

In terms of this field, the total number of electrons in the LLL is $N_e = \int \sum_\sigma \Psi_\sigma^+ \Psi_\sigma(\vec{x}) \, d^2x$. The charge of the Skyrmion is the difference between the number of magnetic fluxes N_ϕ and the number of electrons N_e: $Q_s = N_\phi - N_e$. In other words, it is the number of electrons added or subtracted to the system starting from a situation where the filling factor $\nu = N_e/N_\phi$ is exactly one. In the following we consider the limit N_ϕ, $N_e = \infty$ keeping the charge Q_s fixed.

In [34], it was observed that the zero energy states of the hard-core model Hamiltonian could be completely determined. We consider a closely related short range repulsive Hamiltonian invariant under a particle hole transformation $\Psi \to \Psi^+$ and such that the energy of its ground state coincides with the charge. It is given by

$$H = \frac{1}{\rho} \int (\Psi_\uparrow^+ \Psi_\uparrow - \Psi_\downarrow \Psi_\downarrow^+)^2(\vec{x}) \, d^2x = \frac{2}{\rho} \int (\Psi_\uparrow \Psi_\downarrow)^+ (\Psi_\uparrow \Psi_\downarrow)(\vec{x}) \, d^2x + Q_s \tag{54}$$

where we have used the fact that $\{\Psi_\sigma(\vec{x}), \Psi_{\sigma'}^+(\vec{x})\} = \rho\delta_{\sigma\sigma'}$ to obtain the second equality. Let us for simplicity consider the case where $Q_s > 0$, the other case can be reached using a particle hole transformation. This Hamiltonian is clearly bounded from below by Q_s and the exact eigenstates with energy Q_s are obtained for states $|\Phi\rangle$ such that $\Psi_\uparrow(\vec{x})\Psi_\downarrow(\vec{x})|\Phi\rangle = 0$. In such a state two electrons never occupy the same position and the wave function is blind to the short range potential. This property is precisely guaranteed by the factor $\prod_{i<j}(z_i - z_j)$ in (49).

The states (49) carry a charge $Q_s = k$ where k is the degree of the polynomials f and g. They are Slater determinants:

$$\langle z_1, \ldots, z_{N_e}|\Phi\rangle = \bigwedge_1^{N_e} (f(z_i) \uparrow + g(z_i) \downarrow)\langle z_i|\tilde{l}\rangle \tag{55}$$

where $\langle z|\tilde{l}\rangle$ is basis of orthogonal polynomials for the scalar product:

$$\langle \phi|\phi'\rangle = \int \bar{\phi}(\bar{z})\phi'(z)e^{-q\bar{z}z}(|f|^2 + |g|^2)d^2x. \tag{56}$$

A Slater determinant is fully determined by the matrix expectation value:

$$\langle z|\hat{P}|z\rangle = \rho p = \langle \Phi|, \begin{pmatrix} \Psi_\downarrow^+ \Psi_\downarrow & \Psi_\uparrow^+ \Psi_\downarrow \\ \Psi_\downarrow^+ \Psi_\uparrow & \Psi_\uparrow^+ \Psi_\uparrow \end{pmatrix} |\Phi\rangle \tag{57}$$

and $\hat{P}$ is a projector, $\hat{P}^2 = \hat{P}$. In the case of (55) we can obtain $\hat{P}$ explicitly as follows. The states $|v_l\rangle = (f(b^+)\uparrow + g(b^+)\downarrow)|\tilde{l}\rangle$ can be organized into a vector $V = \sum_l |v_l\rangle\langle l|$. By construction, $V^+V = \mathrm{Id}$, so that $VV^+ = \hat{P}$, where $\hat{P}$ is the projector with Q-symbol p.

To relate the Skyrmion to the classical σ-model, let us evaluate the energy of a Slater determinant $|\Phi\rangle$ using the Wick theorem:

$$\langle\Phi|H|\Phi\rangle = \rho \int (2\det p - \mathrm{tr}\, p + 1)\, d^2x. \tag{58}$$

In the expression above, the determinant is evaluated using the ordinary product. Suppose we replace it with the $*$-product in (58). Using the fact that $p*p = p$ one verifies that the integrand rewrites $(\mathrm{tr}\, p - 1)*(\mathrm{tr}\, p - 1)$. Since $\mathrm{tr}(p-1)$ is $O(\rho^{-1})$, the $*$-square is $O(\rho^{-2})$ and does not contribute to the energy when $\rho \to \infty$. Therefore, the limiting value of (58) is given by the modification induced by the ordinary product at first-order in ρ^{-1}. One obtains from (27):

$$\langle\Phi|H|\Phi\rangle = \frac{1}{\pi} \int \mathrm{tr}\, \partial_z p\, \partial_{\bar{z}} p\, d^2x + O(1/\rho) \tag{59}$$

which is the value of the action (50) and establishes the correspondence between the classical and the quantum problems.

Although the classical action (50) can be obtained straightforwardly from the energy (58) in the limit $\rho \to \infty$, the topological term (51) cannot be related so directly to the charge of the Skyrmion. Nevertheless it is possible to define the topological term at the quantum level and to verify that it coincides with the charge in the present case. For this we need to make the substitutions (28) in (51):

$$p \to \hat{P}, \quad \frac{1}{\pi} \int \cdot\, d^2x \to \frac{1}{q}\mathrm{Tr}\, ., \quad \partial_z. \to q[b, .], \quad \partial_{\bar{z}}. \to -q[b^+, .]\,. \tag{60}$$

The modified expression of K (51) still defines a topological invariant which is a non-commutative analogue of K [37] to which it reduces in the limit $\rho \to \infty$. It can be defined for projectors $\hat{P}$ which do not have a classical limit p. The easiest way to realize a charge $-k$ configuration consists in expelling k electrons from the first $l < k$ angular momentum orbitals in the $\nu = 1$ filled LLL. (Here the spin can be kept fixed and plays no essential role.) The projector which characterizes this configuration is $\hat{P} = \sum_{l \geq k} |l\rangle\langle l|$, equivalently $V = b^{+k}N^{-1} = \sum_l |l + k\rangle\langle l|$. Using the quantum expression one can easily verify that the topological invariant is equal to the charge $K = -k$.

References

[1] The Quantum Hall Effect, Edited by R.E. Prange and S. Girvin, Springer-Verlag, New York, (1990).

[2] Perspectives in Quantum Hall Effects, Edited by S. Das Sarma and A. Pinczuk, Wiley, New-York, (1997).

[3] S.M. Girvin, "The Quantum Hall Effect: Novel Excitations and broken Symmetries", Les Houches Lecture Notes, in Topological Aspects of Low Dimensional Systems, Springer-Verlag, Berlin and Les Editions de Physique, Les Ulis, (1998).

[4] B. Doucot and V. Pasquier, Séminaire Poincaré, novembre 2004,

[5] J. Klauder and B.S. Skagerstam, "Coherent states", Word Scientific (1984)

[6] See the contribution of V. Rivasseau in this volume.

[7] See the contribution of J.M. Maillet in this volume.

[8] L. Susskind, The Quantum Hall Fluid and Non-Commutative Chern Simons Theory, hep-th/0101029

[9] A. Polychronakos, *Quantum Hall states as matrix Chern-Simons theory*, JHEP **0104** (2001) 011, arXiv.org /hep-th/0103013.

[10] J.K. Jain, Phys. Rev. Lett. **63** (1989), 199.

[11] J. Klauder and B.S. Skagerstam, "Coherent states", Word Scientific (1984)

[12] R.L. Willet, M.A. Paalanen, R.R. Ruel, K.W. West, L.N. Pfeiffer and D.J. Bishop, Phys. Rev. Lett. **54** (1990), 112.

[13] B.A. Halperin, P.A. Lee and N. Read, Phys. Rev. B **47** (1993), 7312.

[14] N. Read, Semicon. Sci. Tech. **9** (1994), 1859.

[15] Y.B. Kim, X.G. Wen, P.A. Lee and A. Furusaki, Phys. Rev. B **50** (1994), 17917.

[16] C. Kallin and B. Halperin, Phys. Rev. B **30** (1984), 5655.

[17] I.V. Lerner and Yu. E. Lozovik, Zh. Eksp. Theor. Fiz **78** (1978) , 1167 ; [Sov. Phys. JETP **51** (1980), 588].

[18] S.M. Girvin, A.H. Macdonald and P.M. Platzman, Phys. Rev. B **33** (1986), 2481.

[19] R. Shankar and G. Murthy, Phys. Rev. Lett. **9** (1997), 4437.

[20] R.L. Willet, Surf. Sci. **305** (1994), 76; R.L. Willet et al., Phys. Rev. Lett. **71** (1993), 3846.

[21] W. Kang et al., Phys. Rev. Lett. **71** (1993), 3850; V.J. Goldman et al., Phys. Rev. Lett. **72** (1994), 2065; H.J. Smet et al., Phys. Rev. Lett. **77** (1996), 2272.

[22] R.R. Du, H.L. Stormer, D.C.Tsui, N.L. Pfeiffer, and K.W. West, Phys. Rev. Lett. **70** (1993), 2944. P.T. Colerige, Z.W. Wasilewski, P. Zawadzki, A.S. Sacharajda and H.A. Carmona Phys. Rev. B **52** (1995), 11603.

[23] V. Pasquier "Composite Fermions and confinement" (1996) unpublished, Morion proceedings (1999).

[24] N. Read, http://arxiv.org/abs/cond-mat/9804294.

[25] R. Shankar and G. Murthy, Phys. Rev. Lett. **79** (1997), 4437.

[26] Dung-Hai Lee, Phys. Rev. Lett. **80** (1998), 4547.

[27] A. Stern, B.I. Halperin, F. von Oppen, S. Simon, Phys. Rev. B in press and cond-mat/9812135.

[28] V. Pasquier, F.D.M. Haldane, Nuclear Physics B **516**[FS] (1998), 719.

[29] V.J. Goldman, B. Su and J.K. Jain, Phys. Rev. Lett. **72** (1994), 2065.

[30] F. Von Oppen, B. Halperin, S. Simon and A. Stern, cond-mat/9903430.

[31] S.L. Sondhi, A. Karlhede, S.A. Kivelson and E.H. Rezayi, Phys.Rev.B **47** (1993), 16419.

[32] A.A. Belavin and A.M. Polyakov, JETP Let. **22** (1975), 245.

[33] V. Pasquier, Arxiv preprint hep-th/0007176, 2000.

[34] A.H. MacDonald, H.A. Fertig and L. Brey, Phys. Rev. Lett **76** (1996), 2153.

[35] V. Pasquier, Phys. Lett. B **490** (2000), 258.

[36] F.A. Berezin, Sov. Phys. Usp. **132** (1980), 497.

[37] A. Connes, "Non-commutative Geometry", Academic Press (1994).

[38] Bigatti and Susskind, Arxiv preprint hep-th/9908056.

Vincent Pasquier
Service de Physique Théorique
C.E.A/ Saclay
F-91191 Gif-sur-Yvette
France
e-mail: `vincent.pasquier@cea.fr`

Quantum Spaces, 19–107
© 2007 Birkhäuser Verlag Basel

Non-commutative Renormalization

Vincent Rivasseau

Abstract. A new version of scale analysis and renormalization theory has been found on the non-commutative Moyal space. It could be useful for physics beyond the standard model or for standard physics in strong external fields. The good news is that quantum field theory is better behaved on non-commutative than on ordinary space: indeed it has no Landau ghost. We review this rapidly growing subject.

1. Introduction

The world as we know it today is made of about 61 different scales if we use powers of ten[1]. Indeed there is a fundamental length obtained by combining the three fundamental constants of physics, Newton's gravitation constant G, Planck's constant $\hbar$ and the speed of light c. It is the Planck length $\ell_P = \sqrt{\hbar G/c^3}$, whose value is about $1.6 \cdot 10^{-35}$ meters. Below this length ordinary space-time almost certainly has to be quantized, so that the very notion of scale might be modified. But there is also a maximal observable scale or "horizon" in the universe, not for fundamental but for practical reasons. The current distance from the Earth to the edge of the visible universe is about 46 billion light-years in any direction[2]. This translates into a comoving radius of the visible universe of about $4.4 \cdot 10^{26}$ meters, or more fundamentally $2.7 \cdot 10^{61}$ Planck lengths. Although we do not observe galaxies that far away, the WMAP data indicate that the universe is really at least 80% that big [1]. The geometric mean between the size of the (observable) universe

---

[1]Or of about 140 e-folds if we want to avoid any parochialism due to our ten fingers. What is important is to measure distances on a logarithmic scale.

[2]The age of the universe is only about 13.7 billion years, so one could believe the observable radius would be 13.7 billion light years. This gives already a correct order of magnitude, but in our expanding universe space-time is actually curved so that distances have to be measured in comoving coordinates. The light emitted by matter shortly after the big-bang, that is to say about 13.7 billion years ago, which is reaching us now corresponds to a present distance of that matter to us that has almost trippled, see http://en.wikipedia.org/wiki/Observable_universe.

and the Planck's length stands therefore around 10^{-4} meters, about the size of an (arguably very small) ant. In [2], we proposed to call this the "antropic principle".

Among the roughly sixty scales of the universe, only about ten to eleven were relatively well known to ancient Greeks and Romans two thousand years ago. We have now at least some knowledge of the 45 largest scales from $2 \cdot 10^{-19}$ meters (roughly speaking the scale of 1 Tev, observable at the largest particle colliders on earth) up to the size of the universe. This means that we know about three fourths of all scales. But the sixteen scales between $2 \cdot 10^{-19}$ meters and the Planck length form the last true *terra incognita* of physics. Note that this year the LHC accelerator at Cern with maximum energy of about 10 Tev should start opening a window into a new power of ten. But that truly special treat also will mark the end of an era. The next fifteen scales between $2 \cdot 10^{-20}$ meters and the Planck length may remain largely out of direct reach in the foreseeable future, except for the glimpses which are expected to come from the study of very energetic but rare cosmic rays. Just as the Palomar mountain telescope remained the largest in the world for almost fifty years, we expect the LHC to remain the machine with highest energy for a rather long time until truly new technologies emerge[3]. Therefore we should try to satisfy our understandable curiosity about the *terra incognita* in the coming decades through more and more sophisticated indirect analysis. Here theoretical and mathematical physics have a large part to play because they will help us to better compare and recoup many indirect observations, most of them probably coming from astrophysics and cosmology, and to make better educated guesses.

I would like now to argue both that quantum field theory and renormalization are some of the best tools at our disposal for such educated guesses, but also that very likely we shall also need some generalization of these concepts.

Quantum field theory or QFT provides a quantum description of particles and interactions which is compatible with special relativity [3–6]. It is certainly essential because it lies right at the frontier of the *terra incognita*. It is the accurate formalism at the shortest distances we know, between roughly the atomic scale of 10^{-10} meters, at which relativistic corrections to quantum mechanics start playing a significant role[4], up to the last known scale of a Tev or $2 \cdot 10^{-19}$ meters. Over the years it has evolved into the *standard model* which explains in great detail most experiments in particle physics and is contradicted by none. But it suffers from at least two flaws. First it is not yet compatible with general relativity, that is Einstein's theory of gravitation. Second, the standard model incorporates so many different fermionic matter fields coupled by bosonic gauge fields that it seems more some kind of new Mendeleyev table than a fundamental theory. For these two reasons QFT and the standard model are not supposed to remain valid without

---

[3] New colliders such as the planned linear e^{+}- e^{-} international collider might be built soon. They will be very useful and cleaner than the LHC, but they should remain for a long time with lower total energy.

[4] For instance quantum electrodynamics explains the Lamb shift in the hydrogen atom spectrum.

any changes until the Planck length where gravitation should be quantized. They could in fact become inaccurate much before that scale.

What about renormalization? Nowadays renormalization is considered the heart of QFT, and even much more [7–9]. But initially renormalization was little more than a trick, a quick fix to remove the divergences that plagued the computations of quantum electrodynamics. These divergences were due to summations over exchanges of virtual particles with high momenta. Early renormalization theory succeeded in hiding these divergences into unobservable *bare* parameters of the theory. In this way the physical quantities, when expressed in terms of the *renormalized* parameters at observable scales, no longer showed any divergences. Mathematicians were especially scornful. But many physicists also were not fully satisfied. F. Dyson, one of the founding fathers of that early theory, once told me: "We believed renormalization would not last more than six months, just the time for us to invent something better..."

Surprisingly, renormalization survived and prospered. In the mid 1950s Landau and others found a key difficulty, called the Landau ghost or triviality problem, which plagued simple renormalizable QFT such as the ϕ_4^4 theory or quantum electrodynamics. Roughly speaking Landau showed that the infinities supposedly taken out by renormalization were still there, because the bare coupling corresponding to a nonzero renormalized coupling became infinite at a very small but finite scale. Although his argument was not mathematically fully rigorous, many physicists proclaimed QFT and renormalization dead and looked for a better theory. But in the early 1970s, against all odds, they both made a spectacular comeback. As a double consequence of better experiments but also of better computations, quantum electrodynamics was demoted of its possibly fundamental status and incorporated into the larger electroweak theory of Glashow, Weinberg and Salam. This electroweak theory is still a QFT but with a non-abelian gauge symmetry. Motivated by this work 't Hooft and Veltman proved that renormalization could be extended to non-abelian gauge theories [10]. This difficult technical feat used the new technique of dimensional renormalization to better respect the gauge symmetry. The next and key step was the extraordinary discovery that such non-abelian gauge theories no longer have any Landau ghost. This was done first by 't Hooft in some unpublished work, then by D. Gross, H.D. Politzer and F. Wilczek [11,12]. D. Gross and F. Wilczek then used this discovery to convincingly formulate a non-abelian gauge theory of strong interactions [13], the ones which govern nuclear forces, which they called quantum chromodynamics. Remark that in every key aspect of this striking recovery, renormalization was no longer some kind of trick. It took a life of its own.

But as spectacular as this story might be, something even more important happened to renormalization around that time. In the hands of K. Wilson [14] and others, renormalization theory went out of its QFT cradle. Its scope expanded considerably. Under the alas unfortunate name of renormalization group (RG), it was recognized as the right mathematical technique to move through the different scales of physics. More precisely, over the years it became a completely general paradigm to study changes of scale, whether the relevant physical phenomena are classical or quantum, and whether they are deterministic or statistical. This encompasses in particular the full Boltzmann program to deduce thermodynamics from statistical mechanics and potentially much more. In the hands of Wilson, Kadanoff, Fisher and followers, RG allowed to much better understand phase transitions in statistical mechanics, in particular the universality of critical exponents [15]. The fundamental observation of K. Wilson was that the change from bare to renormalized actions is too complex a phenomenon to be described in a single step. Just like the trajectory of a complicated dynamical system, it must be studied step by step through a *local* evolution equation. To summarize, do not jump over many scales at once!

Let us make a comparison between renormalization and geometry. To describe a manifold, one needs a covering set of maps or atlas with crucial *transition* regions which must appear on different maps and which are glued through transition functions. One can then describe more complicated objects, such as bundles over a base manifold, through connections which allow to parallel transport objects in the fibers when one moves over the base.

Renormalization theory is both somewhat similar and somewhat different. It is some kind of geometry with a very sophisticated infinite-dimensional"bundle" part which loosely speaking describes the effective actions. These actions flow in some infinite-dimensional functional space. But at least until now the "base" part is quite trivial: it is a simple one-dimensional positive real axis, better viewed in fact as a full real axis if we use logarithmic scales. We have indeed both positive and negative scales around a reference scale of observation. The negative or small spatial scales are called *ultraviolet* and the positive or large ones are called *infrared* in reference to the origin of the theory in electrodynamics. An elementary step from one scale to the next is called a renormalization group step. K. Wilson understood that there is an analogy between this step and the elementary evolution step of a dynamical system. This analogy allowed him to bring the techniques of classical dynamical systems into renormalization theory. One can say that he was able to see the classical structure hidden in QFT.

Working in the direction opposite to K. Wilson, G. Gallavotti and collaborators were able to see the quantum field theory structure hidden in classical dynamics. For instance they understood secular averages in celestial mechanics as a kind of renormalization [16, 17]. In classical mechanics, small denominators play the role of high frequencies or ultraviolet divergences in ordinary RG. The interesting physics consists in studying the long time behavior of the classical trajectories, which is the analog of the infrared or large distance effects in statistical mechanics.

At first sight the classical structure discovered by Wilson in QFT and the quantum structure discovered by Gallavotti and collaborators in classical mechanics are both surprising because classical and QFT perturbation theories look very different. Classical perturbation theory, like the inductive solution of any deterministic equation, is indexed by trees, whether QFT perturbation theory is indexed by more complicated "Feynman graphs", which contain the famous "loops" of anti-particles responsible for the ultraviolet divergences[5]. But the classical trees hidden inside QFT were revealed in many steps, starting with Zimmermann (who called them forests...) [18] through Gallavotti and many others, until Kreimer and Connes viewed them as generators of Hopf algebras [19–21]. Roughly speaking the trees were hidden because they are not just subgraphs of the Feynman graphs. They picture abstract inclusion relations of the short distance connected components of the graph within the bigger components at larger scales. Gallavotti and collaborators understood why there is a structure on the trees which index the classical Poincaré-Lindstedt perturbation series similar to Zimmermann's forests in quantum field perturbation theory[6].

Let us make an additional remark which points to another fundamental similarity between renormalization group flow and time evolution. Both seem naturally *oriented* flows. Microscopic laws are expected to determine macroscopic laws, not the converse. Time runs from past to future and entropy increases rather than decreases. This is philosophically at the heart of standard determinism. A key feature of Wilson's RG is to have defined in a mathematically precise way *which* short scale information should be forgotten through coarse graining: it is the part corresponding to the *irrelevant operators* in the action. But coarse graining is also fundamental for the second law in statistical mechanics, which is the only law in classical physics which is "oriented in time" and also the one which can be only understood in terms of change of scales.

Whether this arrow common to RG and to time evolution is of a cosmological origin remains to be further investigated. We remark here simply that in the distant past the big bang has to be explored and understood on a logarithmic time scale. At the beginning of our universe important physics is the one at very short distance. As time passes and the universe evolves, physics at longer distances, lower temperatures and lower momenta becomes literally visible. Hence the history of the universe itself can be summarized as a giant unfolding of the renormalization group.

This unfolding can then be specialized into many different technical versions depending on the particular physical context, and the particular problem at hand. RG has the potential to provide microscopic explanations for many phenomenological theories. Hence it remains today a very active subject, with several important new brands developed in the two last decades at various levels of physical precision and of mathematical rigor.

To name just a few of these brands:

- the RG around extended singularities governs the quantum behavior of condensed matter [23–25]. It should also govern the propagation of wave fronts and the long-distance scattering of particles in Minkowski space. Extended

---

[5] Remember that one can interpret antiparticles as going backwards in time.

[6] In addition Gallavotti also remarked that antimatter loops in Feynman graphs can just be erased by an appropriate choice of non-Hermitian field interactions [22].

singularities alter dramatically the behavior of the renormalization group. For instance because the dimension of the extended singularity of the Fermi surface equals that of the space itself minus one, hence that of space-time minus *two*, local quartic fermionic interactions in condensed matter in *any* dimension have the same power counting than *two*-dimensional fermionic field theories. This means that condensed matter in *any* dimension is similar to *just renormalizable* field theory. Among the main consequences, there is no critical mean field dimension in condensed matter except at infinity, but there is a rigorous way to handle non-perturbative phase transitions such as the BCS formation of superconducting pairs through a dynamical $1/N$ expansion [26].

- the RG trajectories in dimension 2 between conformal theories with different central charges have been pioneered in [27]. Here the theory is less advanced, but again the c-theorem is a very tantalizing analog of Boltzmann's H-theorem.
- the functional RG of [28] governs the behavior of many disordered systems. It might have wide applications from spin glasses to surfaces.

Let us return to our desire to glimpse into the *terra incognita* from currently known physics. We are in the uncomfortable situation of salmons returning to their birthplace, since we are trying to run against the RG flow. Many different bare actions lead to the same effective physics, so that we may be lost in a maze. However the region of *terra incognita* closest to us is still far from the Planck scale. In that region we can expect that any non-renormalizable terms in the action generated at the Planck scale have been washed out by the RG flow and renormalizable theories should still dominate physics. Hence renormalizability remains a guiding principle to lead us into the maze of speculations at the entrance of *terra incognita*. Of course we should also be alert and ready to incorporate possible modifications of QFT as we progress towards the Planck scale, since we know that quantization of gravity at that scale will not happen through standard field theory.

String theory [29] is currently the leading candidate for such a quantum theory of gravitation. Tantalizingly the spectrum of massless particles of the closed string contains particles up to spin 2, hence contains a candidate for the graviton. Open strings only contain spin one massless particles such as gauge bosons. Since closed strings must form out of open strings through interactions, it has been widely argued that string theory provides an explanation for the existence of quantum gravity as a necessary complement to gauge theories. This remains the biggest success of the theory up to now. It is also remarkable that string theory (more precisely membrane theory) allows some microscopic derivations of the Beckenstein-Hawking formula for blackhole entropy [30].

String theory also predicts two crucial features which are unobserved up to now, supersymmetry and six or seven new Kaluza-Klein dimensions of space-time at short distance. Although no superpartner of any real particle has been found yet, there are some indirect indications of supersymmetry, such as the careful study of the flows of the running non-abelian standard model gauge couplings[7]. Extra dimensions might also be welcome, especially if they are significantly larger than the Planck scale, because they might provide an explanation for the puzzling weakness of gravitation with respect to other forces. Roughly speaking gravitation could be weak because in string theory it propagates very naturally into such extra dimensions in contrast with other interactions which may remain stuck to our ordinary four-dimensional universe or "brane".

But there are several difficulties with string theory which cast some doubt on its usefulness to guide us into the first scales of *terra incognita*. First the theory is really a very bold stroke to quantize gravity at the Planck scale, very far from current observations. This giant leap runs directly against the step by step philosophy of the RG. Second the mathematical structure of string theory is complicated up to the point where it may become depressing. For instance great effort is needed to put the string theory at two loops on some rigorous footing [31], and three loops seem almost hopeless. Third, there was for some time the hope that string theory and the phenomenology at lower energies derived from it might be unique. This hope has now vanished with the discovery of a very complicated *landscape* of different possible string vacua and associated long distance phenomenologies.

In view of these difficulties some physicists have started to openly criticize what they consider a disproportionate amount of intellectual resources devoted to the study of string theory compared to other alternatives [32].

I do not share these critics. I think in particular that string theory has been very successful as a brain storming tool. It has lead already to many spectacular insights into pure mathematics and geometry. But my personal bet would be that if somewhere in the mountains near the Planck scale string theory might be useful, or even correct, we should also search for other complementary and more reliable principles to guide us in the maze of waterways at the entrance of *terra incognita*. If these other complementary principles turn out to be compatible with string theory at higher scales, so much the better.

It is a rather natural remark that since gravity alters the very geometry of ordinary space, any quantum theory of gravity should quantize ordinary space, not just the phase space of mechanics, as quantum mechanics does. Hence at some point at or before the Planck scale we should expect the algebra of ordinary co-ordinates or observables to be generalized to a non-commutative algebra. Alain Connes, Michel Dubois-Violette, Ali Chamseddine and others have forcefully advocated that the *classical* Lagrangian of the current standard model arises much more naturally on simple non-commutative geometries than on ordinary commutative Minkowsky space. We refer to Alain's contribution here for these arguments. They remain in the line of Einstein's classical unification of Maxwell's electrodynamics equations through the introduction of a new four-dimensional space-time.

---

[7]The three couplings join better at a single very high scale if supersymmetry is included in the picture. Of course sceptics can remark that this argument requires to continue these flows deep within *terra incognita*, where new physics could occur.

The next logical step seems to be to find the analog of quantum electrodynamics. It should be quantum field theory on non-commutative geometry, or NCQFT. The idea of NCQFT goes back at least to Snyders [33].

A second line of argument ends at the same conclusion. String theorists realized in the late 1990s that NCQFT is an effective theory of strings [34,35]. Roughly this is because in addition to the symmetric tensor $g_{\mu\nu}$ the spectrum of the closed string also contains an antisymmetric tensor $B_{\mu\nu}$. There is no reason for this antisymmetric tensor not to freeze at some lower scale into a classical field, just as $g_{\mu\nu}$ is supposed to freeze into the classical metric of Einstein's general relativity. But such a freeze of $B_{\mu\nu}$ precisely induces an effective non-commutative geometry. In the simplest case of flat Riemannian metric and trivial constant antisymmetric tensor, the geometry is simply of the Moyal type; it reduces to a constant anticommutator between (Euclidean) space-time coordinates. This made NCQFT popular among string theorists. A good review of these ideas can be found in [36].

These two lines of arguments, starting at both ends of *terra incognita* converge to the same conclusion: there should be an intermediate regime between QFT and string theory where NCQFT is the right formalism. The breaking of locality and the appearance of cyclic-symmetric rather than fully symmetric interactions in NCQFT is fully consistent with this intermediate status of NCQFT between fields and strings. The ribbon graphs of NCQFT may be interpreted either as "thicker particle world-lines" or as "simplified open strings world-sheets" in which only the ends of strings appear but not yet their internal oscillations. However until recently a big stumbling block remained. The simplest NCQFT on Moyal space, such as $\phi_4^{\star 4}$, were found not to be renormalizable because of a surprising phenomenon called *uv/ir mixing*. Roughly speaking this $\phi_4^{\star 4}$ theory still has infinitely many ultraviolet divergent graphs but fewer than the ordinary ϕ_4^4 theory. The new "ultraviolet convergent" graphs, such as the non-planar tadpole

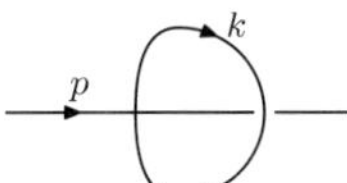

generate completely unexpected infrared divergences which are not of the renormalizable type [37].

However few years ago the solution out of this riddle was found. H. Grosse and R. Wulkenhaar in a brilliant series of papers discovered how to renormalize $\phi_4^{\star 4}$ [38–40]. This "revolution" happened quietly without mediatic fanfare, but it might turn out to develop into a good Ariane's thread at the entrance of the maze. Indeed remember the argument of Wilson: renormalizable theories are the building blocks of physics because they are the ones who survive RG flows ...

It is always very interesting to develop a new brand of RG, but *that* new brand on non-commutative Moyal space is especially exciting. Indeed it changes the very definition of scales in a new and nontrivial way. Therefore it may ultimately change our view of locality and causality, hence our very view of the deterministic relationship from small to large distances. It is fair to say that the same is true of string theory, where T-dualities also change small into large distances and vice-versa. But in contrast with string theory, this new brand of NCQFT is mathematically tractable, not at one or two loops, but as we shall see below, at any number of loops and probably even non-perturbatively! This just means that we can do complicated computations in these NCQFT's with much more ease and confidence than in string theory.

The goal of these lectures is to present this new set of burgeoning ideas.

We start with a blitz introduction to standard renormalization group concepts in QFT: functional integration and Feynman graphs. The system of Feynman graphs of the ϕ_4^4 theory provides the simplest example to play and experiment with the idea of renormalization. It is straightforward to analyze the basic scaling behavior of high energy subgraphs within graphs of lower energy. In this way one discovers relatively easily the most important physical effect under change of the observation scale, namely the flow of the coupling constant. It leads immediately to the fundamental difficulty associated to the short distance behavior of the theory, namely the Landau ghost or triviality problem. That ghost disappears in the "asymptotically free" non-abelian gauge theories [11, 12]. With hindsight this result might perhaps be viewed in a not so distant future as the first glimpse of NCQFT ...

Grosse and Wulkenhaar realized that previous studies of NCQFT had used the wrong propagator! Moyal interactions were noticed to obey a certain Langmann-Szabo duality [41], which exchanges space and momentum variables. Grosse and Wulkenhaar realized that the propagator should be modified to also respect this symmetry [40]. This means that NCQFT on Moyal spaces has to be based on the Mehler kernel, which governs propagation in a harmonic potential, rather than on the heat kernel, which governs ordinary propagation in commutative space. Grosse and Wulkenhaar were able to compute for the first time the Mehler kernel in the *matrix base* which transforms the Moyal product into a matrix product. This is a real *tour de force*! The matrix based Mehler kernel is quasi-diagonal, and they were able to use their computation to prove perturbative renormalizability of the theory, up to some estimates which were finally proven in [42].

By matching correctly propagator and interaction to respect symmetries, Grosse and Wulkenhaar were following one of the main successful thread of quantum field theory. Their renormalizability result is in the direct footsteps of 't Hooft and Veltman, who did the same for non-abelian gauge theories thirty years before. However I have often heard two main critics raised, which I would like to answer here.

The first critic is that it is no wonder that adding a harmonic potential gets rid of the infrared problem. It is naive because the harmonic potential is the only partner of the Laplacian under LS duality. No other infrared regulator would make the theory renormalizable. The theory has infinitely many degrees of freedom, and infinitely many divergent graphs, so the new BPHZ theorem obtained by Grosse and Wulkenhaar is completely nontrivial. In fact now that the RG flow corresponding to these theories is better understood, we understand the former uv/ir mixing just as an ordinary anomaly which signaled a missing marginal term in the Lagrangian under that RG flow.

The second and most serious critic is that since the Mehler kernel is not translation invariant, the Grosse and Wulkenhaar ideas will never be able to describe any mainstream physics in which there should be no preferred origin. This is just *wrong* but for a more subtle reason. We have shown that the Grosse-Wulkenhaar method can be extended to renormalize theories such as the Langmann-Szabo-Zarembo $\bar{\phi} \star \phi \star \bar{\phi} \star \phi$ model [43–45] in four dimensions or the Gross-Neveu model in two dimensions. In these theories the ordinary Mehler kernel is replaced by a related kernel which governs propagation of charged particles in a constant background field. This kernel, which we now propose to call the *covariant Mehler kernel*[8], is still not translation invariant because it depends on non translation-invariant gauge choice. It oscillates rather than decays when particles move away from a preferred origin. But in such theories physical observables, which are gauge invariant, do not feel that preferred origin. That's why translation invariant phenomena can be described!

We proposed to call the whole new class of NCQFT theories built either on the Mehler kernel or on its covariant generalizations *vulcanized* (may be we should have spelled Wulkenized?) because renormalizability means that their structure resists under change of scale[9].

These newly discovered vulcanized theories or NCVQFT and their associated RG flows absolutely deserve a thorough and systematic investigation, not only because they may be relevant for physics beyond the standard model, but also (although this is often less emphasized) because they may provide explanation of nontrivial effective physics in our ordinary standard world whenever strong background gauge fields are present. Many examples come to mind, from various aspects of the quantum Hall effect to the behavior of two-dimensional charged polymers under magnetic fields or even to quark confinement. In such cases appropriate generalizations of the vulcanized RG may be the right tool to show how the correct effective nonlocal interactions emerge out of local interactions.

At the laboratoire de physique théorique at Orsay we have embarked on such a systematic investigation of NCVQFTs and of their RG flows. This program is also actively pursued elsewhere. Let us review briefly the main recent results and open problems.

Multiscale analysis. The initial Grosse-Wulkenhaar breakthrough used sharp cutoffs in matrix space, which like sharp cutoffs in ordinary direct and momentum space are not so well suited to rigorous bounds and multiscale analysis. By replacing these cutoffs by smoother cutoffs which cut directly the Mehler parameter

---

[8]Initially we called such NCQFT theories *critical*, but it was pointed to us that this word may create confusion with critical phenomena, so we suggest now to call them *covariant*.

[9]Vulcanization is a technological operation which adds sulphur to natural rubber to improve its mechanical properties and its resistance to temperature change, and temperature is a scale in imaginary time...

into slices, we could derive rigorously the estimates that were only numerically checked in [40] hence close the last gaps in the BPHZ theorem for vulcanized non-commutative $\phi_4^{\star 4}$ [42]. We could also replace the somewhat cumbersome recursive use of the Polchinski equation [46] by more direct and explicit bounds in a multiscale analysis.

Direct space. Although non translation invariant propagators and nonlocal vertices are unfamiliar, the direct space representation of NCVQFT remains closer to our ordinary intuition than the matrix base. Using direct space methods, we have provided a new proof of the BPHZ theorem for vulcanized non-commutative $\phi_4^{\star 4}$ [47]. We have also extended the Grosse-Wulkenhaar results to the $\bar\phi \star \phi \star \bar\phi \star \phi$ LSZ model [43]. Our proof relies on a multiscale analysis analogous to [42] but in direct space. It allows a more transparent understanding of the *moyality* of the counterterms for *planar* subgraphs at higher scales when seen through external propagators at lower scales. This is the exact analog of the *locality* in ordinary QFT of general subgraphs at higher scales when seen through external propagators at lower scales. Such propagators do not distinguish short distance details, and ordinary locality could be summarized as the obvious remark that from far enough away any object looks roughly like a point. But *moyality* could be summarized as a more surprising fact: viewed from lower RG scales[10], planar higher scale effects, which are the only ones large enough to require renormalization, look like Moyal products.

Fermionic theories. To enlarge the class of renormalizable non-commutative field theories and to attack the quantum Hall effect problem it is essential to extend the results of Grosse-Wulkenhaar to fermionic theories. Vulcanized fermionic propagators have been computed and their scaling properties established, both in matrix base and direct space, in [50]. They seem to be necessarily of the covariant type.

The simplest fermionic NCVQFT theory, corresponding to the two-dimensional ordinary Gross-Neveu model, was then proved renormalizable to all orders in [51]. This was done using the x-space version which seems also the most promising for a complete non-perturbative construction, using Pauli's principle to control the apparent (fake) divergences of perturbation theory.

Ghost Hunting. Grosse and Wulkenhaar made the first nontrivial one loop RG computation in NCVQFT in [52]. Although they did not word it initially in this way, their result means that at this order there is no Landau ghost in NCVQFT! A nontrivial fixed point of the renormalization group develops at high energy, where the Grosse-Wulkenhaar parameter Ω tends to the *self-dual point* $\Omega = 1$, so that Langmann-Szabo duality become exact, and the beta function vanishes. This

---

[10]These scales being defined in the new RG sense, we no longer say "from far away". Although I hate to criticize, I feel a duty here to warn the reader that often cited previous "proofs of moyality" such as [48, 49] should be dismissed. The main theorem in [48], whose proof never appeared, is simply wrong; and even more importantly the analysis in [49] does not lead to any BPHZ theorem nor to any sensible RG flow. This is because using the old definition of RG scales it misses vulcanization.

stops the growth of the bare coupling constant in the ultraviolet regime, hence kills the ghost. So after all NCVQFT is not only as good as QFT with respect to renormalization, it is definitely better! This vindicates, although in a totally unexpected way, the initial intuition of Snyders [33], who like many after him was at least partly motivated by the hope to escape the divergences in QFT which were considered ugly. Remark however that the ghost is not killed because of asymptotic freedom. Both the bare and the renormalized coupling are nonzero. They can be made both small if the renormalized Ω is not too small, in which case perturbation theory is expected to remain valid all along the complete RG trajectory. It is only in the singular limit $\Omega_{\mathrm{ren}} \to 0$ that the ghost begins to reappear.

For mathematical physicists who like me came from the constructive field theory program, the Landau ghost has always been a big frustration. Remember that because non-abelian gauge theories are very complicated and lead to confinement in the infrared regime, there is no good four-dimensional rigorous field theory without unnatural cutoffs up to now[11]. I was therefore from the start very excited by the possibility to build non-perturbatively the $\phi_4^{\star 4}$ theory as the first such rigorous four-dimensional field theory without unnatural cutoffs, even if it lives on the Moyal space which is not the one initially expected, and does not obey the usual axioms of ordinary QFT.

For that happy scenario to happen, two main nontrivial steps are needed. The first one is to extend the vanishing of the beta function at the self-dual point $\Omega = 1$ to all orders of perturbation theory. This has been done in [57,58], using the matrix version of the theory. First the result was checked by brute force computation at two and three loops. Then we devised a general method for all orders. It relies on Ward identities inspired by those of similar theories with quartic interactions in which the beta function vanishes [59–61]. However the relation of these Ward identities to the underlying LS symmetry remains unclear and we would also like to develop again an x-space version of that result to understand better its relation to the LS symmetry.

The second step is to extend in a proper way constructive methods such as cluster and Mayer expansions to build non-perturbatively the connected functions of NCVQFT in a single RG slice. Typically we would like a theorem of Borel summability [62] in the coupling constant for these functions which has to be *uniform* in the slice index. This is in progress. A construction of the model and of its full RG trajectory would then presumably follow from a multiscale analysis similar to that of [63].

$\phi_6^{\star 3}$ **and Kontsevich model.** The non-commutative $\phi^{\star 3}$ model in 6 dimensions has been shown to be renormalizable, asymptotically free, and solvable genus by genus by mapping it to the Kontsevich model, in [64–66]. The running coupling constant has also been computed exactly, and found to decrease more rapidly than predicted

---

[11]We have only renormalizable constructive theories for two-dimensional fermionic theories [53, 54] and for the infrared side of ϕ_4^4 [55,56].

by the one-loop beta function. That model however is not expected to have a non-perturbative definition because it should be unstable at large ϕ.

Gauge theories. A very important and difficult goal is to properly vulcanize gauge theories such as Yang-Mills in four-dimensional Moyal space or Chern-Simons on the two-dimensional Moyal plane plus one additional ordinary commutative time direction. We do not need to look at complicated gauge groups since the $U(1)$ pure gauge theory is nontrivial and interacting on non-commutative geometry even without matter fields. What is not obvious is to find a proper compromise between gauge and Langmann-Szabo symmetries which still has a well-defined perturbation theory around a computable vacuum after gauge invariance has been fixed through appropriate Faddeev-Popov or BRS procedures. We should judge success in my opinion by one main criterion, namely renormalizability. Recently de Goursac, Wallet and Wulkenhaar computed the non-commutative action for gauge fields which can be induced through integration of a scalar renormalizable matter field minimally coupled to the gauge field [67]; the result exhibits both gauge symmetry and LS covariance, hence vulcanization, but the vacuum looks nontrivial so that to check whether the associated perturbative expansion is really renormalizable seems difficult.

Dimensional regularization and renormalization better respect gauge symmetries and they were the key to the initial 't Hooft-Veltman proof of renormalizability of ordinary gauge theories. Therefore no matter what the final word will be on NCV gauge theories, it should be useful to have the corresponding tools ready at hand in the non-commutative context[12]. This requires several steps, the first of which is

Parametric representation. In this compact representation, direct space or momentum variables have been integrated out for each Feynman amplitude. The result is expressed as integrals over the heat kernel parameters of each propagator, and the integrands are the topological polynomials of the graph[13]. These integrals can then be shown analytic in the dimension D of space-time for $\Re D$ small enough. They are in fact meromorphic in the complex plane, and ultraviolet divergences can be extracted through appropriate inductive contour integrations.

The same program can be accomplished in NCVQFT because the Mehler kernel is still quadratic in space variables[14]. The corresponding topological hyperbolic polynomials are richer than in ordinary field theory since they are invariants of the ribbon graph which for instance contain information about the genus of the surface on which these graphs live. They can be computed both for ordinary

---

[12]The Connes-Kreimer works also use abundantly dimensional regularization and renormalization, and this is another motivation.

[13]Mathematicians call these polynomials Kirchhoff polynomials, and physicist call them Symanzik polynomials in the quantum field theory context.

[14]This is true provided "hypermomenta" are introduced to Fourier transform the space conservation at vertices which in Moyal space is the LS dual to ordinary momentum conservation.

NCVQFT [68] and in the more difficult case of covariant theories such as the LSZ model [69].

Dimensional regularization and renormalization. From the parametric representation the corresponding regularization and minimal-dimensional renormalization scheme should follow for NCVQFTs. However appropriate factorization of the leading terms of the new hyperbolic polynomials under rescaling of the parameters of any subgraph is required. This is indeed the analog in the parameter representation of the "moyality" of the counterterms in direct space. This program is under way [70].

Quantum Hall effect. NCQFT and in particular the non-commutative Chern Simons theory has been recognized as effective theory of the quantum Hall effect already for some time [71–73]. We also refer to the contributions of V. Pasquier and of A. Polychronakos in this volume. But the discovery of the vulcanized RG holds promises for a better explanation of how these effective actions are generated from the microscopic level.

In this case there is an interesting reversal of the initial Grosse-Wulkenhaar problematic. In the $\phi_4^{\star 4}$ theory the vertex is given a priori by the Moyal structure, and it is LS invariant. The challenge was to find the right propagator which makes the theory renormalizable, and it turned out to have LS duality.

Now to explain the (fractional) quantum Hall effect, which is a bulk effect whose understanding requires electron interactions, we can almost invert this logic. The propagator is known since it corresponds to non-relativistic electrons in two dimensions in a constant magnetic field. It has LS duality. But the effective theory should be anionic hence not local. Here again we can argue that among all possible nonlocal interactions, a few renormalization group steps should select the only ones which form a renormalizable theory with the corresponding propagator. In the commutative case (i.e., zero magnetic field) local interactions such as those of the Hubbard model are just renormalizable in any dimension because of the extended nature of the Fermi-surface singularity. Since the non-commutative electron propagator (i.e., in nonzero magnetic field) looks very similar to the Grosse-Wulkenhaar propagator (it is in fact a generalization of the Langmann-Szabo-Zarembo propagator) we can conjecture that the renormalizable interaction corresponding to this propagator should be given by a Moyal product. That's why we hope that non-commutative field theory and a suitable generalization of the Grosse-Wulkenhaar RG might be the correct framework for a microscopic *ab initio* understanding of the fractional quantum Hall effect which is currently lacking.

Charged polymers in magnetic fields. Just like the heat kernel governs random motion, the covariant Mehler kernel governs random motion of charged particles in presence of a magnetic field. Ordinary polymers can be studied as random walk with a local self-repelling or self-avoiding interaction. They can be treated by QFT techniques using the $N = 0$ component limit or the supersymmetry trick to erase the unwanted vacuum graphs. Many results, such as various exact critical exponents in two dimensions, approximate ones in three dimensions, and infrared asymptotic freedom in four dimensions have been computed for self-avoiding polymers through renormalization group techniques. In the same way we expect that charged polymers under magnetic field should be studied through the new non-commutative vulcanized RG. The relevant interactions again should be of the Moyal rather than of the local type, and there is no reason that the replica trick could not be extended in this context. Ordinary observables such as N point functions would be only translation *covariant*, but translation invariant physical observables such as density-density correlations should be recovered out of gauge invariant observables. In this way it might be possible to deduce new scaling properties of these systems and their exact critical exponents through the generalizations of the techniques used in the ordinary commutative case [74].

More generally we hope that the conformal invariant two-dimensional theories, the RG flows between them and the c theorem of Zamolodchikov [27] should have appropriate *magnetic generalizations* which should involve vulcanized flows and Moyal interactions.

Quark confinement. It is less clear that NCVQFT gauge theories might shed light on confinement, but this is also possible.

Even for regular commutative field theory such as non-abelian gauge theory, the strong coupling or non-perturbative regimes may be studied fruitfully through their non-commutative (i.e., nonlocal) counterparts. This point of view is forcefully suggested in [35], where a mapping is proposed between ordinary and non-commutative gauge fields which do not preserve the gauge groups but preserve the gauge equivalent classes. Let us further remark that the effective physics of confinement should be governed by a nonlocal interaction, as is the case in effective strings or bags models. The great advantage of NCVQFT over the initial matrix model approach of 't Hooft [75] is that in the latter the planar graphs dominate because a gauge group $SU(N)$ with N large is introduced in an *ad hoc* way instead of the physical $SU(2)$ or $SU(3)$, whether in the former case, there is potentially a perfectly physical explanation for the planar limit, since it should just emerge naturally out of a renormalization group effect. We would like the large N *matrix* limit in NCVQFT's to parallel the large N *vector* limit which allows to understand the formation of Cooper pairs in supraconductivity [26]. In that case N is not arbitrary but is roughly the number of effective quasi particles or sectors around the extended Fermi surface singularity at the superconducting transition temperature. This number is automatically very large if this temperature is very low. This is why we called this phenomenon a *dynamical* large N *vector* limit. NCVQFTs provides us with the first clear example of a *dynamical* large N *matrix* limit. We hope therefore that it should be ultimately useful to understand bound states in ordinary commutative non-abelian gauge theories, hence quark confinement.

Quantum gravity. Although ordinary renormalizable QFTs seem more or less to have NCVQFT analogs on the Moyal space, there is no renormalizable commutative field theory for spin 2 particles, so that the NCVQFTs alone should not allow quantization of gravity. However quantum gravity might enter the picture of NCVQFTs at a later and more advanced stage. Since quantum gravity appears in closed strings, it may have something to do with doubling the ribbons of some NCQFT in an appropriate way. But because there is no reason not to quantize the antisymmetric tensor B which defines the non-commutative geometry as well as the symmetric one g which defines the metric, we should clearly no longer limit ourselves to Moyal spaces. A first step towards a non-commutative approach to quantum gravity along these lines should be to search for the proper analog of vulcanization in more general non-commutative geometries. It might for instance describe physics in the vicinity of a charged rotating black hole generating a strong magnetic field. However we have to admit that any theory of quantum gravity will probably remain highly conjectural for many decades or even centuries ...

I would like to warmly thank all the collaborators who contributed in various ways to the elaboration of this material, in particular M. Disertori, R. Gurau, J. Magnen, A. Tanasa, F. Vignes-Tourneret, J.C. Wallet and R. Wulkenhaar. Special thanks are due to F. Vignes-Tourneret since this review is largely based on our common recent review [76], with introduction and sections added on commutative renormalization, ghost hunting and the parametric representation. I would like also to sincerely apologize to the many people whose work in this area would be worth of citation but has not been cited here: this is because of my lack of time or competence but not out of bad will.

2. Commutative renormalization, a blitz review

This section is a summary of [77] which we include for self-containedness.

2.1. Functional integral

In QFT, particle number is not conserved. Cross sections in scattering experiments contain the physical information of the theory. They are the matrix elements of the diffusion matrix $\mathcal{S}$. Under suitable conditions they are expressed in terms of the Green functions G_N of the theory through so-called "reduction formulae".

Green's functions are time ordered vacuum expectation values of the field ϕ, which is operator valued and acts on the Fock space:

$$G_N(z_1, \ldots, z_N) = \langle \psi_0, T[\phi(z_1) \cdots \phi(z_N)]\psi_0 \rangle . \tag{2.1}$$

Here ψ_0 is the vacuum state and the T-product orders $\phi(z_1) \cdots \phi(z_N)$ according to times.

Consider a Lagrangian field theory, and split the total Lagrangian as the sum of a free plus an interacting piece, $\mathcal{L} = \mathcal{L}_0 + \mathcal{L}_{\text{int}}$. The Gell-Mann-Low formula expresses the Green functions as vacuum expectation values of a similar product of free fields with an $e^{i\mathcal{L}_{\text{int}}}$ insertion:

$$G_N(z_1, \ldots, z_N) = \frac{\left\langle \psi_0, T\left[\phi(z_1)\cdots\phi(z_N)e^{i\int dx\mathcal{L}_{\text{int}}(\phi(x))}\right]\psi_0\right\rangle}{\left\langle \psi_0, T(e^{i\int dx\mathcal{L}_{\text{int}}(\phi(x))})\psi_0\right\rangle}. \tag{2.2}$$

In the functional integral formalism proposed by Feynman [78], the Gell-Mann-Low formula is replaced by a functional integral in terms of an (ill-defined) "integral over histories" which is formally the product of Lebesgue measures over all space-time. The corresponding formula is the Feynman-Kac formula:

$$G_N(z_1, \ldots, z_N) = \frac{\int \prod_j \phi(z_j)e^{i\int \mathcal{L}(\phi(x))dx}D\phi}{\int e^{i\int \mathcal{L}(\phi(x))dx}D\phi}. \tag{2.3}$$

The integrand in (2.3) contains now the full Lagrangian $\mathcal{L} = \mathcal{L}_0 + \mathcal{L}_{int}$ instead of the interacting one. This is interesting to expose symmetries of the theory which may not be separate symmetries of the free and interacting Lagrangians, for instance gauge symmetries. Perturbation theory and the Feynman rules can still be derived as explained in the next subsection. But (2.3) is also well adapted to constrained quantization and to the study of non-perturbative effects. Finally there is a deep analogy between the Feynman-Kac formula and the formula which expresses correlation functions in classical statistical mechanics. For instance, the correlation functions for a lattice Ising model are given by

$$\left\langle \prod_{i=1}^{n} \sigma_{x_i}\right\rangle = \frac{\sum\limits_{\{\sigma_x=\pm 1\}} e^{-L(\sigma)} \prod_i \sigma_{x_i}}{\sum\limits_{\{\sigma_x=\pm 1\}} e^{-L(\sigma)}}, \tag{2.4}$$

where x labels the discrete sites of the lattice, the sum is over configurations $\{\sigma_x = \pm 1\}$ which associate a "spin" with value $+1$ or -1 to each such site and $L(\sigma)$ contains usually nearest neighbor interactions and possibly a magnetic field h:

$$L(\sigma) = \sum_{x,y \text{ nearest neighbors}} J\sigma_x\sigma_y + \sum_x h\sigma_x. \tag{2.5}$$

By analytically continuing (2.3) to imaginary time, or Euclidean space, it is possible to complete the analogy with (2.4), hence to establish a firm contact with statistical mechanics [15, 79, 80].

This idea also allows us to give much better meaning to the path integral, at least for a free bosonic field. Indeed the free Euclidean measure can be defined easily as a Gaussian measure, because in Euclidean space L_0 is a quadratic form of positive type[15].

---

[15]However the functional space that supports this measure is not in general a space of smooth functions, but rather of distributions. This was already true for functional integrals such as those of Brownian motion, which are supported by continuous but not differentiable paths. Therefore "functional integrals" in quantum field theory should more appropriately be called "distributional integrals".

The Green functions continued to Euclidean points are called the Schwinger functions of the model, and are given by the Euclidean Feynman-Kac formula:

$$S_N(z_1, \ldots, z_N) = Z^{-1} \int \prod_{j=1}^{N} \phi(z_j) e^{-\int \mathcal{L}(\phi(x)) dx} D\phi \tag{2.6}$$

$$Z = \int e^{-\int \mathcal{L}(\phi(x)) dx} D\phi. \tag{2.7}$$

The simplest interacting field theory is the theory of a one component scalar bosonic field ϕ with quartic interaction $\lambda \phi^4$ (ϕ^3 which is simpler is unstable). In $\mathbb{R}^d$ it is called the ϕ_d^4 model. For $d = 2, 3$ the model is super-renormalizable and has been built non-perturbatively by constructive field theory. For $d = 4$ it is just renormalizable, and it provides the simplest pedagogical introduction to perturbative renormalization theory. But because of the Landau ghost or triviality problem explained in Subsection 2.5, the model presumably does not exist as a true interacting theory at the non-perturbative level. Its non-commutative version should exist on the Moyal plane, see Section 5.

Formally the Schwinger functions of ϕ_d^4 are the moments of the measure:

$$d\nu = \frac{1}{Z} e^{-\frac{\lambda}{4!} \int \phi^4 - (m^2/2) \int \phi^2 - (a/2) \int (\partial_\mu \phi \partial^\mu \phi)} D\phi, \tag{2.8}$$

where

- λ is the coupling constant, usually assumed positive or complex with positive real part; remark the convenient $1/4!$ factor to take into account the symmetry of permutation of all fields at a local vertex. In the non-commutative version of the theory permutation symmetry becomes the more restricted cyclic symmetry and it is convenient to change the $1/4!$ factor to $1/4$.
- m is the mass, which fixes an energy scale for the theory.
- a is the wave function constant. It can be set to 1 by a rescaling of the field.
- Z is a normalization factor which makes (2.8) a probability measure.
- $D\phi$ is a formal (mathematically ill-defined) product $\prod_{x \in \mathbb{R}^d} d\phi(x)$ of Lebesgue measures at every point of $\mathbb{R}^d$.

The Gaussian part of the measure is

$$d\mu(\phi) = \frac{1}{Z_0} e^{-(m^2/2) \int \phi^2 - (a/2) \int (\partial_\mu \phi \partial^\mu \phi)} D\phi. \tag{2.9}$$

where Z_0 is again the normalization factor which makes (2.9) a probability measure.

More precisely if we consider the translation invariant propagator $C(x, y) \equiv C(x - y)$ (with slight abuse of notation), whose Fourier transform is

$$C(p) = \frac{1}{(2\pi)^d} \frac{1}{p^2 + m^2}, \tag{2.10}$$

we can use Minlos theorem and the general theory of Gaussian processes to define $d\mu(\phi)$ as the centered Gaussian measure on the Schwartz space of tempered distributions $S'(\mathbb{R}^d)$ whose covariance is C. A Gaussian measure is uniquely defined by its moments, or the integral of polynomials of fields. Explicitly this integral is zero for a monomial of odd degree, and for $n = 2p$ even it is equal to

$$\int \phi(x_1) \cdots \phi(x_n) d\mu(\phi) = \sum_{\gamma} \prod_{\ell \in \gamma} C(x_{i_\ell}, x_{j_\ell}), \qquad (2.11)$$

where the sum runs over all the $2p!! = (2p-1)(2p-3)\cdots 5 \cdot 3 \cdot 1$ pairings γ of the $2p$ arguments into p disjoint pairs $\ell = (i_\ell, j_\ell)$.

Note that since for $d \geq 2$, $C(p)$ is not integrable, $C(x, y)$ must be understood as a distribution. It is therefore convenient to also use regularized kernels, for instance

$$C_\kappa(p) = \frac{1}{(2\pi)^d} \frac{e^{-\kappa(p^2+m^2)}}{p^2+m^2} = \int_\kappa^\infty e^{-\alpha(p^2+m^2)} d\alpha \qquad (2.12)$$

whose Fourier transform $C_\kappa(x, y)$ is now a smooth function and not a distribution:

$$C_\kappa(x, y) = \int_\kappa^\infty e^{-\alpha m^2 - (x-y)^2/4\alpha} \frac{d\alpha}{\alpha^{D/2}}. \qquad (2.13)$$

$\alpha^{-D/2} e^{-(x-y)^2/4\alpha}$ is the *heat kernel* and therefore this α-representation has also an interpretation in terms of Brownian motion:

$$C_\kappa(x, y) = \int_\kappa^\infty d\alpha \exp(-m^2\alpha) \, P(x, y; \alpha) \qquad (2.14)$$

where $P(x, y; \alpha) = (4\pi\alpha)^{-d/2} \exp(-|x-y|^2/4\alpha)$ is the Gaussian probability distribution of a Brownian path going from x to y in time α.

Such a regulator κ is called an ultraviolet cutoff, and we have (in the distribution sense) $\lim_{\kappa \to 0} C_\kappa(x, y) = C(x, y)$. Remark that due to the nonzero m^2 mass term, the kernel $C_\kappa(x, y)$ decays exponentially at large $|x - y|$ with rate m. For some constant K and $d > 2$ we have:

$$|C_\kappa(x, y)| \leq K \kappa^{1-d/2} e^{-m|x-y|}. \qquad (2.15)$$

It is a standard useful construction to build from the Schwinger functions the connected Schwinger functions, given by:

$$C_N(z_1, \ldots, z_N) = \sum_{\substack{P_1 \cup \cdots \cup P_k = \{1, \ldots, N\} \\ P_i \cap P_j = 0}} (-1)^{k+1}(k-1)! \prod_{i=1}^k S_{p_i}(z_{j_1}, \ldots, z_{j_{p_i}}), \qquad (2.16)$$

where the sum is performed over all distinct partitions of $\{1, \ldots, N\}$ into k subsets $P_1, \ldots, P_k$, P_i being made of p_i elements called $j_1, \ldots, j_{p_i}$. For instance in the ϕ^4 theory, where all odd Schwinger functions vanish due to the unbroken $\phi \to -\phi$

38 V. Rivasseau

symmetry, the connected 4-point function is simply:

$$C_4(z_1,\ldots,z_4) = S_4(z_1,\ldots,z_4) - S_2(z_1,z_2)S_2(z_3,z_4) \qquad (2.17)$$
$$- S_2(z_1,z_3)S_2(z_2,z_4) - S_2(z_1,z_4)S_2(z_2,z_3).$$

2.2. Feynman rules

The full interacting measure may now be defined as the multiplication of the Gaussian measure $d\mu(\phi)$ by the interaction factor:

$$d\nu = \frac{1}{Z}e^{-\frac{\lambda}{4!}\int \phi^4(x)dx}d\mu(\phi) \qquad (2.18)$$

and the Schwinger functions are the normalized moments of this measure:

$$S_N(z_1,\ldots,z_N) = \int \phi(z_1)\cdots\phi(z_N)d\nu(\phi). \qquad (2.19)$$

Expanding the exponential as a power series in the coupling constant λ, one obtains a formal expansion for the Schwinger functions:

$$S_N(z_1,\ldots,z_N) = \frac{1}{Z}\sum_{n=0}^{\infty}\frac{(-\lambda)^n}{n!}\int\Big[\int\frac{\phi^4(x)dx}{4!}\Big]^n\phi(z_1)\cdots\phi(z_N)d\mu(\phi) \qquad (2.20)$$

It is now possible to perform explicitly the functional integral of the corresponding polynomial. The result gives at any order n a sum over $(4n + N - 1)!!$ "Wick contractions schemes $\mathcal{W}$", i.e., ways of pairing together $4n+N$ fields into $2n+N/2$ pairs. At order n the result of this perturbation scheme is therefore simply the sum over all these schemes $\mathcal{W}$ of the spatial integrals over $x_1,\ldots,x_n$ of the integrand $\prod_{\ell\in\mathcal{W}} C(x_{i_\ell},x_{j_\ell})$ times the factor $\frac{1}{n!}(\frac{-\lambda}{4!})^n$. These integrals are then functions (in fact distributions) of the external positions $z_1,\ldots,z_N$. But they may diverge either because they are integrals over all of $\mathbb{R}^4$ (no volume cutoff) or because of the singularities in the propagator C at coinciding points.

Labeling the n dummy integration variables in (2.20) as $x_1,\ldots,x_n$, we draw a line ℓ for each contraction of two fields. Each position $x_1,\ldots,x_n$ is then associated to a four-legged vertex and each external source z_i to a one-legged vertex, as shown in Figure 1.

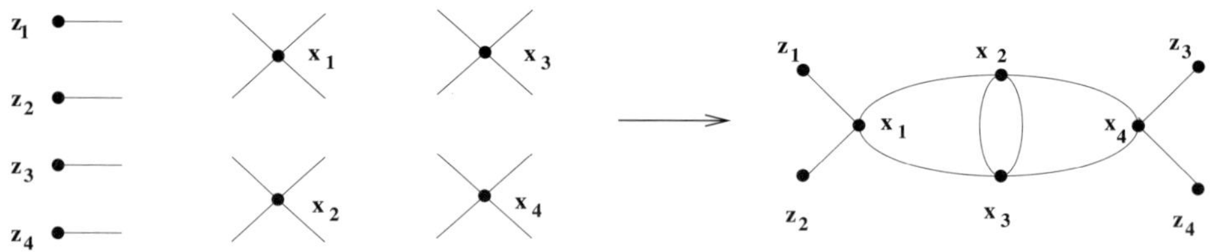

Figure 1: A possible contraction scheme with $n = N = 4$.

For practical computations, it is obviously more convenient to gather all the contractions which lead to the same drawing, hence to the same integral. This leads to the notion of Feynman graphs. To any such graph is associated a contribution or amplitude, which is the sum of the contributions associated with the corresponding set of Wick contractions. The "Feynman rules" summarize how to compute this amplitude with its correct combinatoric factor.

We always use the following notation for a graph G:

- $n(G)$ or simply n is the number of internal vertices of G, or the order of the graph.
- $l(G)$ or l is the number of internal lines of G, i.e., lines hooked at both ends to an internal vertex of G.
- $N(G)$ or N is the number of external vertices of G; it corresponds to the order of the Schwinger function one is looking at. When $N = 0$ the graph is a vacuum graph, otherwise it is called an N-point graph.
- $c(G)$ or c is the number of connected components of G.
- $L(G)$ or L is the number of independent loops of G.

For a *regular* ϕ^4 graph, i.e., a graph which has no line hooked at both ends to external vertices, we have the relations

$$l(G) = 2n(G) - N(G)/2, \tag{2.21}$$

$$L(G) = l(G) - n(G) + c(G) = n(G) + 1 - N(G)/2, \tag{2.22}$$

where in the last equality we assume connectedness of G, hence $c(G) = 1$.

A *subgraph* F of a graph G is a subset of internal lines of G, together with the corresponding attached vertices. Lines in the subset defining F are the internal lines of F, and their number is simply $l(F)$, as before. Similarly all the vertices of G hooked to at least one of these internal lines of F are called the internal vertices of F and considered to be in F; their number by definition is $n(F)$. Finally a good convention is to call external half-line of F every half-line of G which is not in F but which is hooked to a vertex of F; it is then the number of such external half-lines which we call $N(F)$. With these conventions one has for ϕ^4 subgraphs the same relation (2.21) as for regular ϕ^4 graphs.

To compute the amplitude associated to a ϕ^4 graph, we have to add the contributions of the corresponding contraction schemes.

This is summarized by the "Feynman rules":

- To each line ℓ with end vertices at positions x_ℓ and y_ℓ, associate a propagator $C(x_j, y_j)$.
- To each internal vertex, associate $(-\lambda)/4!$.
- Count all the contraction schemes giving this diagram. The number should be of the form $(4!)^n n!/S(G)$ where $S(G)$ is an integer called the symmetry factor of the diagram. The 4! represents the permutation of the fields hooked to an internal vertex.
- Multiply all these factors, divide by $n!$ and sum over the position of all internal vertices.

The formula for the bare amplitude of a graph is therefore, as a distribution in $z_1, \ldots, z_N$:

$$A_G(z_1, \ldots, z_N) \equiv \int \prod_{i=1}^{n} dx_i \prod_{\ell \in G} C(x_\ell, y_\ell). \tag{2.23}$$

This is the "direct" or "x-space" representation of a Feynman integral. As stated above, this integral suffers of possible divergences. But the corresponding quantities with both volume cutoff and ultraviolet cutoff κ are well defined. They are:

$$A_{G,\Lambda}^{\kappa}(z_1, \ldots, z_N) \equiv \int_{\Lambda^n} \prod_{i=1}^{n} dx_i \prod_{\ell \in G} C_\kappa(x_\ell, y_\ell). \tag{2.24}$$

The integrand is indeed bounded and the integration domain is a compact box Λ.

The *unnormalized* Schwinger functions are therefore formally given by the sum over all graphs with the right number of external lines of the corresponding Feynman amplitudes:

$$ZS_N = \sum_{\phi^4 \text{ graphs } G \text{ with } N(G)=N} \frac{(-\lambda)^{n(G)}}{S(G)} A_G . \tag{2.25}$$

Z itself, the normalization, is given by the sum of all vacuum amplitudes:

$$Z = \sum_{\phi^4 \text{ graphs } G \text{ with } N(G)=0} \frac{(-\lambda)^{n(G)}}{S(G)} A_G. \tag{2.26}$$

Let us remark that since the total number of Feynman graphs is $(4n + N)!!$, taking into account Stirling's formula and the symmetry factor $1/n!$ from the exponential we expect perturbation theory at large order to behave as $K^n n!$ for some constant K. Indeed at order n the amplitude of a Feynman graph is a $4n$-dimensional integral. It is reasonable to expect that in average it should behave as c^n for some constant c. But this means that one should expect zero radius of convergence for the series (2.25). This is not too surprising. Even the one-dimensional integral

$$F(g) = \int_{-\infty}^{+\infty} e^{-x^2/2 - \lambda x^4/4!} dx \tag{2.27}$$

is well defined only for $\lambda \geq 0$. We cannot hope infinite-dimensional functional integrals of the same kind to behave better than this one-dimensional integral. In mathematically precise terms, F is not analytic near $\lambda = 0$, but only Borel summable [62]. Borel summability is therefore the best we can hope for the ϕ^4 theory, and it has indeed been proved for the theory in dimensions 2 and 3 [81,82].

From translation invariance, we do not expect $A_{G,\Lambda}^{\kappa}$ to have a limit as $\Lambda \to \infty$ if there are vacuum subgraphs in G. But we can remark that an amplitude factorizes as the product of the amplitudes of its connected components.

With simple combinatoric verification at the level of contraction schemes we can factorize the sum over all vacuum graphs in the expansion of unnormalized

Schwinger functions, hence get for the normalized functions a formula analog to (2.25):

$$S_N = \sum_{\substack{\phi^4 \text{ graphs } G \text{ with } N(G)=N \\ G \text{ without any vacuum subgraph}}} \frac{(-\lambda)^{n(G)}}{S(G)} A_G. \qquad (2.28)$$

Now in (2.28) it is possible to pass to the thermodynamic limit (in the sense of formal power series) because using the exponential decrease of the propagator, each individual graph has a limit at fixed external arguments. There is of course no need to divide by the volume for that because each connected component in (2.28) is tied to at least one external source, and they provide the necessary breaking of translation invariance.

Finally one can find the perturbative expansions for the connected Schwinger functions and the vertex functions. As expected, the connected Schwinger functions are given by sums over connected amplitudes:

$$C_N = \sum_{\phi^4 \text{ connected graphs } G \text{ with } N(G)=N} \frac{(-\lambda)^{n(G)}}{S(G)} A_G \qquad (2.29)$$

and the vertex functions are the sums of the *amputated* amplitudes for proper graphs, also called one-particle-irreducible. They are the graphs which remain connected even after removal of any given internal line. The amputated amplitudes are defined in momentum space by omitting the Fourier transform of the propagators of the external lines. It is therefore convenient to write these amplitudes in the so-called momentum representation:

$$\Gamma_N(z_1,\ldots,z_N) = \sum_{\phi^4 \text{ proper graphs } G \text{ with } N(G)=N} \frac{(-\lambda)^{n(G)}}{S(G)} A_G^T(z_1,\ldots,z_N), \qquad (2.30)$$

$$A_G^T(z_1,\ldots,z_N) \equiv \frac{1}{(2\pi)^{dN/2}} \int dp_1 \ldots dp_N e^{i \sum p_i z_i} A_G(p_1,\ldots,p_N), \qquad (2.31)$$

$$A_G(p_1,\ldots,p_N) = \int \prod_{\ell \text{ internal line of } G} \frac{d^d p_\ell}{p_\ell^2 + m^2} \prod_{v \in G} \delta(\sum_\ell \epsilon_{v,\ell}\, p_\ell). \qquad (2.32)$$

Remark in (2.32) the δ functions which ensure momentum conservation at each internal vertex v; the sum inside is over both internal and external momenta; each internal line is oriented in an arbitrary way and each external line is oriented towards the inside of the graph. The incidence matrix $\epsilon(v,\ell)$ is 1 if the line ℓ arrives at v, -1 if it starts from v and 0 otherwise. Remark also that there is an overall momentum conservation rule $\delta(p_1 + \cdots + p_N)$ hidden in (2.32). The drawback of the momentum representation lies in the necessity for practical computations to eliminate the δ functions by a "momentum routing" prescription, and there is no canonical choice for that. Although this is rarely explicitly explained in the quantum field theory literature, such a choice of a momentum routing is equivalent to the choice of a particular spanning tree of the graph.

2.3. Scale analysis and renormalization

In order to analyze the ultraviolet or short distance limit according to the renormalization group method, we can cut the propagator C into slices C_i so that $C = \sum_{i=0}^{\infty} C_i$. This can be done conveniently within the parametric representation, since α in this representation roughly corresponds to $1/p^2$. So we can define the propagator within a slice as

$$C_0 = \int_1^{\infty} e^{-m^2\alpha - \frac{|x-y|^2}{4\alpha}} \frac{d\alpha}{\alpha^{d/2}} \ , \quad C_i = \int_{M^{-2i}}^{M^{-2(i-1)}} e^{-m^2\alpha - \frac{|x-y|^2}{4\alpha}} \frac{d\alpha}{\alpha^{d/2}} \quad \text{for } i \geq 1. \tag{2.33}$$

where M is a fixed number, for instance 10, or 2, or e (see footnote 1 in the Introduction). We can intuitively imagine C_i as the piece of the field oscillating with Fourier momenta essentially of size M^i. In fact it is easy to prove the bound (for $d > 2$)

$$|C_i(x,y)| \leq K.M^{(d-2)i} e^{-M^i|x-y|} \tag{2.34}$$

where K is some constant.

Now the full propagator with ultraviolet cutoff M^ρ, ρ being a large integer, may be viewed as a sum of slices:

$$C_{\leq \rho} = \sum_{i=0}^{\rho} C_i \tag{2.35}$$

Then the basic renormalization group step is made of two main operations:

- a functional integration;
- the computation of a logarithm.

Indeed decomposing a covariance in a Gaussian process corresponds to a decomposition of the field into independent Gaussian random variables ϕ^i, each distributed with a measure $d\mu_i$ of covariance C_i. Let us introduce

$$\Phi_i = \sum_{j=0}^{i} \phi_j. \tag{2.36}$$

This is the "low-momentum" field for all frequencies lower than i. The RG idea is that starting from scale ρ and performing $\rho - i$ steps, one arrives at an effective action for the remaining field Φ_i. Then, writing $\Phi_i = \phi_i + \Phi_{i-1}$, one splits the field into a "fluctuation" field ϕ_i and a "background" field Φ_{i-1}. The first step, functional integration, is performed solely on the fluctuation field, so it computes

$$Z_{i-1}(\Phi_{i-1}) = \int d\mu_i(\phi_i) e^{-S_i(\phi_i + \Phi_{i-1})}. \tag{2.37}$$

Then the second step rewrites this quantity as the exponential of an effective action, hence simply computes

$$S_{i-1}(\Phi_{i-1}) = -\log[Z_{i-1}(\Phi_{i-1})]. \tag{2.38}$$

Now $Z_{i-1} = e^{-S_{i-1}}$ and one can iterate! The flow from the initial bare action $S = S_\rho$ for the full field to an effective renormalized action S_0 for the last "slowly varying" component ϕ_0 of the field is similar to the flow of a dynamical system. Its evolution is decomposed into a sequence of discrete steps from S_i to S_{i-1}.

This renormalization group strategy can be best understood on the system of Feynman graphs which represent the perturbative expansion of the theory. The first step, functional integration over fluctuation fields, means that we have to consider subgraphs with all their internal lines in higher slices than any of their external lines. The second step, taking the logarithm, means that we have to consider only *connected* such subgraphs. We call such connected subgraphs *quasi-local*. Renormalizability is then a nontrivial result that combines locality and power counting for these quasi-local subgraphs.

Locality simply means that *quasi-local* subgraphs S look *local* when seen through their external lines. Indeed since they are connected and since their internal lines have scale say $\geq i$, all the internal vertices are roughly at distance M^{-i}. But the external lines have scales $\leq i - 1$, which only distinguish details larger than $M^{-(i-1)}$. Therefore they cannot distinguish the internal vertices of S one from the other. Hence quasi-local subgraphs look like "fat dots" when seen through their external lines, see Figure 2. Obviously this locality principle is completely independent of dimension.

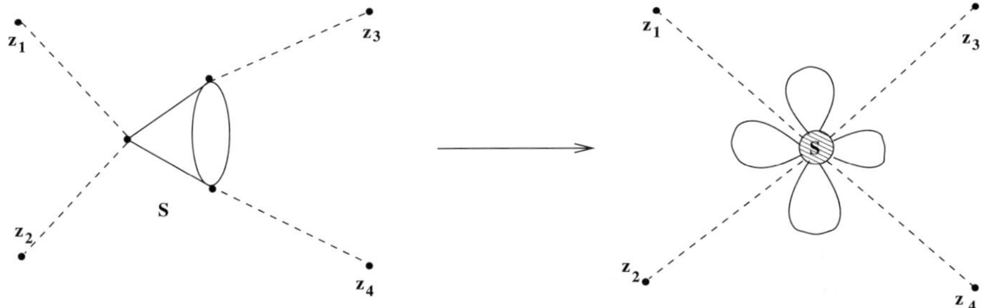

Figure 2: A high energy subgraph **S** seen from lower energies looks quasi-local.

Power counting is a rough estimate which compares the size of a fat dot such as S in Figure 2 with N external legs to the coupling constant that would be in front of an *exactly local* $\int \phi^N(x)dx$ interaction term if it were in the Lagrangian. To simplify we now assume that the internal scales are all equal to i, the external scales are $O(1)$, and we do not care about constants and so on, but only about the dependence in i as i gets large. We must first save one internal position such as the barycentre of the fat dot or the position of a particular internal vertex to represent the $\int dx$ integration in $\int \phi^N(x)dx$. Then we must integrate over the positions of all internal vertices of the subgraph *save that one*. This brings about a weight $M^{-di(n-1)}$, because since S is connected we can use the decay of the internal lines to evaluate these $n - 1$ integrals. Finally we should not forget the prefactor $M^{(D-2)li}$ coming from (2.34), for the l internal lines. Multiplying these two factors and using relation (2.21)–(2.22) we obtain that the "coupling constant" or factor in front of the fat dot is of order $M^{-di(n-1)+2i(2n-N/2)} = M^{\omega(G)}$, if we define the superficial degree of divergence of a ϕ_d^4 connected graph as

$$\omega(G) = (d-4)n(G) + d - \frac{d-2}{2}N(G). \tag{2.39}$$

So power counting, in contrast with locality, depends on the space-time dimension.

Let us return to the concrete example of Figure 2. A 4-point subgraph made of three vertices and four internal lines at a high slice i index. If we suppose the four external dashed lines have much lower index, say of order unity, the subgraph looks almost local, like a fat dot at this unit scale. We have to save one vertex integration for the position of the fat dot. Hence the coupling constant of this fat dot is made of two vertex integrations and the four weights of the internal lines (in order not to forget these internal line factors we kept internal lines apparent as four tadpoles attached to the fat dot in the right of Figure 2). In dimension 4 this total weight turns out to be independent of the scale.

At lower scales propagators can branch either through the initial bare coupling or through any such fat dot in all possible ways because of the combinatorial rules of functional integration. Hence they feel effectively a new coupling which is the sum of the bare coupling plus all the fat dot corrections coming from higher scales. To compute these new couplings only graphs with $\omega(G) \geq 0$, which are called primitively divergent, really matter because their weight does not decrease as the gap i increases.

- If $d = 2$, we find $\omega(G) = 2 - 2n$, so the only primitively divergent graphs have $n = 1$, and $N = 0$ or $N = 2$. The only divergence is due to the "tadpole" loop $\int \frac{d^2p}{(p^2+m^2)}$ which is logarithmically divergent.
- If $d = 3$, we find $\omega(G) = 3 - n - N/2$, so the only primitively divergent graphs have $n \leq 3$, $N = 0$, or $n \leq 2$ and $N = 2$. Such a theory with only a finite number of "primitively divergent" subgraphs is called super-renormalizable.
- If $d = 4$, $\omega(G) = 4 - N$. Every two-point graph is quadratically divergent and every four-point graph is logarithmically divergent. This is in agreement with the superficial degree of these graphs being respectively 2 and 0. The couplings that do not decay with i all correspond to terms that were already present in the Lagrangian, namely $\int \phi^4$, $\int \phi^2$ and $\int (\nabla\phi).(\nabla\phi)$ [16]. Hence the structure of the Lagrangian resists under change of scale, although the values of the coefficients can change. The theory is called just renormalizable.

---

[16] Because the graphs with $N = 2$ are quadratically divergent we must Taylor expand the quasi local fat dots until we get convergent effects. Using parity and rotational symmetry, this generates only a logarithmically divergent $\int (\nabla\phi).(\nabla\phi)$ term beyond the quadratically divergent $\int \phi^2$. Furthermore this term starts only at $n = 2$ or two loops, because the first tadpole graph at $N = 2$, $n = 1$ is *exactly* local.

– Finally for $d > 4$ we have infinitely many primitively divergent graphs with arbitrarily large number of external legs, and the theory is called non-renormalizable, because fat dots with N larger than 4 are important and they correspond to new couplings generated by the renormalization group which are not present in the initial bare Lagrangian.

To summarize:

- Locality means that quasi-local subgraphs look local when seen through their external lines. It holds in any dimension.
- Power counting gives the rough size of the new couplings associated to these subgraphs as a function of their number N of external legs, of their order n and of the dimension of space-time d.
- Renormalizability (in the ultraviolet regime) holds if the structure of the Lagrangian resists under change of scale, although the values of the coefficients or coupling constants may change. For ϕ^4 it occurs if $d \leq 4$, with $d = 4$ the most interesting case.

2.4. The BPHZ theorem

The BPHZ theorem is both a brilliant historic piece of mathematical physics which gives precise mathematical meaning to the notion of renormalizability, using the mathematics of formal power series, but it is also ultimately a bad way to understand and express renormalization. Let us try to explain both statements.

For the massive Euclidean ϕ_4^4 theory we could for instance state the following normalization conditions on the connected functions in momentum space at zero momenta:

$$C^4(0,0,0,0) = -\lambda_{\mathrm{ren}}, \tag{2.40}$$

$$C^2(p^2 = 0) = \frac{1}{m_{\mathrm{ren}}^2}, \tag{2.41}$$

$$\frac{d}{dp^2}C^2|_{p^2=0} = -\frac{a_{\mathrm{ren}}}{m_{\mathrm{ren}}^4}. \tag{2.42}$$

Usually one puts $a_{\mathrm{ren}} = 1$ by rescaling the field ϕ.

Using the inversion theorem on formal power series for any *fixed ultraviolet cutoff* κ it is possible to reexpress any formal power series in λ_{bare} with bare propagators $1/(a_{\mathrm{bare}}p^2 + m_{\mathrm{bare}}^2)$ for any Schwinger functions as a formal power series in λ_{ren} with renormalized propagators $1/(a_{\mathrm{ren}}p^2 + m_{\mathrm{ren}}^2)$. The BPHZ theorem then states that the formal perturbative formal power series has finite coefficients order by order when the ultraviolet cutoff κ is lifted. The first proof by Hepp relied on the inductive Bogoliubov's recursion scheme. Then a completely explicit expression for the coefficients of the renormalized series was written by Zimmermann and many followers. The coefficients of that renormalized series can be written as sums of renormalized Feynman amplitudes. They are similar to Feynman integrals but with additional subtractions indexed by Zimmermann's forests. Returning to an inductive rather than explicit scheme, Polchinski remarked that it is possible to also deduce the BPHZ theorem from a renormalization group equation and inductive bounds which does not decompose each order of perturbation theory into Feynman graphs [46]. This method was clarified and applied by C. Kopper and coworkers, see [83].

The solution of the difficult "overlapping" divergence problem through Bogoliubov's or Polchinski's recursions and Zimmermann's forests becomes particularly clear in the parametric representation using Hepp's sectors. A Hepp sector is simply a complete ordering of the α parameters for all the lines of the graph. In each sector there is a different classification of forests into packets so that each packet gives a finite integral [84, 85].

But from the physical point of view we cannot conceal the fact that purely perturbative renormalization theory is not very satisfying. At least two facts hint at a better theory which lies behind:

- The forest formula seems unnecessarily complicated, with too many terms. For instance in any given Hepp sector only one particular packet of forests is really necessary to make the renormalized amplitude finite, the one which corresponds to the quasi-local divergent subgraphs of *that* sector. The other packets seem useless, a little bit like "junk DNA". They are there just because they are necessary for other sectors. This does not look optimal.
- The theory makes renormalized amplitudes finite, but at tremendous cost! The size of some of these renormalized amplitudes becomes unreasonably large as the size of the graph increases. This phenomenon is called the "renormalon problem". For instance it is easy to check that the renormalized amplitude (at 0 external momenta) of the graphs P_n with 6 external legs and $n + 2$ internal vertices in Figure 3 becomes as large as $c^n n!$ when $n \to \infty$. Indeed at large q the renormalized amplitude $A_{G_2}^R$ in Figure 5 grows like $\log |q|$. Therefore the chain of n such graphs in Figure 3 behaves as $[\log |q|]^n$, and the total amplitude of P_n behaves as

$$\int [\log |q|]^n \frac{d^4 q}{[q^2 + m^2]^3} \simeq_{n \to \infty} c^n n! \ . \tag{2.43}$$

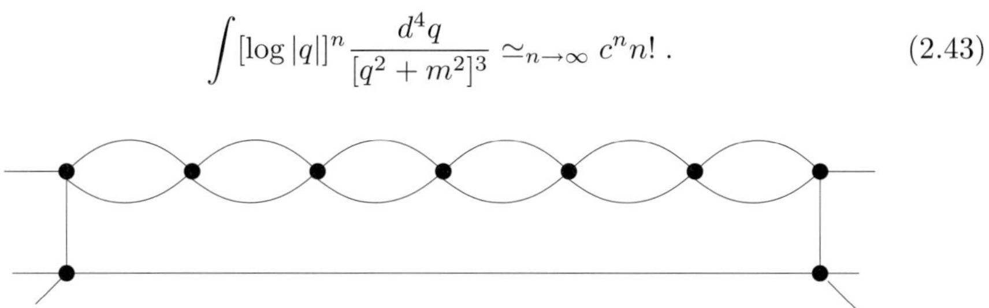

Figure 3: A family of graphs P_n producing a renormalon.

So after renormalization some families of graphs acquire so large values that they cannot be resumed! Physically this is just as bad as if infinities were still there. These two hints are in fact linked. As their name indicates, renormalons are due to renormalization. Families of completely convergent graphs such as the

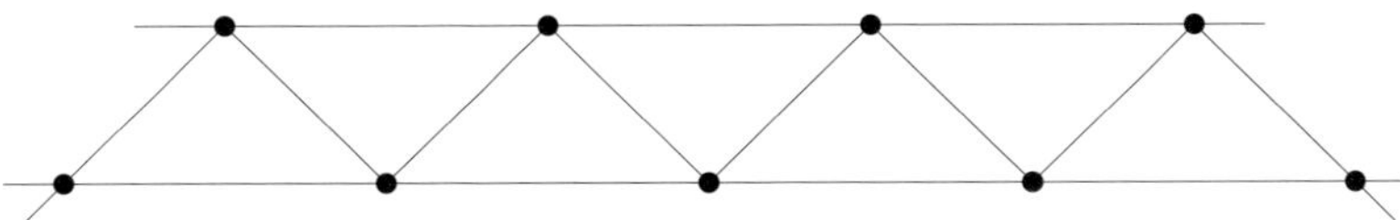

Figure 4: A family of convergent graphs Q_n, that do not produce any renormalon.

graphs Q_n of Figure 4, are bounded by c^n, and produce no renormalons. Studying more carefully renormalization in the α parametric representation one can check that renormalons are solely due to the forests packets that we compared to "junk DNA". Renormalons are due to subtractions that are not necessary to ensure convergence, just like the strange $\log|q|$ growth of $A_{G_0}^R$ at large q is solely due to the counterterm in the region where this counterterm is not necessary to make the amplitude finite.

We can therefore conclude that subtractions are not organized in an optimal way by the Bogoliubov recursion. What is wrong from a physical point of view in the BPHZ theorem is to use the size of the graph as the relevant parameter to organize Bogoliubov's induction. It is rather the size of the line momenta that should be used to better organize the renormalization subtractions.

This leads to the point of view advocated in [9]: neither the bare nor the renormalized series are optimal. Perturbation should be organized as a power series in an infinite set of effective expansions, which are related through the RG flow equation. In the end exactly the same contributions are resumed than in the bare or in the renormalized series, but they are regrouped in a much better way.

2.5. The Landau ghost and asymptotic freedom

In the case of ϕ_4^4 only the flow of the coupling constants really matters, because the flow of m and of a for different reasons are not very important in the ultraviolet limit:

- the flow of m is governed at leading order by the tadpole. The bare mass m_i^2 corresponding to a finite positive physical mass $m_{\rm ren}^2$ is negative and grows as λM^{2i} with the slice index i. But since p^2 in the ith slice is also of order M^{2i} but without the λ, as long as the coupling λ remains small it remains much larger than m_i^2. Hence the mass term plays no significant role in the higher slices. It was remarked in [9] that because there are no overlapping problems associated to 1PI two-point subgraphs, there is in fact no inconvenience to use the full renormalized $m_{\rm ren}$ all the way from the bare to renormalized scales, with subtractions on 1PI two-point subgraphs independent of their scale.

- the flow of a is also not very important. Indeed it really starts at two loops because the tadpole is exactly local. So this flow is in fact bounded, and generates no renormalons. In fact as again remarked in [9] for theories of the ϕ_4^4 type one might as well use the bare value $a_{\rm bare}$ all the way from bare to

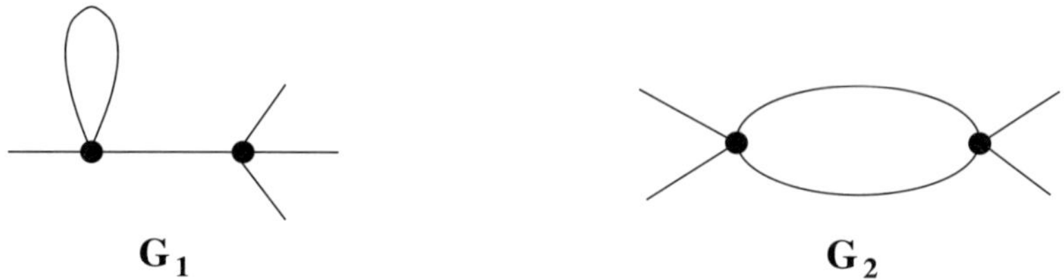

Figure 5: The ϕ^4 connected graphs with $n = 2$, $N = 4$.

renormalized scales and perform no second Taylor subtraction on any 1PI two-point subgraphs.

But the physics of ϕ_4^4 in the ultraviolet limit really depends of the flow of λ. By a simple second-order computation there are only 2 connected graphs with $n = 2$ and $N = 4$ pictured in Figure 5. They govern at leading order the flow of the coupling constant.

In the commutative ϕ_4^4 theory the graph G_1 does not contribute to the coupling constant flow. This can be seen in many ways, for instance after mass renormalization the graph G_1 vanishes exactly because it contains a tadpole which is not quasi-local but *exactly* local. One can also remark that the graph is one particle reducible. In ordinary translation-invariant, hence momentum-conserving theories, one-particle-reducible quasi-local graphs never contribute significantly to RG flows. Indeed they become very small when the gap i between internal and external scales grows. This is because by momentum conservation the momentum of any one-particle-reducible line ℓ has to be the sum of a finite set of external momenta on one of its sides. But a finite sum of small momenta remains small and this clashes directly with the fact that ℓ being internal its momentum should grow as the gap i grows. Remark that this is no longer true in non-commutative vulcanized $\phi_4^{\star 4}$, because that theory is not translation invariant, and that's why it will ultimately escape the Landau ghost curse.

So in ϕ_4^4 the flow is intimately linked to the sign of the graph G_2 of Figure 5. More precisely, we find that at second order the relation between λ_i and λ_{i-1} is

$$\lambda_{i-1} \quad \simeq \quad \lambda_i - \beta\lambda_i^2 \tag{2.44}$$

(remember the minus sign in the exponential of the action), where β is a constant, namely the asymptotic value of $\sum_{j,j'/\inf(j,j')=i} \int d^4y\, C_j(x,y) C_{j'}(x,y)$ when $i \to \infty$. Clearly this constant is positive. So for the normal stable ϕ_4^4 theory, the relation (2.44) inverts into

$$\lambda_i \simeq \lambda_{i-1} + \beta\lambda_{i-1}^2, \tag{2.45}$$

so that fixing the renormalized coupling seems to lead at finite i to a large, diverging bare coupling, incompatible with perturbation theory. This is the Landau ghost problem, which affects both the ϕ_4^4 theory and electrodynamics. Equivalently if one keeps λ_i finite as i gets large, $\lambda_0 = \lambda_{\rm ren}$ tends to zero and the final effective theory is "trivial" which means it is a free theory without interaction, in contradiction with the physical observation, e.g., of a coupling constant of about $1/137$ in electrodynamics.

But in non-abelian gauge theories an extra minus sign is created by the algebra of the Lie brackets. This surprising discovery has deep consequences. The flow relation becomes approximately

$$\lambda_i \simeq \lambda_{i-1} - \beta \lambda_i \lambda_{i-1}, \tag{2.46}$$

with $\beta > 0$, or, dividing by $\lambda_i \lambda_{i-1}$,

$$1/\lambda_i \simeq 1/\lambda_{i-1} + \beta, \tag{2.47}$$

with solution $\lambda_i \simeq \frac{\lambda_0}{1+\lambda_0 \beta i}$. A more precise computation to third order in fact leads to

$$\lambda_i \simeq \frac{\lambda_0}{1 + \lambda_0(\beta i + \gamma \log i + O(1))}. \tag{2.48}$$

Such a theory is called asymptotically free (in the ultraviolet limit) because the effective coupling tends to 0 with the cutoff for a finite fixed small renormalized coupling. Physically the interaction is turned off at small distances. This theory is in agreement with scattering experiments which see a collection of almost free particles (quarks and gluons) inside the hadrons at very high energy. This was the main initial argument to adopt quantum chromodynamics, a non-abelian gauge theory with $SU(3)$ gauge group, as the theory of strong interactions [13].

Remark that in such asymptotically free theories which form the backbone of today's standard model, the running coupling constants remain bounded between far ultraviolet "bare" scales and the lower energy scale where renormalized couplings are measured. Ironically the point of view on early renormalization theory as a trick to hide the ultraviolet divergences of QFT into infinite unobservable bare parameters could not turn out to be more wrong than in the standard model. Indeed the bare coupling constants tend to 0 with the ultraviolet cutoff, and what can be farther from infinity than 0?

3. Non-commutative field theory

3.1. Field theory on Moyal space

The recent progresses concerning the renormalization of non-commutative field theory have been obtained on a very simple non-commutative space namely the Moyal space. From the point of view of quantum field theory, it is certainly the most studied space. Let us start with its precise definition.

3.1.1. The Moyal space $\mathbb{R}_\theta^D$. Let us define $E = \{x^\mu, \ \mu \in [\![1, D]\!]\}$ and $\mathbb{C}\langle E \rangle$ the free algebra generated by E. Let Θ a $D \times D$ non-degenerate skew-symmetric matrix (which requires D even) and I the ideal of $\mathbb{C}\langle E \rangle$ generated by the elements $x^\mu x^\nu - x^\nu x^\mu - \imath\Theta^{\mu\nu}$. The Moyal algebra $\mathcal{A}_\Theta$ is the quotient $\mathbb{C}\langle E \rangle / I$. Each element in $\mathcal{A}_\Theta$ is a formal power series in the x^μ's for which the relation $[x^\mu, x^\nu] = \imath\Theta^{\mu\nu}$ holds.

Usually, one puts the matrix Θ into its canonical form:

$$\Theta = \begin{pmatrix} \begin{matrix} 0 & \theta_1 \\ -\theta_1 & 0 \end{matrix} & & (0) \\ & \ddots & \\ (0) & & \begin{matrix} 0 & \theta_{D/2} \\ -\theta_{D/2} & 0 \end{matrix} \end{pmatrix}. \tag{3.1}$$

Sometimes one even set $\theta = \theta_1 = \cdots = \theta_{D/2}$. The preceding algebraic definition whereas short and precise may be too abstract to perform real computations. One then needs a more analytical definition. A representation of the algebra $\mathcal{A}_\Theta$ is given by some set of functions on $\mathbb{R}^d$ equipped with a non-commutative product: the *Groenwald-Moyal* product. What follows is based on [86].

The algebra $\mathcal{A}_\Theta$. The Moyal algebra $\mathcal{A}_\Theta$ is the linear space of smooth and rapidly decreasing functions $\mathcal{S}(\mathbb{R}^D)$ equipped with the non-commutative product defined by: $\forall f, g \in \mathcal{S}_D \overset{\text{def}}{=} \mathcal{S}(\mathbb{R}^D)$,

$$(f \star_\Theta g)(x) = \int_{\mathbb{R}^D} \frac{d^D k}{(2\pi)^D} d^D y \, f(x + \tfrac{1}{2}\Theta \cdot k) g(x + y) e^{\imath k \cdot y} \tag{3.2}$$

$$= \frac{1}{\pi^D |\det \Theta|} \int_{\mathbb{R}^D} d^D y \, d^D z \, f(x+y) g(x+z) e^{-2\imath y \Theta^{-1} z} \, . \tag{3.3}$$

This algebra may be considered as the "functions on the Moyal space $\mathbb{R}^D_\theta$". In the following we will write $f \star g$ instead of $f \star_\Theta g$ and use: $\forall f, g \in \mathcal{S}_D$, $\forall j \in [\![1, 2N]\!]$,

$$(\mathscr{F}f)(x) = \int f(t) e^{-\imath t x} dt \tag{3.4}$$

for the Fourier transform and

$$(f \diamond g)(x) = \int f(x - t) g(t) e^{2\imath x \Theta^{-1} t} dt \tag{3.5}$$

for the twisted convolution. As on $\mathbb{R}^D$, the Fourier transform exchanges product and convolution:

$$\mathscr{F}(f \star g) = \mathscr{F}(f) \diamond \mathscr{F}(g) \tag{3.6}$$

$$\mathscr{F}(f \diamond g) = \mathscr{F}(f) \star \mathscr{F}(g). \tag{3.7}$$

One also shows that the Moyal product and the twisted convolution are *associative*:

$$((f \diamond g) \diamond h)(x) = \int f(x - t - s) g(s) h(t) e^{2\imath(x\Theta^{-1}t + (x-t)\Theta^{-1}s)} ds \, dt \tag{3.8}$$

$$= \int f(u - v) g(v - t) h(t) e^{2\imath(x\Theta^{-1}v - t\Theta^{-1}v)} dt \, dv$$

$$= (f \diamond (g \diamond h))(x). \tag{3.9}$$

Using (3.7), we show the associativity of the $\star$-product. The complex conjugation is *involutive* in $\mathcal{A}_\Theta$,

$$\overline{f \star_\Theta g} = \bar{g} \star_\Theta \bar{f}. \tag{3.10}$$

One also has

$$f \star_\Theta g = g \star_{-\Theta} f. \tag{3.11}$$

Proposition 3.1 (Trace). *For all* $f, g \in \mathcal{S}_D$,

$$\int dx \, (f \star g)(x) = \int dx \, f(x)g(x) = \int dx \, (g \star f)(x) \,. \tag{3.12}$$

Proof.

$$\int dx \, (f \star g)(x) = \mathscr{F}(f \star g)(0) = (\mathscr{F}f \diamond \mathscr{F}g)(0) \tag{3.13}$$

$$= \int \mathscr{F}f(-t)\mathscr{F}g(t)dt = (\mathscr{F}f \ast \mathscr{F}g)(0) = \mathscr{F}(fg)(0) = \int f(x)g(x)dx$$

where $\ast$ is the ordinary convolution. $\qquad\square$

In the following sections, we will need Lemma 3.2 to compute the interaction terms for the $\Phi_4^{\star 4}$ and Gross-Neveu models. We write $x \wedge y \overset{\text{def}}{=} 2x\Theta^{-1}y$.

Lemma 3.2. *For all* $j \in [\![1, 2n+1]\!]$, *let* $f_j \in \mathcal{A}_\Theta$. *Then*

$$(f_1 \star_\Theta \cdots \star_\Theta f_{2n})(x) = \frac{1}{\pi^{2D} \det^2 \Theta} \int \prod_{j=1}^{2n} dx_j f_j(x_j) \, e^{-\imath x \wedge \sum_{i=1}^{2n} (-1)^{i+1} x_i} \, e^{-\imath \varphi_{2n}},$$

$$\tag{3.14}$$

$$(f_1 \star_\Theta \cdots \star_\Theta f_{2n+1})(x)$$

$$= \frac{1}{\pi^D \det \Theta} \int \prod_{j=1}^{2n+1} dx_j f_j(x_j) \, \delta\Big(x - \sum_{i=1}^{2n+1} (-1)^{i+1} x_i\Big) e^{-\imath \varphi_{2n+1}}, \tag{3.15}$$

$$\forall p \in \mathbb{N}, \ \varphi_p = \sum_{i<j=1}^{p} (-1)^{i+j+1} x_i \wedge x_j. \tag{3.16}$$

Corollary 3.3. *For all* $j \in [\![1, 2n+1]\!]$, *let* $f_j \in \mathcal{A}_\Theta$. *Then*

$$\int dx \, (f_1 \star_\Theta \cdots \star_\Theta f_{2n})(x) = \frac{1}{\pi^D \det \Theta} \int \prod_{j=1}^{2n} dx_j f_j(x_j) \, \delta\Big(\sum_{i=1}^{2n} (-1)^{i+1} x_i\Big) e^{-\imath \varphi_{2n}},$$

$$\tag{3.17}$$

$$\int dx \, (f_1 \star_\Theta \cdots \star_\Theta f_{2n+1})(x) = \frac{1}{\pi^D \det \Theta} \int \prod_{j=1}^{2n+1} dx_j f_j(x_j) \, e^{-\imath \varphi_{2n+1}}, \tag{3.18}$$

$$\forall p \in \mathbb{N}, \ \varphi_p = \sum_{i<j=1}^{p} (-1)^{i+j+1} x_i \wedge x_j. \tag{3.19}$$

The cyclicity of the product, inherited from Proposition 3.1 implies: $\forall f, g, h \in \mathcal{S}_D$,

$$\langle f \star g, h \rangle = \langle f, g \star h \rangle = \langle g, h \star f \rangle \tag{3.20}$$

and allows to extend the Moyal algebra by duality into an algebra of tempered distributions.

Extension by Duality. Let us first consider the product of a tempered distribution with a Schwartz-class function. Let $T \in \mathcal{S}'_D$ and $h \in \mathcal{S}_D$. We define $\langle T, h \rangle \overset{\text{def}}{=} T(h)$ and $\langle T^*, h \rangle = \overline{\langle T, \overline{h} \rangle}$.

Definition 3.1. *Let* $T \in \mathcal{S}'_D$, $f, h \in \mathcal{S}_D$, *we define* $T \star f$ *and* $f \star T$ *by*

$$\langle T \star f, h \rangle = \langle T, f \star h \rangle, \tag{3.21}$$
$$\langle f \star T, h \rangle = \langle T, h \star f \rangle. \tag{3.22}$$

For example, the identity $\mathbb{1}$ as an element of $\mathcal{S}'_D$ is the unity for the $\star$-product: $\forall f, h \in \mathcal{S}_D$,

$$\langle \mathbb{1} \star f, h \rangle = \langle \mathbb{1}, f \star h \rangle \tag{3.23}$$
$$= \int (f \star h)(x) dx = \int f(x) h(x) dx$$
$$= \langle f, h \rangle.$$

We are now ready to define the linear space $\mathcal{M}$ as the intersection of two sub-spaces $\mathcal{M}_L$ and $\mathcal{M}_R$ of $\mathcal{S}'_D$.

Definition 3.2 (Multipliers algebra).

$$\mathcal{M}_L = \left\{ S \in \mathcal{S}'_D : \forall f \in \mathcal{S}_D, S \star f \in \mathcal{S}_D \right\}, \tag{3.24}$$
$$\mathcal{M}_R = \left\{ R \in \mathcal{S}'_D : \forall f \in \mathcal{S}_D, f \star R \in \mathcal{S}_D \right\}, \tag{3.25}$$
$$\mathcal{M} = \mathcal{M}_L \cap \mathcal{M}_R. \tag{3.26}$$

One can show that $\mathcal{M}$ is an associative $*$-algebra. It contains, among others, the identity, the polynomials, the δ distribution and its derivatives. Then the relation

$$[x^\mu, x^\nu] = \imath \Theta^{\mu\nu}, \tag{3.27}$$

often given as a definition of the Moyal space, holds in $\mathcal{M}$ (but not in $\mathcal{A}_\Theta$).

3.1.2. The $\Phi^{\star 4}$ theory on $\mathbb{R}^4_\theta$ Moyal space. The simplest non-commutative model one may consider is the $\Phi^{\star 4}$ theory on the four-dimensional Moyal space. Its Lagrangian is the usual (commutative) one where the pointwise product is replaced by the Moyal one:

$$S[\phi] = \int d^4x \left(-\frac{1}{2} \partial_\mu \phi \star \partial^\mu \phi + \frac{1}{2} m^2 \, \phi \star \phi + \frac{\lambda}{4} \phi \star \phi \star \phi \star \phi \right)(x). \tag{3.28}$$

Thanks to the formula (3.3), this action can be explicitly computed. The interaction part is given by the Corollary 3.3:

$$\int dx\, \phi^{\star 4}(x) = \int \prod_{i=1}^{4} dx_i\, \phi(x_i)\, \delta(x_1 - x_2 + x_3 - x_4) e^{\iota\varphi}, \qquad (3.29)$$

$$\varphi = \sum_{i<j=1}^{4} (-1)^{i+j+1} x_i \wedge x_j.$$

The most obvious characteristic of the Moyal product is its non-locality. But its non-commutativity implies that the vertex of the model (3.28) is only invariant under cyclic permutation of the fields. This restricted invariance incites to represent the associated Feynman graphs with ribbon propagators. One can then make a clear distinction between planar and non-planar graphs. This will be detailed in Section 4.

Thanks to the delta function in (3.29), the oscillation may be written in different ways:

$$\delta(x_1 - x_2 + x_3 - x_4) e^{\iota\varphi} = \delta(x_1 - x_2 + x_3 - x_4) e^{\iota x_1 \wedge x_2 + \iota x_3 \wedge x_4} \qquad (3.30\text{a})$$

$$= \delta(x_1 - x_2 + x_3 - x_4) e^{\iota x_4 \wedge x_1 + \iota x_2 \wedge x_3} \qquad (3.30\text{b})$$

$$= \delta(x_1 - x_2 + x_3 - x_4) \exp \iota(x_1 - x_2) \wedge (x_2 - x_3). \qquad (3.30\text{c})$$

The interaction is real and positive[17]:

$$\int \prod_{i=1}^{4} dx_i \phi(x_i)\, \delta(x_1 - x_2 + x_3 - x_4) e^{\iota\varphi}$$

$$= \int dk \left(\int dx dy\, \phi(x)\phi(y) e^{\iota k(x-y) + \iota x \wedge y} \right)^2 \in \mathbb{R}_+. \qquad (3.31)$$

It is also translation invariant as shows equation (3.30c).

Property 3.1 implies that the propagator is the usual one: $\hat{C}(p) = 1/(p^2 + m^2)$.

3.1.3. UV/IR mixing. In the article [87], Filk computed the Feynman rules corresponding to (3.28). He showed that the planar amplitudes equal the commutative ones whereas the non-planar ones give rise to oscillations coupling the internal and external legs. Hence contrary perhaps to overoptimistic initial expectations, non-commutative geometry alone does not eliminate the ultraviolet divergences of QFT. Since there are infinitely many planar graphs with four external legs, the model (3.28) might at best be just renormalizable in the ultraviolet regime, as ordinary ϕ_4^4.

In fact it is not. Minwalla, Van Raamsdonk and Seiberg discovered that the model (3.28) exhibits a new type of divergences making it non-renormalizable [37].

---

[17] Another way to prove it is from (3.10), $\overline{\phi^{\star 4}} = \phi^{\star 4}$.

A typical example is the non-planar tadpole:

$$
\frac{\lambda}{12} \int \frac{d^4k}{(2\pi)^4} \frac{e^{ip_\mu k_\nu \Theta^{\mu\nu}}}{k^2 + m^2}
$$

$$
= \frac{\lambda}{48\pi^2} \sqrt{\frac{m^2}{(\Theta p)^2}} K_1\left(\sqrt{m^2(\Theta p)^2}\right) \underset{p\to 0}{\simeq} p^{-2}. \tag{3.32}
$$

If $p \neq 0$, this amplitude is finite but, for small p, it diverges like p^{-2}. In other words, if we put an ultraviolet cut-off Λ to the k-integral, the two limits $\Lambda \to \infty$ and $p \to 0$ do not commute. This is the UV/IR mixing phenomenon. A chain of non-planar tadpoles, inserted in bigger graphs, makes divergent any function (with six points or more for example). But this divergence is not local and can't be absorbed in a mass redefinition. This is what makes the model non-renormalizable. We will see in sections 6.4 and 7 that the UV/IR mixing results in a coupling of the different scales of the theory. We will also note that we should distinguish different types of mixing.

The UV/IR mixing was studied by several groups. First, Chepelev and Roiban [48] gave a power counting for different scalar models. They were able to identify the divergent graphs and to classify the divergences of the theories thanks to the topological data of the graphs. Then V. Gayral [88] showed that UV/IR mixing is present on all isospectral deformations (they consist in curved generalizations of the Moyal space and of the non-commutative torus). For this, he considered a scalar model (3.28) and discovered contributions to the effective action which diverge when the external momenta vanish. The UV/IR mixing is then a general characteristic of the non-commutative theories, at least on these deformations.

3.2. The Grosse-Wulkenhaar breakthrough

The situation remained unchanged until H. Grosse and R. Wulkenhaar discovered a way to define a renormalizable non-commutative model. We will detail their result in Section 4 but the main message is the following. By adding an harmonic term to the Lagrangian (3.28),

$$
S[\phi] = \int d^4x \left(-\frac{1}{2}\partial_\mu \phi \star \partial^\mu \phi + \frac{\Omega^2}{2}(\tilde{x}_\mu \phi) \star (\tilde{x}^\mu \phi) + \frac{1}{2}m^2 \phi \star \phi + \frac{\lambda}{4}\phi \star \phi \star \phi \star \phi \right)(x) \tag{3.33}
$$

where $\tilde{x} = 2\Theta^{-1}x$ and the metric is Euclidean, the model, in four dimensions, is renormalizable at all orders of perturbation [40]. We will see in Section 7 that this additional term gives rise to an infrared cut-off and allows to decouple the different scales of the theory. The new model (3.33), which we call vulcanized $\Phi_4^{\star 4}$, does not exhibit any mixing. This result is very important because it opens the way towards other non-commutative field theories. Remember that we call *vulcanization* the procedure consisting in adding a new term to a Lagrangian of a non-commutative theory in order to make it renormalizable, see footnote 9.

The propagator C of this Φ^4 theory is the kernel of the inverse operator $-\Delta + \Omega^2 \widetilde{x}^2 + m^2$. It is known as the Mehler kernel [50, 89]:

$$C(x, y) = \frac{\Omega^2}{\theta^2 \pi^2} \int_0^\infty \frac{dt}{\sinh^2(2\widetilde{\Omega}t)} \, e^{-\frac{\widetilde{\Omega}}{2} \coth(2\widetilde{\Omega}t)(x-y)^2 - \frac{\widetilde{\Omega}}{2} \tanh(2\widetilde{\Omega}t)(x+y)^2 - m^2 t}. \tag{3.34}$$

Langmann and Szabo remarked that the quartic interaction with Moyal product is invariant under a duality transformation. It is a symmetry between momentum and direct space. The interaction part of the model (3.33) is (see equation (3.17))

$$S_{\text{int}}[\phi] = \int d^4 x \, \frac{\lambda}{4} (\phi \star \phi \star \phi \star \phi)(x) \tag{3.35}$$

$$= \int \prod_{a=1}^4 d^4 x_a \, \phi(x_a) \, V(x_1, x_2, x_3, x_4) \tag{3.36}$$

$$= \int \prod_{a=1}^4 \frac{d^4 p_a}{(2\pi)^4} \, \hat{\phi}(p_a) \, \hat{V}(p_1, p_2, p_3, p_4) \tag{3.37}$$

with

$$V(x_1, x_2, x_3, x_4) = \frac{\lambda}{4} \frac{1}{\pi^4 \det \Theta} \delta(x_1 - x_2 + x_3 - x_4) \cos(2(\Theta^{-1})_{\mu\nu}(x_1^\mu x_2^\nu + x_3^\mu x_4^\nu)),$$

$$\hat{V}(p_1, p_2, p_3, p_4) = \frac{\lambda}{4} (2\pi)^4 \delta(p_1 - p_2 + p_3 - p_4) \cos(\frac{1}{2} \Theta^{\mu\nu}(p_{1,\mu} p_{2,\nu} + p_{3,\mu} p_{4,\nu}))$$

where we used a *cyclic* Fourier transform: $\hat{\phi}(p_a) = \int dx \, e^{(-1)^a \imath p_a x_a} \phi(x_a)$. The transformation

$$\hat{\phi}(p) \leftrightarrow \pi^2 \sqrt{|\det \Theta|} \, \phi(x), \qquad p_\mu \leftrightarrow \widetilde{x}_\mu \tag{3.38}$$

exchanges (3.36) and (3.37). In addition, the free part of the model (3.28) isn't co-variant under this duality. The vulcanization adds a term to the Lagrangian which restores the symmetry. The theory (3.33) is then covariant under the Langmann-Szabo duality:

$$S[\phi; m, \lambda, \Omega] \mapsto \Omega^2 \, S[\phi; \frac{m}{\Omega}, \frac{\lambda}{\Omega^2}, \frac{1}{\Omega}]. \tag{3.39}$$

By symmetry, the parameter Ω is confined in $[0, 1]$. Let us note that for $\Omega = 1$, the model is invariant.

The interpretation of that harmonic term is not yet clear. But the vulcanization procedure already allowed to prove the renormalizability of several other models on Moyal spaces such that $\Phi_2^{\star 4}$ [39], $\phi_{2,4}^3$ [64,65] and the LSZ models [43–45]. These last ones are of the type

$$S[\phi] = \int d^n x \left(\frac{1}{2} \bar{\phi} \star (-\partial_\mu + \widetilde{x}_\mu + m)^2 \phi + \frac{\lambda}{4} \bar{\phi} \star \phi \star \bar{\phi} \star \phi \right)(x). \tag{3.40}$$

By comparison with (3.33), one notes that here the additional term is formally equivalent to a fixed magnetic background. Therefore such a model is invariant under magnetic translations which combine a translation and a phase shift on the field. This model is invariant under the above duality and is exactly soluble. Let us remark that the complex interaction in (3.40) makes the Langmann-Szabo duality more natural. It doesn't need a cyclic Fourier transform. The $\phi^{\star 3}$ model at $\Omega = 1$ also exhibits a soluble structure [64–66].

3.3. The non-commutative Gross-Neveu model

Apart from the $\Phi_4^{\star 4}$, the modified bosonic LSZ model [47] and supersymmetric theories, we now know several renormalizable non-commutative field theories. Nevertheless they either are super-renormalizable ($\Phi_2^{\star 4}$ [39]) or (and) studied at a special point in the parameter space where they are solvable ($\Phi_2^{\star 3}$, $\Phi_4^{\star 3}$ [64, 65], the LSZ models [43–45]). Although only logarithmically divergent for parity reasons, the non-commutative Gross-Neveu model is a just renormalizable quantum field theory as $\Phi_4^{\star 4}$. One of its main interesting features is that it can be interpreted as a nonlocal fermionic field theory in a constant magnetic background. Then apart from strengthening the "vulcanization" procedure to get renormalizable non-commutative field theories, the Gross-Neveu model may also be useful for the study of the quantum Hall effect. It is also a good first candidate for a constructive study [9] of a non-commutative field theory as fermionic models are usually easier to construct. Moreover its commutative counterpart being asymptotically free and exhibiting dynamical mass generation [90–92], a study of the physics of this model would be interesting.

The non-commutative Gross-Neveu model (GN_Θ^2) is a fermionic quartically interacting quantum field theory on the Moyal plane $\mathbb{R}_\theta^2$. The skew-symmetric matrix Θ is

$$\Theta = \begin{pmatrix} 0 & -\theta \\ \theta & 0 \end{pmatrix}. \tag{3.41}$$

The action is

$$S[\bar{\psi}, \psi] = \int dx \left(\bar{\psi} \left(-\imath \partial\!\!\!/ + \Omega \tilde{x}\!\!\!/ + m + \mu\,\gamma_5 \right) \psi + V_{\mathrm{o}}(\bar{\psi}, \psi) + V_{\mathrm{no}}(\bar{\psi}, \psi) \right) (x) \tag{3.42}$$

where $\tilde{x} = 2\Theta^{-1}x$, $\gamma_5 = \imath\gamma^0\gamma^1$ and $V = V_{\mathrm{o}} + V_{\mathrm{no}}$ is the interaction part given hereafter. The μ-term appears at two-loop order. We use an Euclidean metric and the Feynman convention $a\!\!\!/ = \gamma^\mu a_\mu$. The γ^0 and γ^1 matrices form a two-dimensional representation of the Clifford algebra $\{\gamma^\mu, \gamma^\nu\} = -2\delta^{\mu\nu}$. Let us remark that the γ^μ's are then skew-Hermitian: $\gamma^{\mu\dagger} = -\gamma^\mu$.

Propagator. The propagator corresponding to the action (3.42) is given by the following lemma:

Lemma 3.4 (Propagator [50]). *The propagator of the Gross-Neveu model is*

$$C(x,y) = \int d\mu_C(\bar{\psi}, \psi)\,\psi(x)\bar{\psi}(y) = \left(-\imath \partial\!\!\!/ + \Omega \tilde{x}\!\!\!/ + m \right)^{-1} (x, y) \tag{3.43}$$

$$= \int_0^\infty dt\, C(t; x, y),$$

$$C(t;x,y) = -\frac{\Omega}{\theta\pi}\frac{e^{-tm^2}}{\sinh(2\widetilde{\Omega}t)}\,e^{-\frac{\widetilde{\Omega}}{2}\coth(2\widetilde{\Omega}t)(x-y)^2 + \imath\Omega x\wedge y} \tag{3.44}$$

$$\times\left\{\imath\widetilde{\Omega}\coth(2\widetilde{\Omega}t)(\not{x}-\not{y}) + \Omega(\widetilde{\not{x}}-\widetilde{\not{y}}) - m\right\}e^{-2\imath\Omega t\gamma\Theta^{-1}\gamma}$$

with $\widetilde{\Omega} = \frac{2\Omega}{\theta}$ and $x\wedge y = 2x\Theta^{-1}y$.

We also have $e^{-2\imath\Omega t\gamma\Theta^{-1}\gamma} = \cosh(2\widetilde{\Omega}t)\mathbb{1}_2 - \imath\frac{\theta}{2}\sinh(2\widetilde{\Omega}t)\gamma\Theta^{-1}\gamma$.

If we want to study a *N-color* model, we can consider a propagator diagonal in these color indices.

Interactions. Concerning the interaction part V, recall that (see Corollary 3.3) for any f_1, f_2, f_3, f_4 in $\mathcal{A}_\Theta$,

$$\int dx\,(f_1\star f_2\star f_3\star f_4)\,(x) = \frac{1}{\pi^2\det\Theta}\int\prod_{j=1}^{4}dx_j\,f_j(x_j)\,\delta(x_1 - x_2 + x_3 - x_4)e^{-\imath\varphi}, \tag{3.45}$$

$$\varphi = \sum_{i<j=1}^{4}(-1)^{i+j+1}x_i\wedge x_j. \tag{3.46}$$

This product is nonlocal and only invariant under cyclic permutations of the fields. Then, contrary to the commutative Gross-Neveu model, for which there exists only one spinorial interaction, the GN_Θ^2 model has, at least, six different interactions: the *orientable* ones

$$V_{\mathrm{o}} = \frac{\lambda_1}{4}\int dx\,\left(\bar{\psi}\star\psi\star\bar{\psi}\star\psi\right)(x) \tag{3.47a}$$

$$+\frac{\lambda_2}{4}\int dx\,\left(\bar{\psi}\star\gamma^\mu\psi\star\bar{\psi}\star\gamma_\mu\psi\right)(x) \tag{3.47b}$$

$$+\frac{\lambda_3}{4}\int dx\,\left(\bar{\psi}\star\gamma_5\psi\star\bar{\psi}\star\gamma_5\psi\right)(x), \tag{3.47c}$$

where ψ's and $\bar{\psi}$'s alternate and the *non-orientable* ones

$$V_{\mathrm{no}} = \frac{\lambda_4}{4}\int dx\,\left(\psi\star\bar{\psi}\star\bar{\psi}\star\psi\right)(x) \tag{3.48a}$$

$$+\frac{\lambda_5}{4}\int dx\,\left(\psi\star\gamma^\mu\bar{\psi}\star\bar{\psi}\star\gamma_\mu\psi\right)(x) \tag{3.48b}$$

$$+\frac{\lambda_6}{4}\int dx\,\left(\psi\star\gamma_5\bar{\psi}\star\bar{\psi}\star\gamma_5\psi\right)(x). \tag{3.48c}$$

All these interactions have the same x kernel thanks to the equation (3.45). The reason for which we call these interactions orientable or not will be clear in Section 7.

4. Multi-scale analysis in the matrix basis

The matrix basis is a basis for Schwartz-class functions. In this basis, the Moyal product becomes a simple matrix product. Each field is then represented by an infinite matrix [39, 86, 93].

4.1. A dynamical matrix model

4.1.1. From the direct space to the matrix basis.
In the matrix basis, the action (3.33) takes the form:

$$S[\phi] = (2\pi)^{D/2}\sqrt{\det \Theta}\Big(\frac{1}{2}\phi\Delta\phi + \frac{\lambda}{4}\operatorname{Tr}\phi^4\Big) \tag{4.1}$$

where $\phi = \phi_{mn}$, $m, n \in \mathbb{N}^{D/2}$ and

$$\Delta_{mn,kl} = \sum_{i=1}^{D/2}\Big(\mu_0^2 + \frac{2}{\theta}(m_i + n_i + 1)\Big)\delta_{ml}\delta_{nk} - \frac{2}{\theta}(1-\Omega^2) \tag{4.2}$$

$$\times \Big(\sqrt{(m_i+1)(n_i+1)}\,\delta_{m_i+1,l_i}\delta_{n_i+1,k_i} + \sqrt{m_i n_i}\,\delta_{m_i-1,l_i}\delta_{n_i-1,k_i}\Big)\prod_{j\neq i}\delta_{m_j l_j}\delta_{n_j k_j}.$$

The (four-dimensional) matrix Δ represents the quadratic part of the Lagrangian. The first difficulty to study the matrix model (4.1) is the computation of its propagator G defined as the inverse of Δ:

$$\sum_{r,s\in\mathbb{N}^{D/2}}\Delta_{mn;rs}G_{sr;kl} = \sum_{r,s\in\mathbb{N}^{D/2}}G_{mn;rs}\Delta_{sr;kl} = \delta_{ml}\delta_{nk}. \tag{4.3}$$

Fortunately, the action is invariant under $SO(2)^{D/2}$ thanks to the form (3.1) of the Θ matrix. It implies a conservation law

$$\Delta_{mn,kl} = 0 \iff m + k \neq n + l. \tag{4.4}$$

The result is [39, 40]

$$G_{m,m+h;l+h,l} = \frac{\theta}{8\Omega}\int_0^1 d\alpha\,\frac{(1-\alpha)^{\frac{\mu_0^2\theta}{8\Omega}+(\frac{D}{4}-1)}}{(1+C\alpha)^{\frac{D}{2}}}\prod_{s=1}^{\frac{D}{2}}G^{(\alpha)}_{m^s,m^s+h^s;l^s+h^s,l^s}, \tag{4.5}$$

$$G^{(\alpha)}_{m,m+h;l+h,l} = \Big(\frac{\sqrt{1-\alpha}}{1+C\alpha}\Big)^{m+l+h}$$

$$\times \sum_{u=\max(0,-h)}^{\min(m,l)} A(m,l,h,u)\Big(\frac{C\alpha(1+\Omega)}{\sqrt{1-\alpha}(1-\Omega)}\Big)^{m+l-2u},$$

where $A(m,l,h,u) = \sqrt{\binom{m}{m-u}\binom{m+h}{m-u}\binom{l}{l-u}\binom{l+h}{l-u}}$ and C is a function in Ω: $C(\Omega) = \frac{(1-\Omega)^2}{4\Omega}$. The main advantage of the matrix basis is that it simplifies the interaction part: $\phi^{\star 4}$ becomes $\operatorname{Tr}\phi^4$. But the propagator becomes very complicated.

Let us remark that the matrix model (4.1) is *dynamical*: its quadratic part is not trivial. Usually, matrix models are *local*.

Definition 4.1. *A matrix model is called* local *if* $G_{mn;kl} = G(m,n)\delta_{ml}\delta_{nk}$ *and* nonlocal *otherwise.*

In the matrix theories, the Feynman graphs are ribbon graphs. The propagator $G_{mn;kl}$ is then represented by Figure 6. In a local matrix model, the propagator

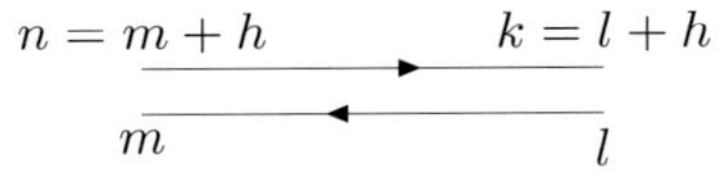

$$n = m + h \qquad\qquad k = l + h$$
$$m \qquad\qquad l$$

Figure 6: Matrix propagator.

preserves the index values along the trajectories (simple lines).

4.1.2. Topology of ribbon graphs. The power counting of a matrix model depends on the topological data of its graphs. Figure 7 gives two examples of ribbon graphs. Each ribbon graph may be drawn on a two-dimensional manifold. Actually each

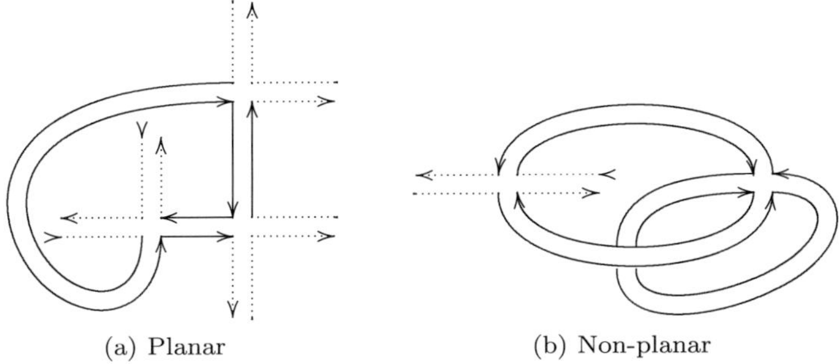

(a) Planar   (b) Non-planar

Figure 7: Ribbon graphs.

graph defines a surface on which it is drawn. Let a graph G with V vertices, I internal propagators (double lines) and F faces (made of simple lines). The Euler characteristic

$$\chi = 2 - 2g = V - I + F \tag{4.6}$$

gives the genus g of the manifold. One can make this clear by passing to the *dual graph*. The dual of a given graph G is obtained by exchanging faces and vertices. The dual graphs of the $\Phi^{\star 4}$ theory are tesselations of the surfaces on which they are drawn. Moreover each direct face broken by external legs becomes, in the dual graph, a *puncture*. If among the F faces of a graph, B are broken, this graph may be drawn on a surface of genus $g = 1 - \frac{1}{2}(V - I + F)$ with B punctures. Figure 8 gives the topological data of the graphs of Figure 7.

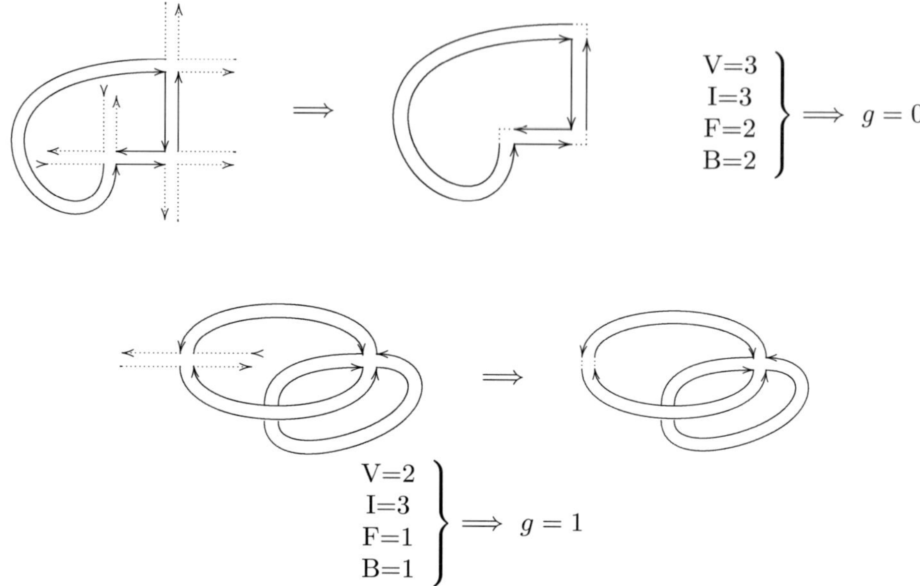

Figure 8: Topological data of ribbon graphs.

4.2. Multi-scale analysis

In [42], a multi-scale analysis was introduced to complete the rigorous study of the power counting of the non-commutative $\Phi^{\star 4}$ theory.

4.2.1. Bounds on the propagator. Let G a ribbon graph of the $\Phi_4^{\star 4}$ theory with N external legs, V vertices, I internal lines and F faces. Its genus is then $g = 1 - \frac{1}{2}(V - I + F)$. Four indices $\{m, n; k, l\} \in \mathbb{N}^2$ are associated to each internal line of the graph and two indices to each external line, that is to say $4I + 2N = 8V$ indices. But, at each vertex, the left index of a ribbon equals the right one of the neighbor ribbon. This gives rise to $4V$ independent identifications which allows to write each index in terms of a set $\mathcal{I}$ made of $4V$ indices, four per vertex, for example the left index of each half-ribbon.

The graph amplitude is then

$$A_G = \sum_{\mathcal{I}} \prod_{\delta \in G} G_{m_\delta(\mathcal{I}), n_\delta(\mathcal{I}); k_\delta(\mathcal{I}), l_\delta(\mathcal{I})} \; \delta_{m_\delta - l_\delta, n_\delta - k_\delta} \, , \tag{4.7}$$

where the four indices of the propagator G of the line δ are functions of $\mathcal{I}$ and written $\{m_\delta(\mathcal{I}), n_\delta(\mathcal{I}); k_\delta(\mathcal{I}), l_\delta(\mathcal{I})\}$. We decompose each propagator, given by (4.5):

$$G = \sum_{i=0}^{\infty} G^i \qquad \text{thanks to} \quad \int_0^1 d\alpha = \sum_{i=1}^{\infty} \int_{M^{-2i}}^{M^{-2(i-1)}} d\alpha, \; M > 1. \tag{4.8}$$

We have an associated decomposition for each amplitude

$$A_G = \sum_{\mu} A_{G,\mu} \, , \tag{4.9}$$

$$A_{G,\mu} = \sum_{\mathcal{I}} \prod_{\delta \in G} G^{i_\delta}_{m_\delta(\mathcal{I}),n_\delta(\mathcal{I});k_\delta(\mathcal{I}),l_\delta(\mathcal{I})} \, \delta_{m_\delta(\mathcal{I})-l_\delta(\mathcal{I}),n_\delta(\mathcal{I})-k_\delta(\mathcal{I})} \, , \tag{4.10}$$

where $\mu = \{i_\delta\}$ runs over all the possible assignments of a positive integer i_δ to each line δ. We proved the following four propositions:

Proposition 4.1. *For M large enough, there exists a constant K such that, for $\Omega \in [0.5, 1]$, we have the uniform bound*

$$G^{i}_{m,m+h;l+h,l} \leqslant K M^{-2i} e^{-\frac{\Omega}{3} M^{-2i} \|m+l+h\|}. \tag{4.11}$$

Proposition 4.2. *For M large enough, there exist two constants K and K_1 such that, for $\Omega \in [0.5, 1]$, we have the uniform bound*

$$G^{i}_{m,m+h;l+h,l} \leqslant K M^{-2i} e^{-\frac{\Omega}{4} M^{-2i} \|m+l+h\|}$$

$$\times \prod_{s=1}^{\frac{D}{2}} \min \left(1, \left(\frac{K_1 \min(m^s, l^s, m^s + h^s, l^s + h^s)}{M^{2i}} \right)^{\frac{|m^s - l^s|}{2}} \right). \tag{4.12}$$

This bound allows to prove that the only diverging graphs have either a constant index along the trajectories or a total jump of 2.

Proposition 4.3. *For M large enough, there exists a constant K such that, for $\Omega \in [\frac{2}{3}, 1]$, we have the uniform bound*

$$\sum_{l=-m}^{p} G^{i}_{m,p-l,p,m+l} \leqslant K M^{-2i} e^{-\frac{\Omega}{4} M^{-2i}(\|p\| + \|m\|)} \, . \tag{4.13}$$

This bound shows that the propagator is almost local in the following sense: with m fixed, the sum over l doesn't cost anything (see Figure 6). Nevertheless the sums we'll have to perform are entangled (a given index may enter different propagators) so that we need the following proposition.

Proposition 4.4. *For M large enough, there exists a constant K such that, for $\Omega \in [\frac{2}{3}, 1]$, we have the uniform bound*

$$\sum_{l=-m}^{\infty} \max_{p \geqslant \max(l,0)} G^{i}_{m,p-l;p,m+l} \leqslant K M^{-2i} e^{-\frac{\Omega}{36} M^{-2i} \|m\|} \, . \tag{4.14}$$

We refer to [42] for the proofs of these four propositions.

4.2.2. Power counting. About half of the $4V$ indices initially associated to a graph is determined by the external indices and the delta functions in (4.7). The other indices are summation indices. The power counting consists in finding which sums cost M^{2i} and which cost $\mathcal{O}(1)$ thanks to (4.13). The M^{2i} factor comes from (4.11) after a summation over an index[18] $m \in \mathbb{N}^2$,

$$\sum_{m^1,m^2=0}^{\infty} e^{-cM^{-2i}(m^1+m^2)} = \frac{1}{(1-e^{-cM^{-2i}})^2} = \frac{M^{4i}}{c^2}(1+\mathcal{O}(M^{-2i})). \qquad (4.15)$$

We first use the delta functions as much as possible to reduce the set $\mathcal{I}$ to a true minimal set $\mathcal{I}'$ of independent indices. For this, it is convenient to use the dual graphs where the resolution of the delta functions is equivalent to a usual momentum routing.

The dual graph is made of the same propagators than the direct graph except the position of their indices. Whereas in the original graph we have $G_{mn;kl} = $ the position of the indices in a dual propagator is

$$G_{mn;kl} = \qquad . \qquad (4.16)$$

The conservation $\delta_{l-m,-(n-k)}$ in (4.7) implies that the difference $l-m$ is conserved along the propagator. These differences behave like *angular momenta* and the conservation of the differences $\ell = l - m$ and $-\ell = n - k$ is nothing else than the conservation of the angular momentum thanks to the symmetry $SO(2) \times SO(2)$ of the action (4.1):

$$l = m + \ell \ , \ n = k + (-\ell). \qquad (4.17)$$

The cyclicity of the vertices implies the vanishing of the sum of the angular momenta entering a vertex. Thus the angular momentum in the dual graph behaves exactly like the usual momentum in ordinary Feynman graphs.

We know that the number of independent momenta is exactly the number L' $(= I - V' + 1$ for a connected graph) of loops in the dual graph. Each index at a (dual) vertex is then given by a unique *reference index* and a sum of momenta. If the dual vertex under consideration is an external one, we choose an external index for the reference index. The reference indices in the dual graph correspond to the loop indices in the direct graph. The number of summation indices is then $V' - B + L' = I + (1 - B)$ where $B \geqslant 0$ is the number of broken faces of the direct graph or the number of external vertices in the dual graph.

By using a well-chosen order on the lines, an optimized tree and an $L^1 - L^\infty$ bound, one can prove that the summation over the angular momenta does not cost anything thanks to (4.13). Recall that a connected component is a subgraph for which all internal lines have indices greater than all its external ones. The power

---

[18]Recall that each index is in fact made of two indices, one for each symplectic pair of $\mathbb{R}^4_\theta$.

counting is then:

$$A_G \leqslant K'^V \sum_{\mu} \prod_{i,k} M^{\omega(G_k^i)} \tag{4.18}$$

$$\text{with } \omega(G_k^i) = 4(V'_{i,k} - B_{i,k}) - 2I_{i,k} = 4(F_{i,k} - B_{i,k}) - 2I_{i,k} \tag{4.19}$$

$$= (4 - N_{i,k}) - 4(2g_{i,k} + B_{i,k} - 1)$$

where $N_{i,k}$, $V_{i,k}$, $I_{i,k} = 2V_{i,k} - \frac{N_{i,k}}{2}$, $F_{i,k}$ and $B_{i,k}$ are respectively the numbers of external legs, of vertices, of (internal) propagators, of faces and broken faces of the connected component G_k^i ; $g_{i,k} = 1 - \frac{1}{2}(V_{i,k} - I_{i,k} + F_{i,k})$ is its genus. We have

Theorem 4.5. *The sum over the scales attributions μ converges if $\forall i, k, \omega(G_k^i) < 0$.*

We recover the power counting obtained in [38].

From this point on, renormalizability of $\Phi_4^{\star 4}$ can proceed (however remark that it remains limited to $\Omega \in [0.5, 1]$ by the technical estimates such as (4.11); this limitation is overcome in the direct space method below).

The multiscale analysis allows to define the so-called effective expansion, in between the bare and the renormalized expansion, which is optimal, both for physical and for constructive purposes [9]. In this effective expansion only the subcontributions with all *internal* scales higher than all *external* scales have to be renormalized by counterterms of the form of the initial Lagrangian.

In fact only planar such subcontributions with a single external face must be renormalized by such counterterms. This follows simply from the Grosse-Wulkenhaar moves defined in [38]. These moves translate the external legs along the outer border of the planar graph, up to irrelevant corrections, until they all merge together into a term of the proper Moyal form, which is then absorbed in the effective constants definition. This requires only the estimates (4.11)–(4.14), which were checked numerically in [38].

In this way the relevant and marginal counterterms can be shown to be of the Moyal type, namely renormalize the parameters λ, m and Ω^{19}.

Notice that in the multiscale analysis there is no need for the relatively complicated use of Polchinski's equation [46] made in [38]. Polchinski's method, although undoubtedly very elegant for proving perturbative renormalizability does not seem directly suited to constructive purposes, even in the case of simple fermionic models such as the commutative Gross Neveu model, see, e.g., [94].

The BPHZ theorem itself for the renormalized expansion follows from finiteness of the effective expansion by developing the counterterms still hidden in the effective couplings. Its own finiteness can be checked, e.g., through the standard classification of forests [9]. Let us however recall once again that in our opinion the effective expansion, not the renormalized one is the more fundamental object, both to describe the physics and to attack deeper mathematical problems, such as those of constructive theory [9, 77].

---

[19]The wave function renormalization, i.e., renormalization of the $\partial_\mu \phi \star \partial^\mu \phi$ term can be absorbed in a rescaling of the field, called "field strength renormalization".

5. Hunting the Landau ghost

The matrix base simplifies very much at $\Omega = 1$, where the matrix propagator becomes diagonal, i.e., conserves exact indices. This property has been used for the general proof that the beta function of the theory vanishes in the ultraviolet regime [58]. At the moment this is the only concrete result that shows that NCVQFT is definitely *better* behaved than QFT. It also opens the perspective of a full non-perturbative construction of the model.

We summarize now the sequence of three papers [52,57,58] which lead to this exciting result, using the simpler notations of [58].

5.1. One loop

The propagator in the matrix base at $\Omega = 1$ is

$$C_{mn;kl} = G_{mn}\delta_{ml}\delta_{nk} \; ; \; G_{mn} = \frac{1}{A + m + n} \, , \tag{5.1}$$

where $A = 2 + \mu^2/4$, $m, n \in \mathbb{N}^2$ (μ being the mass) and we use the notation

$$\delta_{ml} = \delta_{m_1 l_1}\delta_{m_2 l_2} \, , \qquad m + m = m_1 + m_2 + n_1 + n_2 \, . \tag{5.2}$$

We focus on the complex $\bar{\phi} \star \phi \star \bar{\phi} \star \phi$ theory, since the result for the real case is similar [57]. The generating functional is:

$$Z(\eta, \bar{\eta}) = \int d\phi d\bar{\phi} \; e^{-S(\bar{\phi},\phi)+F(\bar{\eta},\eta,;\bar{\phi},\phi)}$$

$$F(\bar{\eta}, \eta; \bar{\phi}, \phi) = \bar{\phi}\eta + \bar{\eta}\phi$$

$$S(\bar{\phi}, \phi) = \bar{\phi}X\phi + \phi X\bar{\phi} + A\bar{\phi}\phi + \frac{\lambda}{2}\phi\bar{\phi}\phi\bar{\phi} \tag{5.3}$$

where traces are implicit and the matrix X_{mn} stands for $m\delta_{mn}$. S is the action and F the external sources.

We denote $\Gamma^4(0,0,0,0)$ the amputated one particle irreducible four-point function and $\Sigma(0,0)$ the amputated one particle irreducible two-point function with external indices set to zero. The wave function renormalization is $\partial_L\Sigma = \partial_R\Sigma = \Sigma(1,0) - \Sigma(0,0)$ [57], and the corresponding field strength renormalization is $Z = (1 - \partial_L\Sigma(0,0)) = (1 - \partial_R\Sigma(0,0))$ The main result to prove is that *after field strength renormalization*[20] the effective coupling is asymptotically constant, hence:

Theorem 5.1. *The equation:*

$$\Gamma^4(0,0,0,0) = \lambda Z^2 \tag{5.4}$$

holds up to irrelevant terms to all *orders of perturbation, either as a bare equation with fixed ultraviolet cutoff, or as an equation for the renormalized theory. In the*

---

[20] We recall that in the ordinary commutative ϕ_4^4 field theory there is no one loop wave-function renormalization, hence the Landau ghost can be seen directly on the four-point function renormalization at one loop.

latter case λ should still be understood as the bare constant, but reexpressed as a series in powers of λ_{ren}.

The field strength renormalization at one loop is

$$Z = 1 - a\lambda \tag{5.5}$$

where we can keep in a only the coefficient of the logarithmic divergence, as the rest does not contribute but to finite irrelevant corrections.

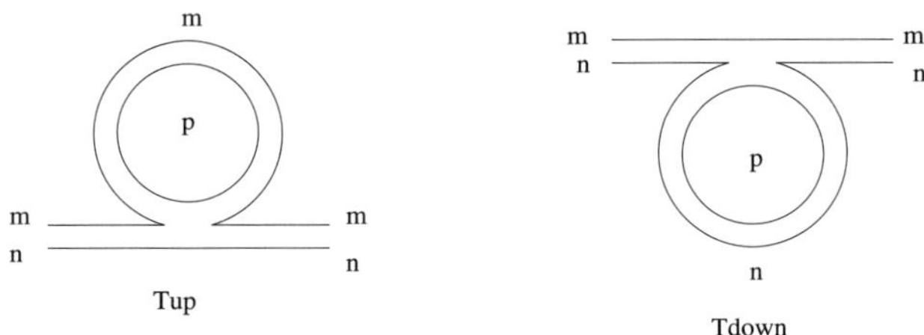

Figure 9: Two-point graphs at one loops: The up and down tadpoles.

To compute a we should add the wave function renormalization for the two tadpoles T *up* and T *down* of Figure 9. These two graphs have both a coupling constant $-\lambda/2$, and a combinatorial factor 2 for choosing to which leg of the vertex the external $\bar{\phi}$ contracts. Then the logarithmic divergence of T *up* is

$$\sum_p \left(\frac{1}{m+p+A} - \frac{1}{p+A} \right) = -\sum_p \left[\frac{m}{(m+p+A)(p+A)} \right] \tag{5.6}$$

so it corresponds to the renormalization of the coefficient of the m factor in $G_{m,n}$ in 5.1, with logarithmic divergence $\lambda \sum_p [\frac{1}{p^2}]$. Similarly the logarithmic divergence of T *down* gives the same renormalization but for the n factor in $G_{m,n}$ in 5.1.

Altogether we find therefore that

$$a = + \sum_p \left[\frac{1}{p^2} \right]. \tag{5.7}$$

In the real case we have a combinatoric factor 4 instead of 2, but the coupling constant is $\lambda/4$, so a is the same. The four-point function perturbative expansion

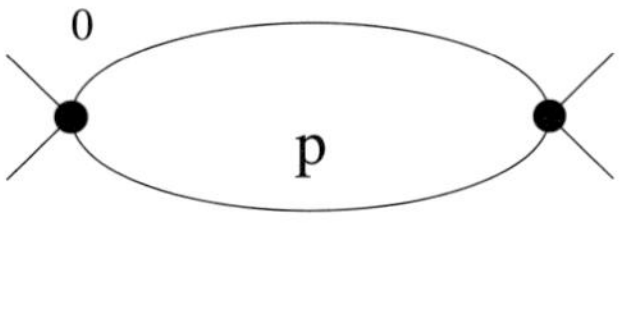

Figure 10: Four-point graph at one loop.

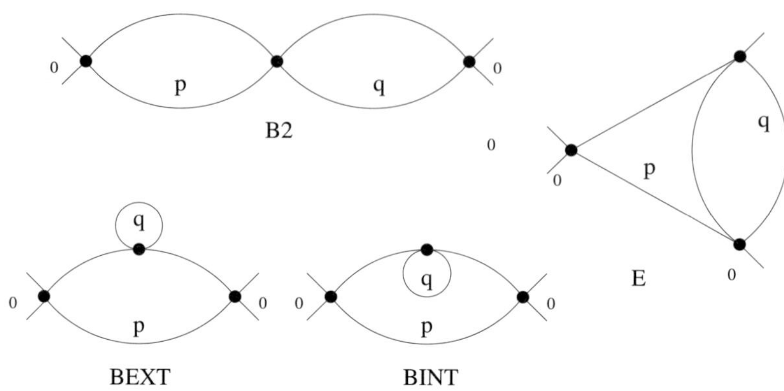

Figure 11: Four-point graphs at two loops.

at one loop is

$$\Gamma_4(0,0,0,0) = -\lambda[1 - a'\lambda]. \qquad (5.8)$$

Only the graph B1 of Figure 10 contributes to a'. It has a prefactor $\frac{1}{2!}(\lambda/2)^2$ and a combinatoric factor 2^4 for contractions, since there is a factor 2 to choose whether the bubble is "vertical or horizontal", i.e., if the horizontal bubble of Figure 10 is of $\bar{\phi} \star \phi \ \star \bar{\phi} \star \phi$ or of $\bar{\phi} \star \phi \ \star \bar{\phi} \star \phi$ type, then a factor 2 to choose to which vertex the first external; $\bar{\phi}$ contracts, then a factor 2 for the leg to which it contracts in that vertex and finally another factor 2 for the leg to which the other external $\bar{\phi}$ contracts.

The corresponding sum gives

$$a' = (2^4\lambda/8)\sum_p \frac{1}{p^2} = 2a \qquad (B1). \qquad (5.9)$$

so that at one loop equation (5.4) holds. In the real case we have a combinatoric factor 4^3 instead of 2^4, but the coupling constant is $\lambda/4$, so a is the same and (5.4) holds.

5.2. Two and three loops

This computation was extended to two and three loops in [57]. The results were given in the form of tables for the discrete divergent sums and combinatoric weights of all planar regular graphs which appears at two and three loops in Γ_4 and Z. Equation (5.4) holds again, both in the real and complex cases.

Here we simply reproduce the list of contributing Feynman graphs. Indeed it is interesting to notice that although at large order there are less planar regular graphs than the general graphs of the commutative theory, the effect is opposite at small orders.

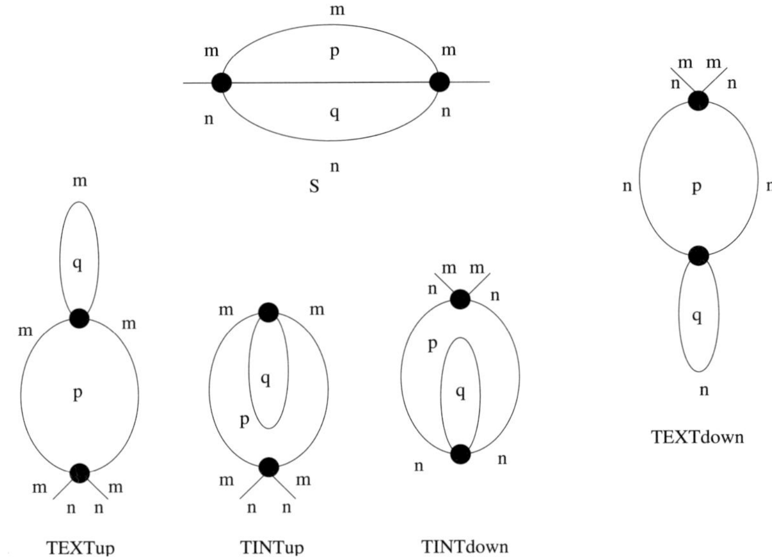

Figure 12: Two-point graphs at two loops.

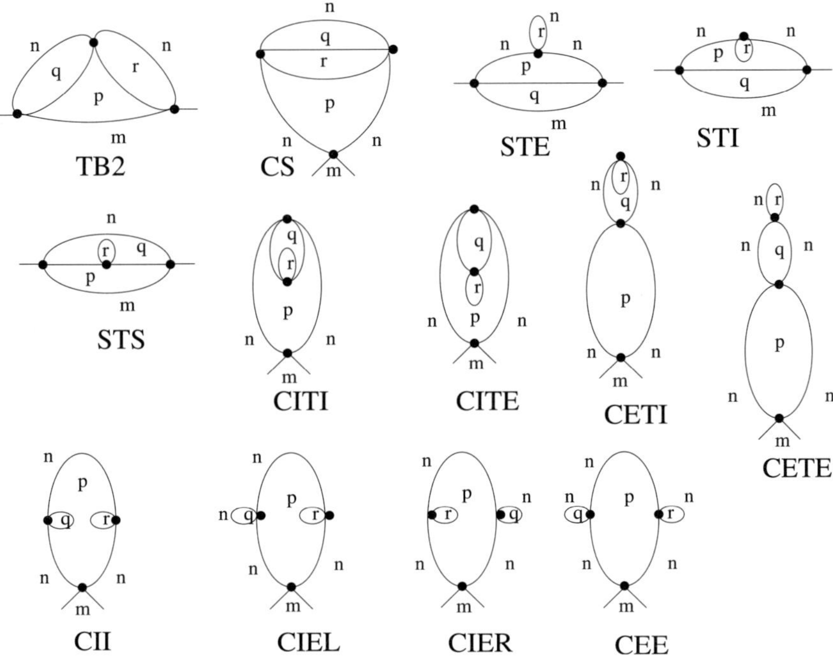

Figure 13: Two-point graphs at three loops.

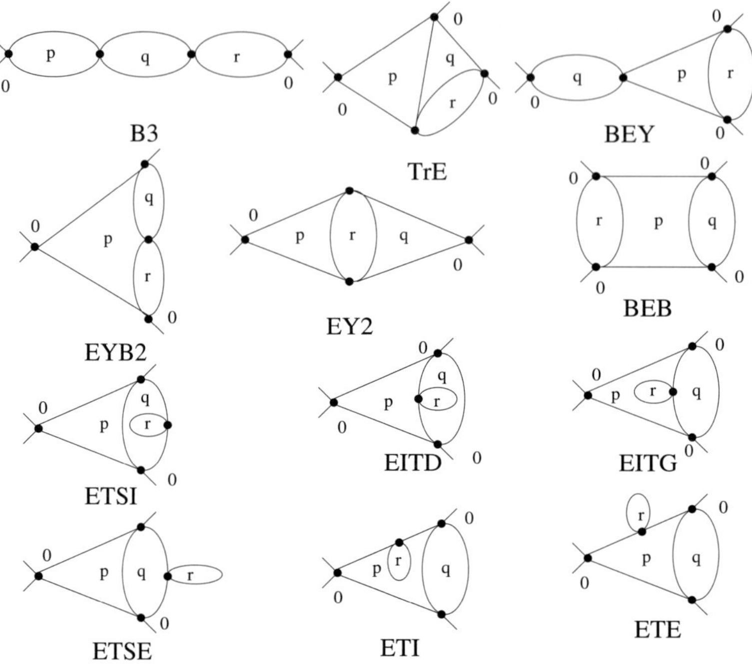

Figure 14: Four-point graphs at three loops, part I.

5.3. The general Ward identity

In this section, essentially reproduced from [58], we prove a general Ward identity which allows to check that Theorem 5.1 continues to hold at any order in perturbation theory.

We orient the propagators from a $\bar{\phi}$ to a ϕ. For a field $\bar{\phi}_{ab}$ we call the index a a *left index* and the index, b a *right index*. The first (second) index of a $\bar{\phi}$ *always* contracts with the second (first) index of a ϕ. Consequently for ϕ_{cd}, c is a *right index* and d is a *left index*.

Let $U = e^{\imath B}$ with B a small hermitian matrix. We consider the "left" (as it acts only on the left indices) change of variables:

$$\phi^U = \phi U; \bar{\phi}^U = U^\dagger \bar{\phi} \, . \tag{5.10}$$

There is a similar "right" change of variables. The variation of the action is, at first order,

$$\begin{aligned} \delta S &= \phi U X U^\dagger \bar{\phi} - \phi X \bar{\phi} \approx \imath \left(\phi B X \bar{\phi} - \phi X B \bar{\phi} \right) \\ &= \imath B \left(X \bar{\phi} \phi - \bar{\phi} \phi X \right) \end{aligned} \tag{5.11}$$

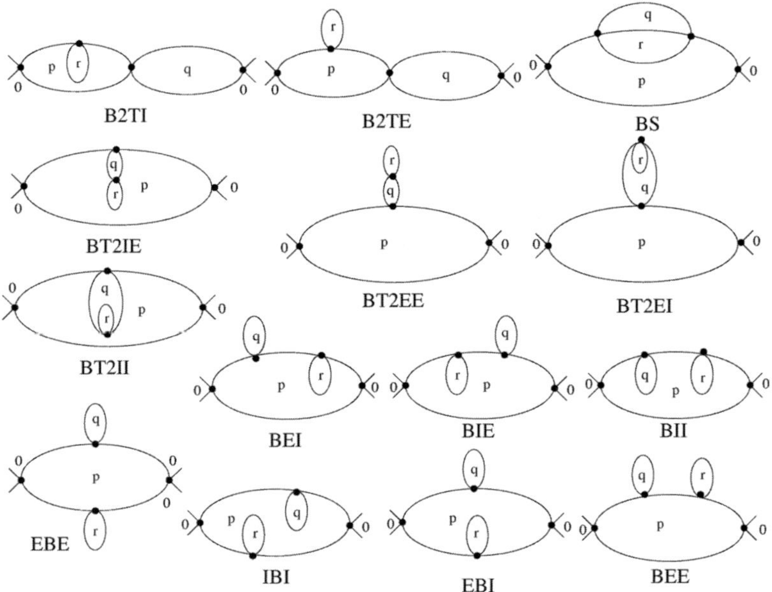

Figure 15: Four-point graphs at three loops, part II.

and the variation of the external sources is

$$\begin{aligned} \delta F &= U^\dagger \bar\phi \eta - \bar\phi \eta + \bar\eta \phi U - \bar\eta \phi \approx -\imath B \bar\phi \eta + \imath \bar\eta \phi B \\ &= \imath B \big(-\bar\phi \eta + \bar\eta \phi \big). \end{aligned} \qquad (5.12)$$

We obviously have

$$\frac{\delta \ln Z}{\delta B_{ba}} = 0 = \frac{1}{Z(\bar\eta, \eta)} \int d\bar\phi d\phi \Big(-\frac{\delta S}{\delta B_{ba}} + \frac{\delta F}{\delta B_{ba}} \Big) e^{-S+F}$$

$$= \frac{1}{Z(\bar\eta, \eta)} \int d\bar\phi d\phi \, e^{-S+F} \Big(-[X\bar\phi\phi - \bar\phi\phi X]_{ab} + [-\bar\phi\eta + \bar\eta\phi]_{ab} \Big) . \qquad (5.13)$$

We now apply $\partial_\eta \partial_{\bar\eta}|_{\eta=\bar\eta=0}$ on the above expression. As we have at most two insertions, we get only the connected components of the correlation functions.

$$0 = \Big\langle \partial_\eta \partial_{\bar\eta} \big(-[X\bar\phi\phi - \bar\phi\phi X]_{ab} + [-\bar\phi\eta + \bar\eta\phi]_{ab} \big) e^{F(\bar\eta,\eta)}|_0 \Big\rangle_c , \qquad (5.14)$$

which gives

$$\Big\langle \frac{\partial(\bar\eta\phi)_{ab}}{\partial\bar\eta} \frac{\partial(\bar\phi\eta)}{\partial\eta} - \frac{\partial(\bar\phi\eta)_{ab}}{\partial\eta} \frac{\partial(\bar\eta\phi)}{\partial\bar\eta} - [X\bar\phi\phi - \bar\phi\phi X]_{ab} \frac{\partial(\bar\eta\phi)}{\partial\bar\eta} \frac{\partial(\bar\phi\eta)}{\partial\eta} \Big\rangle_c = 0. \qquad (5.15)$$

Using the explicit form of X we get

$$(a-b) \Big\langle [\bar\phi\phi]_{ab} \frac{\partial(\bar\eta\phi)}{\partial\bar\eta} \frac{\partial(\bar\phi\eta)}{\partial\eta} \Big\rangle_c = \Big\langle \frac{\partial(\bar\eta\phi)_{ab}}{\partial\bar\eta} \frac{\partial(\bar\phi\eta)}{\partial\eta} \Big\rangle_c - \Big\langle \frac{\partial(\bar\phi\eta)_{ab}}{\partial\eta} \frac{\partial(\bar\eta\phi)}{\partial\bar\eta} \Big\rangle ,$$

and for $\bar{\eta}_{\beta\alpha}\eta_{\nu\mu}$ we get

$$(a-b)\left\langle[\bar{\phi}\phi]_{ab}\phi_{\alpha\beta}\bar{\phi}_{\mu\nu}\right\rangle_c = \left\langle\delta_{a\beta}\phi_{ab}\bar{\phi}_{\mu\nu}\right\rangle_c - \left\langle\delta_{b\mu}\bar{\phi}_{a\nu}\phi_{\alpha\beta}\right\rangle_c . \qquad (5.16)$$

We restrict to terms in the above expressions which are planar with a single external face, as all others are irrelevant. Such terms have $\alpha = \nu$, $a = \beta$ and $b = \mu$. The Ward identity for the two-point function reads

$$(a-b)\left\langle[\bar{\phi}\phi]_{ab}\phi_{\nu a}\bar{\phi}_{b\nu}\right\rangle_c = \left\langle\phi_{\nu b}\bar{\phi}_{b\nu}\right\rangle_c - \left\langle\bar{\phi}_{a\nu}\phi_{\nu a}\right\rangle_c \qquad (5.17)$$

(repeated indices are not summed up).

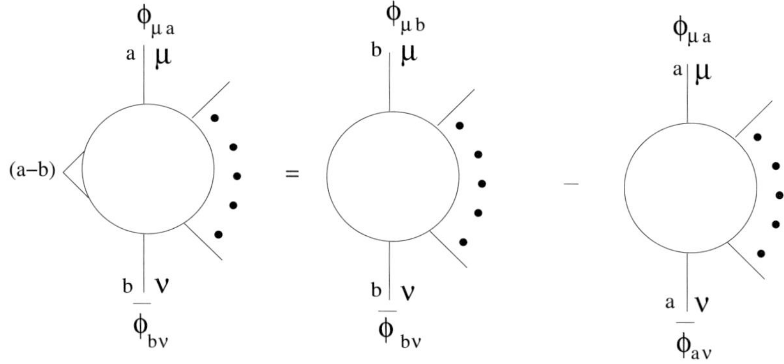

Figure 16: The Ward identity for a 2p-point function with insertion on the left face.

Derivating further we get

$$(a-b)\left\langle[\bar{\phi}\phi]_{ab}\partial_{\bar{\eta}_1}(\bar{\eta}\phi)\partial_{\eta_1}(\bar{\phi}\eta)\partial_{\bar{\eta}_2}(\bar{\eta}\phi)\partial_{\eta_2}(\bar{\phi}\eta)\right\rangle_c \qquad (5.18)$$
$$= \left\langle\partial_{\bar{\eta}_1}(\bar{\eta}\phi)\partial_{\eta_1}(\bar{\phi}\eta)\left[\partial_{\bar{\eta}_2}(\bar{\eta}\phi)_{ab}\partial_{\eta_2}(\bar{\phi}\eta) - \partial_{\eta_2}(\bar{\phi}\eta)_{ab}\partial_{\bar{\eta}_2}(\bar{\eta}\phi)\right]\right\rangle_c + 1 \leftrightarrow 2 .$$

Take $\bar{\eta}_1{}_{\beta\alpha}$, $\eta_1{}_{\nu\mu}$, $\bar{\eta}_2{}_{\delta\gamma}$ and $\eta_2{}_{\sigma\rho}$. We get

$$(a-b)\left\langle[\bar{\phi}\phi]_{ab}\phi_{\alpha\beta}\bar{\phi}_{\mu\nu}\phi_{\gamma\delta}\bar{\phi}_{\rho\sigma}\right\rangle_c \qquad (5.19)$$
$$= \left\langle\phi_{\alpha\beta}\bar{\phi}_{\mu\nu}\delta_{a\delta}\phi_{\gamma b}\bar{\phi}_{\rho\sigma}\right\rangle_c - \left\langle\phi_{\alpha\beta}\bar{\phi}_{\mu\nu}\phi_{\gamma\delta}\bar{\phi}_{a\sigma}\delta_{b\rho}\right\rangle_c +$$
$$+ \left\langle\phi_{\gamma\delta}\bar{\phi}_{\rho\sigma}\delta_{a\beta}\phi_{ab}\bar{\phi}_{\mu\nu}\right\rangle_c - \left\langle\phi_{\gamma\delta}\bar{\phi}_{\rho\sigma}\phi_{\alpha\beta}\bar{\phi}_{a\nu}\delta_{b\mu}\right\rangle_c .$$

Again neglecting all terms which are not planar with a single external face leads to

$$(a-b)\left\langle\phi_{\alpha a}[\bar{\phi}\phi]_{ab}\bar{\phi}_{b\nu}\phi_{\nu\delta}\bar{\phi}_{\delta\alpha}\right\rangle_c = \left\langle\phi_{ab}\bar{\phi}_{b\nu}\phi_{\nu\delta}\bar{\phi}_{\delta\alpha}\right\rangle_c - \left\langle\phi_{\alpha a}\bar{\phi}_{a\nu}\phi_{\nu\delta}\bar{\phi}_{\delta\alpha}\right\rangle_c . \qquad (5.20)$$

Clearly there are similar identities for 2p-point functions for any p.

The indices a and b are left indices, so that we have the Ward identity with an insertion on a left face as represented in Figure 16. There is a similar Ward identity obtained with the "right" transformation, consequently with the insertion on a right face.

5.3.1. Proof of Theorem 5.1. We start this section by some definitions: we will denote $G^4(m, n, k, l)$ the connected four-point function restricted to the planar one broken face case, where m, n, k, l are the indices of the external face in the correct cyclic order. The first index m always represents a left index.

Similarly, $G^2(m, n)$ is the connected planar one broken face two-point function with m, n the indices on the external face (also called the *dressed* propagator, see Figure 17). $G^2(m, n)$ and $\Sigma(m, n)$ are related by

$$G^2(m, n) = \frac{C_{mn}}{1 - C_{mn}\Sigma(m, n)} = \frac{1}{C_{mn}^{-1} - \Sigma(m, n)}. \tag{5.21}$$

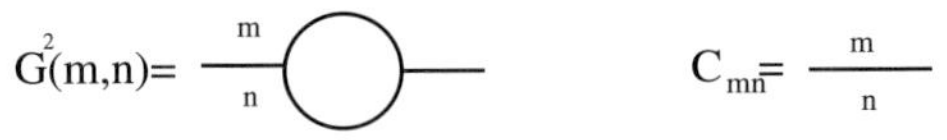

Figure 17: The *dressed* and the bare propagators.

$G_{\mathrm{ins}}(a, b; \dots)$ will denote the planar one broken face connected function with one insertion on the left border where the matrix index jumps from a to b. With this notation the Ward identity (5.17) writes:

$$(a - b)\, G_{\mathrm{ins}}^2(a, b; \nu) = G^2(b, \nu) - G^2(a, \nu). \tag{5.22}$$

All the identities we use, either Ward identities or the Dyson equation of motion can be written either for the bare theory or for the theory with complete mass renormalization, which is the one considered in [57]. In the first case the parameter A in (5.1) is the bare one, A_{bare} and there is no mass subtraction. In the second case the parameter A in (5.1) is $A_{\mathrm{ren}} = A_{\mathrm{bare}} - \Sigma(0, 0)$, and every two-point 1PI subgraph is subtracted at 0 external indices[21]. ∂_L denotes the derivative with respect to a left index and ∂_R the one with respect to a right index. When the two derivatives are equal we use the generic notation ∂.

Let us prove first the theorem in the mass-renormalized case, then in the next subsection in the bare case. Indeed the mass-renormalized theory used is free from any quadratic divergences. Remaining logarithmic subdivergences in the ultra violet cutoff can be removed easily by passing to the effective series as explained in [57].

We analyze a four-point connected function $G^4(0, m, 0, m)$ with index $m \neq 0$ on the right borders. This explicit break of left-right symmetry is adapted to our problem.

Consider a $\bar{\phi}$ external line and the first vertex hooked to it. Turning right on the m border at this vertex we meet a new line (the slashed line in Figure 18). The slashed line either separates the graph into two disconnected components ($G^4_{(1)}$ and $G^4_{(2)}$ in Figure 18) or not ($G^4_{(3)}$ in Figure 18). Furthermore, if the slashed line separates the graph into two disconnected components the first vertex may

---

[21] These mass subtractions need not be rearranged into forests since 1PI two-point subgraphs never overlap non trivially.

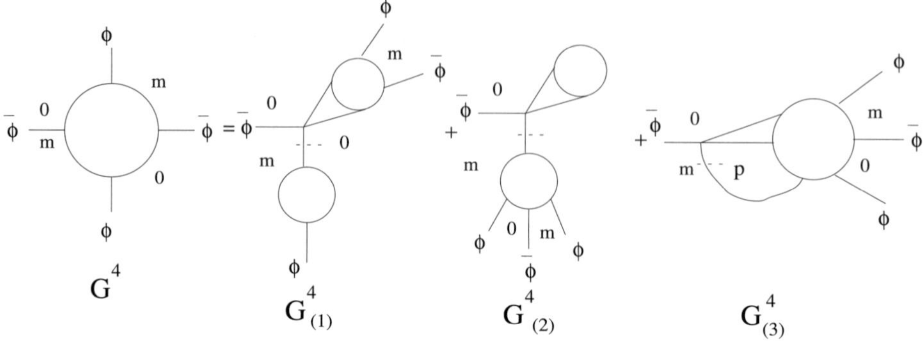

Figure 18: The Dyson equation.

either belong to the four-point component ($G^4_{(1)}$ in Figure 18) or to the two-point component ($G^4_{(2)}$ in Figure 18).

We stress that this is a *classification* of graphs: the different components depicted in Figure 18 take into account all the combinatoric factors. Furthermore, the setting of the external indices to 0 on the left borders and m on the right borders distinguishes the $G^4_{(1)}$ and $G^4_{(2)}$ from their counterparts "pointing upwards": indeed, the latter are classified in $G^4_{(3)}$!

We have thus the Dyson equation:

$$G^4(0, m, 0, m) = G^4_{(1)}(0, m, 0, m) + G^4_{(2)}(0, m, 0, m) + G^4_{(3)}(0, m, 0, m)\,. \quad (5.23)$$

The second term, $G^4_{(2)}$, is zero. Indeed the mass renormalized two-point insertion is zero, as it has the external left index set to zero. Note that this is an insertion exclusively on the left border. The simplest case of such an insertion is a (left) tadpole. We will (naturally) call a general insertion touching only the left border a "generalized left tadpole".

We will prove that $G^4_{(1)} + G^4_{(3)}$ yields $\Gamma^4 = \lambda(1 - \partial\Sigma)^2$ after amputation of the four external propagators.

We start with $G^4_{(1)}$. It is of the form

$$G^4_{(1)}(0, m, 0, m) = \lambda C_{0m} G^2(0, m) G^2_{\text{ins}}(0, 0; m)\,. \quad (5.24)$$

By the Ward identity we have

$$G^2_{\text{ins}}(0, 0; m) = \lim_{\zeta \to 0} G^2_{\text{ins}}(\zeta, 0; m) = \lim_{\zeta \to 0} \frac{G^2(0, m) - G^2(\zeta, m)}{\zeta}$$

$$= -\partial_L G^2(0, m)\,. \quad (5.25)$$

Using the explicit form of the bare propagator we have $\partial_L C_{ab}^{-1} = \partial_R C_{ab}^{-1} = \partial C_{ab}^{-1} = 1$. Reexpressing $G^2(0,m)$ by equation (5.21) we conclude that:

$$G^4_{(1)}(0,m,0,m) = \lambda C_{0m} \frac{C_{0m} C_{0m}^2 [1 - \partial_L \Sigma(0,m)]}{[1 - C_{0m}\Sigma(0,m)](1 - C_{0m}\Sigma(0,m))^2}$$

$$= \lambda [G^2(0,m)]^4 \frac{C_{0m}}{G^2(0,m)} [1 - \partial_L \Sigma(0,m)]. \tag{5.26}$$

The self energy is (again up to irrelevant terms [40])

$$\Sigma(m,n) = \Sigma(0,0) + (m+n)\partial\Sigma(0,0). \tag{5.27}$$

Therefore up to irrelevant terms $(C_{0m}^{-1} = m + A_{\mathrm{ren}})$ we have

$$G^2(0,m) = \frac{1}{m + A_{\mathrm{bare}} - \Sigma(0,m)} = \frac{1}{m[1 - \partial\Sigma(0,0)] + A_{\mathrm{ren}}}, \tag{5.28}$$

and

$$\frac{C_{0m}}{G^2(0,m)} = 1 - \partial\Sigma(0,0) + \frac{A_{\mathrm{ren}}}{m + A_{\mathrm{ren}}}\partial\Sigma(0,0). \tag{5.29}$$

Inserting equation (5.29) in equation (5.26) gives

$$G^4_{(1)}(0,m,0,m) = \lambda [G^2(0,m)]^4 \left(1 - \partial\Sigma(0,0) + \frac{A_{\mathrm{ren}}}{m + A_{\mathrm{ren}}}\partial\Sigma(0,0)\right)$$

$$[1 - \partial_L \Sigma(0,m)]. \tag{5.30}$$

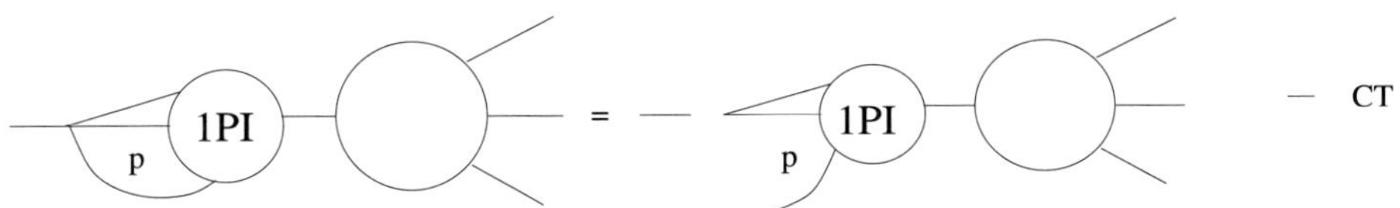

Figure 19: Two-point insertion and opening of the loop with index p.

For the $G^4_{(3)}(0,m,0,m)$ one starts by "opening" the face which is "first on the right". The summed index of this face is called p (see Figure 18). For bare Green functions this reads

$$G^{4,\mathrm{bare}}_{(3)}(0,m,0,m) = C_{0m} \sum_p G^{4,\mathrm{bare}}_{\mathrm{ins}}(p,0;m,0,m). \tag{5.31}$$

When passing to mass renormalized Green functions one must be cautious. It is possible that the face p belonged to a 1PI two-point insertion in $G^4_{(3)}$ (see the left-hand side in Figure 19). Upon opening the face p this two-point insertion disappears (see right-hand side of Figure 19)! When renormalizing, the counterterm corresponding to this kind of two-point insertion will be subtracted on the left-hand side of equation (5.31), but not on the right-hand side. In the equation for $G^4_{(3)}(0, m, 0, m)$ one must therefore *add its missing counterterm*, so that

$$G^4_{(3)}(0, m, 0, m) = C_{0m} \sum_p G^4_{\text{ins}}(0, p; m, 0, m)$$

$$- C_{0m}(CT_{lost})G^4(0, m, 0, m) . \tag{5.32}$$

It is clear that not all 1PI two-point insertions on the left-hand side of Figure 19 will be "lost" on the right-hand side. If the insertion is a "generalized left tadpole" it is not "lost" by opening the face p (imagine a tadpole pointing upwards in Figure 19: clearly it will not be opened by opening the line). We will call the two-point 1PI insertions "lost" on the right-hand side $\Sigma^R(m, n)$. Denoting the generalized left tadpole T^L we can write (see Fig .20):

$$\Sigma(m, n) = T^L(m, n) + \Sigma^R(m, n) . \tag{5.33}$$

Note that as $T^L(m, n)$ is an insertion exclusively on the left border, it does not depend upon the right index n. We therefore have $\partial\Sigma(m, n) = \partial_R\Sigma(m, n) = \partial_R\Sigma^R(m, n)$.

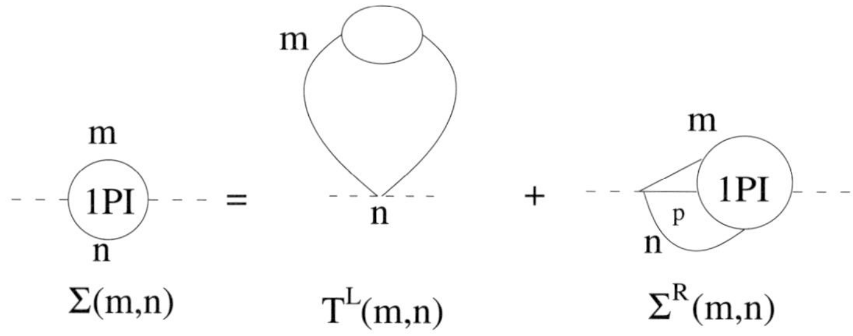

Figure 20: The self energy.

The missing mass counterterm writes

$$CT_{lost} = \Sigma^R(0, 0) = \Sigma(0, 0) - T^L . \tag{5.34}$$

In order to evaluate $\Sigma^R(0, 0)$ we procede by opening its face p and using the Ward identity (5.17), to obtain

$$\Sigma^R(0, 0) = \frac{1}{G^2(0, 0)} \sum_p G^2_{\text{ins}}(0, p; 0)$$

$$= \frac{1}{G^2(0, 0)} \sum_p \frac{1}{p}[G^2(0, 0) - G^2(p, 0)]$$

$$= \sum_p \frac{1}{p}\left(1 - \frac{G^2(p, 0)}{G^2(0, 0)}\right) . \tag{5.35}$$

Using equation (5.32) and equation (5.35) we have

$$G^4_{(3)}(0,m,0,m) = C_{0m} \sum_p G^4_{\text{ins}}(0,p;m,0,m)$$
$$- C_{0m} G^4(0,m,0,m) \sum_p \frac{1}{p}\left(1 - \frac{G^2(p,0)}{G^2(0,0)}\right). \tag{5.36}$$

But by the Ward identity (5.20)

$$C_{0m} \sum_p G^4_{\text{ins}}(0,p;m,0,m) = C_{0m} \sum_p \frac{1}{p}\left(G^4(0,m,0,m) - G^4(p,m,0,m)\right). \tag{5.37}$$

The second term in equation (5.37), having at least three denominators linear in p, is irrelevant[22] . Substituting equation (5.37) in equation (5.36) we have

$$G^4_{(3)}(0,m,0,m) = C_{0m} \frac{G^4(0,m,0,m)}{G^2(0,0)} \sum_p \frac{G^2(p,0)}{p}. \tag{5.38}$$

To conclude we must evaluate the sum in equation (5.38). Using equation (5.28) we have

$$\sum_p \frac{G^2(p,0)}{p} = \sum_p \frac{G^2(p,0)}{p}\left(\frac{1}{G^2(0,1)} - \frac{1}{G^2(0,0)}\right)\frac{1}{1-\partial\Sigma(0,0)}. \tag{5.39}$$

In order to interpret the two terms in the above equation we start by performing the same manipulations as in equation (5.35) for $\Sigma^R(0,1)$. We get

$$\Sigma^R(0,1) = \sum_p \frac{1}{p}\left(1 - \frac{G^2(p,1)}{G^2(0,1)}\right) = \sum_p \frac{1}{p}\left(1 - \frac{G^2(p,0)}{G^2(0,1)}\right). \tag{5.40}$$

where in the second equality we have neglected an irrelevant term.

Substituting equation (5.35) and equation (5.40) in equation (5.39) we get

$$\sum_p \frac{G^2(p,0)}{p} = \frac{\Sigma^R(0,0) - \Sigma^R(0,1)}{1-\partial\Sigma(0,0)} = -\frac{\partial_R\Sigma^R(0,0)}{1-\partial\Sigma(0,0)} = -\frac{\partial\Sigma(0,0)}{1-\partial\Sigma(0,0)}, \tag{5.41}$$

as $\partial_R\Sigma^R = \partial\Sigma$. Hence:

$$G^4_{(3)}(0,m,0,m;p) = -C_{0m} G^4(0,m,0,m)\frac{1}{G^2(0,0)}\frac{\partial\Sigma(0,0)}{1-\partial\Sigma(0,0)}$$
$$= -G^4(0,m,0,m)\frac{A_{\text{ren}}\,\partial\Sigma(0,0)}{(m + A_{\text{ren}})[1 - \partial\Sigma(0,0)]}. \tag{5.42}$$

---

[22] Any perturbation order of $G^4(p,m,0,m)$ is a polynomial in $\ln(p)$ divided by p^2. Therefore the sums over p above are always convergent.

Using (5.30) and (5.42), equation (5.23) rewrites as

$$G^4(0, m, 0, m)\left(1 + \frac{A_{\text{ren}}\,\partial\Sigma(0,0)}{(m + A_{\text{ren}})\,[1 - \partial\Sigma(0,0)]}\right) \tag{5.43}$$

$$= \lambda_{\text{bare}}(G^2(0,m))^4\left(1 - \partial\Sigma(0,0) + \frac{A_{\text{ren}}}{m + A_{\text{ren}}}\partial\Sigma(0,0)\right)[1 - \partial_L\Sigma(0,m)]\,.$$

We multiply (5.43) by $[1 - \partial\Sigma(0,0)]$ and amputate four times. As the differences $\Gamma^4(0, m, 0, m,) - \Gamma^4(0,0,0,0)$ and $\partial_L\Sigma(0,m) - \partial_L\Sigma(0,0)$ are irrelevant we get:

$$\Gamma^4(0,0,0,0) = \lambda(1 - \partial\Sigma(0,0))^2\,. \tag{5.44}$$

□

5.3.2. Bare identity.

5.3.2. Bare identity. Let us explain now why the main theorem is also true as an identity between bare functions, without any renormalization, but with ultraviolet cutoff.

Using the same Ward identities, all the equations go through with only few differences:

- we should no longer add the lost mass counterterm in (5.34);
- the term $G^4_{(2)}$ is no longer zero;
- equation (5.29) and all propagators now involve the bare A parameter.

But these effects compensate. Indeed the bare $G^4_{(2)}$ term is the left generalized tadpole $\Sigma - \Sigma^R$, hence

$$G^4_{(2)}(0, m, 0, m) = C_{0,m}\big(\Sigma(0,m) - \Sigma^R(0,m)\big)G^4(0, m, 0, m)\,. \tag{5.45}$$

Equation (5.29) becomes up to irrelevant terms

$$\frac{C^{\text{bare}}_{0m}}{G^{2,\text{bare}}(0,m)} = 1 - \partial_L\Sigma(0,0) + \frac{A_{\text{bare}}}{m + A_{\text{bare}}}\partial_L\Sigma(0,0) - \frac{1}{m + A_{\text{bare}}}\Sigma(0,0)\,. \tag{5.46}$$

The first term proportional to $\Sigma(0, m)$ in (5.45) combines with the new term in (5.46), and the second term proportional to $\Sigma^R(0, m)$ in (5.45) is exactly the former "lost counterterm" (5.34). This proves (5.4) in the bare case.

5.4. The RG flow

It remains to understand better the meaning of the Langmann-Szabo symmetry which certainly lies behind this Ward identity. Of course we also need to develop a non-perturbative or constructive analysis of the theory to fully confirm the absence of the Landau ghost. If this constructive analysis confirms the perturbative picture the expected non-perturbative flow for the effective parameters λ and Ω should be:

$$\frac{d\lambda_i}{di} \simeq a(1 - \Omega_i)F(\lambda_i)\,, \tag{5.47}$$

$$\frac{d\Omega_i}{di} \simeq b(1 - \Omega_i)G(\lambda_i)\,, \tag{5.48}$$

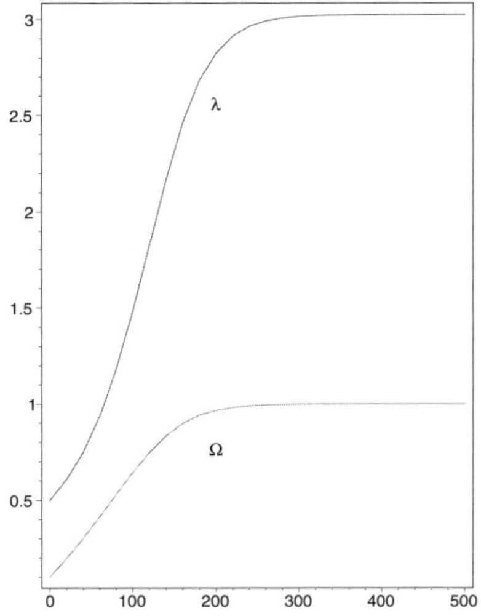

Figure 21: Numerical flow for λ and Ω.

where $F(\lambda_i) = \lambda_i^2 + O(\lambda_i^3)$, $G(\lambda_i) = \lambda_i + O(\lambda_i^2)$ and $a, b \in \mathbb{R}$ are two constants. The behavior of this system is qualitatively the same as the simpler system

$$\frac{d\lambda_i}{di} \simeq a(1 - \Omega_i)\lambda_i^2 \ , \tag{5.49}$$

$$\frac{d\Omega_i}{di} \simeq b(1 - \Omega_i)\lambda_i \ , \tag{5.50}$$

whose solution is

$$\lambda_i = \lambda_0 e^{\frac{a}{b}(\Omega_i - \Omega_0)} \ , \tag{5.51}$$

with Ω_i solution of

$$b\, i\, \lambda_0 = \int_{1-\Omega_i}^{1-\Omega_0} e^{\frac{au}{b}} \frac{du}{u} \ , \tag{5.52}$$

hence going exponentially fast to 1 as i goes to infinity. The corresponding numerical flow is drawn on Figure 21.

Of course to establish fully rigorously this picture is beyond the reach of perturbative theorems and requires a constructive analysis.

6. Propagators on non-commutative space

We give here the results we derived in [50]. In this article, we computed the x-space and matrix basis kernels of operators which generalize the Mehler kernel (3.34). Then we proceeded to a study of the scaling behaviors of these kernels in the matrix basis. This work is useful to study the non-commutative Gross-Neveu model in the matrix basis.

6.1. Bosonic kernel

The following lemma generalizes the Mehler kernel [89]:

Lemma 6.1. *Let H be the operator*

$$H = \frac{1}{2}\Big(-\Delta + \Omega^2 x^2 - 2\imath B(x_0\partial_1 - x_1\partial_0)\Big). \tag{6.1}$$

The x-space kernel of e^{-tH} is

$$e^{-tH}(x,x') = \frac{\Omega}{2\pi\sinh\Omega t}\,e^{-A}, \tag{6.2}$$

$$A = \frac{\Omega\cosh\Omega t}{2\sinh\Omega t}(x^2 + x'^2) - \frac{\Omega\cosh Bt}{\sinh\Omega t}x\cdot x' - \imath\frac{\Omega\sinh Bt}{\sinh\Omega t}x\wedge x'. \tag{6.3}$$

Remark. *The Mehler kernel corresponds to $B = 0$. The limit $\Omega = B \to 0$ gives the usual heat kernel.*

Lemma 6.2. *Let H be given by (6.1) with $\Omega(B) \to 2\Omega/\theta(2B\theta)$. Its inverse in the matrix basis is*

$$H^{-1}_{m,m+h;l+h,l} = \frac{\theta}{8\Omega}\int_0^1 d\alpha\,\frac{(1-\alpha)^{\frac{\mu_0^2\theta}{8\Omega}+(\frac{D}{4}-1)}}{(1+C\alpha)^{\frac{D}{2}}}(1-\alpha)^{-\frac{4B}{8\Omega}h}\prod_{s=1}^{\frac{D}{2}}G^{(\alpha)}_{m^s,m^s+h^s;l^s+h^s,l^s},$$

$$G^{(\alpha)}_{m,m+h;l+h,l} = \left(\frac{\sqrt{1-\alpha}}{1+C\alpha}\right)^{m+l+h} \tag{6.4}$$

$$\times \sum_{u=\max(0,-h)}^{\min(m,l)} \mathcal{A}(m,l,h,u)\left(\frac{C\alpha(1+\Omega)}{\sqrt{1-\alpha}(1-\Omega)}\right)^{m+l-2u},$$

where $\mathcal{A}(m,l,h,u) = \sqrt{\binom{m}{m-u}\binom{m+h}{m-u}\binom{l}{l-u}\binom{l+h}{l-u}}$ and C is a function of Ω: $C(\Omega) = \frac{(1-\Omega)^2}{4\Omega}$.

6.2. Fermionic kernel

On the Moyal space, we modified the commutative Gross-Neveu model by adding a $\tilde{\slashed{x}}$ term (see Lemma 3.4). We have

$$G(x,y) = -\frac{\Omega}{\theta\pi}\int_0^\infty \frac{dt}{\sinh(2\tilde{\Omega}t)}e^{-\frac{\tilde{\Omega}}{2}\coth(2\tilde{\Omega}t)(x-y)^2+\imath\tilde{\Omega}x\wedge y} \tag{6.5}$$

$$\times \left\{\imath\tilde{\Omega}\coth(2\tilde{\Omega}t)(\slashed{x}-\slashed{y}) + \Omega(\tilde{\slashed{x}}-\tilde{\slashed{y}}) - \mu\right\}e^{-2\imath\tilde{\Omega}t\gamma^0\gamma^1}e^{-t\mu^2}.$$

It will be useful to express G in terms of commutators:

$$G(x,y) = -\frac{\Omega}{\theta\pi}\int_0^\infty dt\,\Big\{\imath\tilde{\Omega}\coth(2\tilde{\Omega}t)\left[\slashed{x},\Gamma^t\right](x,y)$$

$$+\Omega\left[\tilde{\slashed{x}},\Gamma^t\right](x,y) - \mu\Gamma^t(x,y)\Big\}e^{-2\imath\tilde{\Omega}t\gamma^0\gamma^1}e^{-t\mu^2}, \tag{6.6}$$

where

$$\Gamma^t(x,y) = \frac{1}{\sinh(2\widetilde{\Omega}t)} \, e^{-\frac{\widetilde{\Omega}}{2}\coth(2\widetilde{\Omega}t)(x-y)^2 + i\widetilde{\Omega}x\wedge y} \tag{6.7}$$

with $\widetilde{\Omega} = \frac{2\Omega}{\theta}$ and $x \wedge y = x^0 y^1 - x^1 y^0$.

We now give the expression of the fermionic kernel (6.6) in the matrix basis. The inverse of the quadratic form

$$\Delta = p^2 + \mu^2 + \frac{4\Omega^2}{\theta^2}x^2 + \frac{4B}{\theta}L_2 \tag{6.8}$$

is given by (6.4) in the preceding section:

$$\Gamma_{m,m+h;l+h,l} = \frac{\theta}{8\Omega}\int_0^1 d\alpha \, \frac{(1-\alpha)^{\frac{\mu^2\theta}{8\Omega}-\frac{1}{2}}}{(1+C\alpha)}\Gamma^\alpha_{m,m+h;l+h,l}\,, \tag{6.9}$$

$$\Gamma^{(\alpha)}_{m,m+h;l+h,l} = \left(\frac{\sqrt{1-\alpha}}{1+C\alpha}\right)^{m+l+h}(1-\alpha)^{-\frac{Bh}{2\Omega}} \tag{6.10}$$

$$\times \sum_{u=0}^{\min(m,l)} A(m,l,h,u)\left(\frac{C\alpha(1+\Omega)}{\sqrt{1-\alpha}\,(1-\Omega)}\right)^{m+l-2u}.$$

The fermionic propagator G (6.6) in the matrix basis may be deduced from the kernel (6.9). We just set $B = \Omega$, add the missing term with $\gamma^0\gamma^1$ and compute the action of $-\not{p} - \Omega\not{\tilde{x}} + \mu$ on Γ. We must then evaluate $[x^\nu, \Gamma]$ in the matrix basis:

$$\begin{aligned}\left[x^0, \Gamma\right]_{m,n;k,l} =&\, 2\pi\theta\sqrt{\frac{\theta}{8}}\Big\{\sqrt{m+1}\Gamma_{m+1,n;k,l} - \sqrt{l}\Gamma_{m,n;k,l-1} + \sqrt{m}\Gamma_{m-1,n;k,l} \\ &- \sqrt{l+1}\Gamma_{m,n;k,l+1} + \sqrt{n+1}\Gamma_{m,n+1;k,l} - \sqrt{k}\Gamma_{m,n;k-1,l} \\ &+ \sqrt{n}\Gamma_{m,n-1;k,l} - \sqrt{k+1}\Gamma_{m,n;k+1,l}\Big\}, \end{aligned} \tag{6.11}$$

$$\begin{aligned}\left[x^1, \Gamma\right]_{m,n;k,l} =&\, 2i\pi\theta\sqrt{\frac{\theta}{8}}\Big\{\sqrt{m+1}\Gamma_{m+1,n;k,l} - \sqrt{l}\Gamma_{m,n;k,l-1} - \sqrt{m}\Gamma_{m-1,n;k,l} \\ &+ \sqrt{l+1}\Gamma_{m,n;k,l+1} - \sqrt{n+1}\Gamma_{m,n+1;k,l} + \sqrt{k}\Gamma_{m,n;k-1,l} \\ &+ \sqrt{n}\Gamma_{m,n-1;k,l} - \sqrt{k+1}\Gamma_{m,n;k+1,l}\Big\}. \end{aligned} \tag{6.12}$$

This allows us to prove:

Lemma 6.3. *Let $G_{m,n;k,l}$ be the kernel, in the matrix basis, of the operator $\left(\not{p} + \Omega\not{\tilde{x}} + \mu\right)^{-1}$. We have:*

$$G_{m,n;k,l} = -\frac{2\Omega}{\theta^2\pi^2}\int_0^1 d\alpha \, G^\alpha_{m,n;k,l}, \tag{6.13}$$

$$G^{\alpha}_{m,n;k,l} = \left(i\widetilde{\Omega}\frac{2-\alpha}{\alpha}[\not{x},\Gamma^{\alpha}]_{m,n;k,l} + \Omega\left[\not{\tilde{x}},\Gamma^{\alpha}\right]_{m,n;k,l} - \mu\,\Gamma^{\alpha}_{m,n;k,l}\right)$$
$$\times\left(\frac{2-\alpha}{2\sqrt{1-\alpha}}\mathbb{1}_2 - i\frac{\alpha}{2\sqrt{1-\alpha}}\gamma^0\gamma^1\right),\tag{6.14}$$

where Γ^{α} is given by (6.10) and the commutators by the formulas (6.11) and (6.12).

The first two terms in the equation (6.14) contain commutators and will be gathered under the name $G^{\alpha,\mathrm{comm}}_{m,n;k,l}$. The last term will be called $G^{\alpha,\mathrm{mass}}_{m,n;k,l}$:

$$G^{\alpha,\mathrm{comm}}_{m,n;k,l} = \left(i\widetilde{\Omega}\frac{2-\alpha}{\alpha}[\not{x},\Gamma^{\alpha}]_{m,n;k,l} + \Omega\left[\not{\tilde{x}},\Gamma^{\alpha}\right]_{m,n;k,l}\right)$$
$$\times\left(\frac{2-\alpha}{2\sqrt{1-\alpha}}\mathbb{1}_2 - i\frac{\alpha}{2\sqrt{1-\alpha}}\gamma^0\gamma^1\right),\tag{6.15}$$

$$G^{\alpha,\mathrm{mass}}_{m,n;k,l} = -\mu\,\Gamma^{\alpha}_{m,n;k,l}\times\left(\frac{2-\alpha}{2\sqrt{1-\alpha}}\mathbb{1}_2 - i\frac{\alpha}{2\sqrt{1-\alpha}}\gamma^0\gamma^1\right).\tag{6.16}$$

6.3. Bounds

We use the multi-scale analysis to study the behavior of the propagator (6.14) and revisit more finely the bounds (4.11) to (4.14). In a slice i, the propagator is

$$\Gamma^{i}_{m,m+h,l+h,l} = \frac{\theta}{8\Omega}\int_{M^{-2i}}^{M^{-2(i-1)}} d\alpha\,\frac{(1-\alpha)^{\frac{\mu_0^2\theta}{8\Omega}-\frac{1}{2}}}{(1+C\alpha)}\Gamma^{(\alpha)}_{m,m+h;l+h,l}\,,\tag{6.17}$$

$$G_{m,n;k,l} = \sum_{i=1}^{\infty}G^{i}_{m,n;k,l}\;;\;\;G^{i}_{m,n;k,l} = -\frac{2\Omega}{\theta^2\pi^2}\int_{M^{-2i}}^{M^{-2(i-1)}} d\alpha\,G^{\alpha}_{m,n;k,l}\,.\tag{6.18}$$

Let $h = n-m$ and $p = l-m$. Without loss of generality, we assume $h \geqslant 0$ and $p \geqslant 0$. Then the smallest index among m,n,k,l is m and the biggest is $k = m+h+p$. We have:

Theorem 6.4. *Under the assumptions $h = n - m \geqslant 0$ and $p = l - m \geqslant 0$, there exist $K, c \in \mathbb{R}_+$ (c depends on Ω) such that the propagator of the non-commutative Gross-Neveu model in a slice i obeys the bound*

$$|G^{i,\mathrm{comm}}_{m,n;k,l}| \leqslant KM^{-i}\left(\chi(\alpha k > 1)\frac{\exp\{-\frac{cp^2}{1+kM^{-2i}} - \frac{cM^{-2i}}{1+k}(h-\frac{k}{1+C})^2\}}{(1+\sqrt{kM^{-2i}})}\right.$$
$$\left.+ \min(1,(\alpha k)^p)e^{-ckM^{-2i}-cp}\right).\tag{6.19}$$

The mass term is slightly different:

$$|G^{i,\mathrm{mass}}_{m,n;k,l}| \leqslant KM^{-2i}\left(\chi(\alpha k > 1)\frac{\exp\{-\frac{cp^2}{1+kM^{-2i}} - \frac{cM^{-2i}}{1+k}(h-\frac{k}{1+C})^2\}}{1+\sqrt{kM^{-2i}}}\right.$$
$$\left.+ \min(1,(\alpha k)^p)e^{-ckM^{-2i}-cp}\right).\tag{6.20}$$

Remark. We can redo the same analysis for the Φ^4 propagator and get

$$G^i_{m,n;k,l} \leqslant KM^{-2i} \min\left(1, (\alpha k)^p\right) e^{-c(M^{-2i}k+p)} \tag{6.21}$$

which allows to recover the bounds (4.11) to (4.14).

6.4. Propagators and renormalizability

Let us consider the propagator (6.13) of the non-commutative Gross-Neveu model. We saw in Section 6.3 that there exist two regions in the space of indices where the propagator behaves very differently. In one of them it behaves as the Φ^4 propagator and leads then to the same power counting. In the critical region, we have

$$G^i \leqslant K \frac{M^{-i}}{1+\sqrt{kM^{-2i}}}\, e^{-\frac{cp^2}{1+kM^{-2i}} - \frac{cM^{-2i}}{1+k}\left(h-\frac{k}{1+C}\right)^2}. \tag{6.22}$$

The point is that such a propagator does not allow to sum two reference indices with a unique line. This fact was useful in the proof of the power counting of the Φ^4 model. This leads to a *renormalizable* UV/IR mixing.

Let us consider the graph in Figure 22b where the two external lines bear an index $i \gg 1$ and the internal one an index $j < i$. The propagator (6.13) obeys the bound in Proposition 4.13 which means that it is almost local. We only have to sum over one index per internal face.

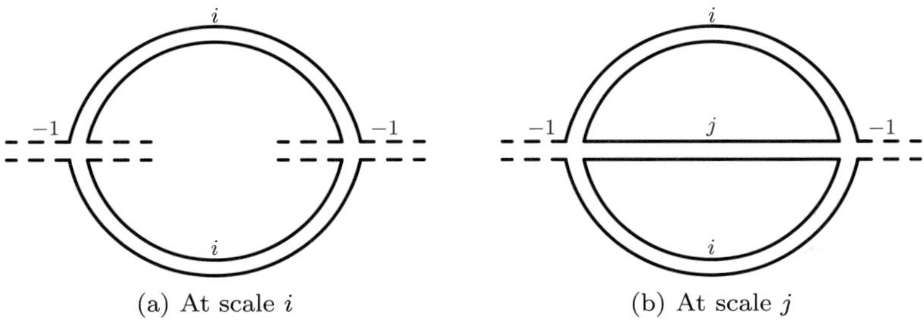

(a) At scale i (b) At scale j

Figure 22: Sunset graph.

On the graph of Figure 22a, if the two lines inside are true external ones, the graph has two broken faces and there is no index to sum over. Then by using Proposition 4.11 we get $A_G \leqslant M^{-2i}$. The sum over i converges and we have the same behavior as the Φ^4 theory, that is to say the graphs with $B \geqslant 2$ broken faces are finite. But if these two lines belongs to a line of scale $j < i$ (see Figure 22b), the result is different. Indeed, at scale i, we recover the graph of Figure 22a. To maintain the previous result (M^{-2i}), we should sum up the two indices corresponding to the internal faces with the propagator of scale j. This is not possible. Instead we have:

$$\sum_{k,h} M^{-2i-j}\, e^{-M^{-2i}k}\, \frac{e^{-\frac{cM^{-2j}}{1+k}\left(h-\frac{k}{1+C}\right)^2}}{1+\sqrt{kM^{-2j}}} \leqslant KM^j. \tag{6.23}$$

The sum over i diverges logarithmically. The graph of Figure 22a converges if it is linked to true external legs and diverges if it is a subgraph of a graph at a lower scale. The power counting depends on the scales lower than the lowest scale of the graph. It can't then be factorized into the connected components: this is UV/IR mixing.

Let's remark that the graph of Figure 22a is not renormalizable by a counterterm in the Lagrangian. Its logarithmic divergence cannot be absorbed in a redefinition of a coupling constant. Fortunately the renormalization of the two-point graph of Figure 22b makes the four-point subdivergence finite [51]. This makes the non-commutative Gross-Neveu model renormalizable.

7. Direct space

We now want to explain how the power counting analysis can be performed in direct space, and the "moyality" of the necessary counterterms can be checked by a Taylor expansion which is a generalization of the one used in direct commutative space.

In the commutative case there is translation invariance, hence each propagator depends on a single difference variable which is short in the ultraviolet regime; in the non-commutative case the propagator depends on both the difference of end positions, which is again short in the uv regime, but also on the sum which is long in the uv regime, considering the explicit form (3.34) of the Mehler kernel.

This distinction between short and long variables is at the basis of the power counting analysis in direct space.

7.1. Short and long variables

Let G be an arbitrary connected graph. The amplitude associated with this graph is in direct space (with hopefully self-explaining notation):

$$A_G = \int \prod_{v,i=1,\dots,4} dx_{v,i} \prod_l dt_l \tag{7.1}$$

$$\prod_v \left[\delta(x_{v,1} - x_{v,2} + x_{v,3} - x_{v,4}) e^{\imath \sum_{i<j}(-1)^{i+j+1} x_{v,i}\theta^{-1} x_{v,j}} \right] \prod_l C_l \,,$$

$$C_l = \frac{\Omega^2}{[2\pi \sinh(\Omega t_l)]^2} e^{-\frac{\Omega}{2}\coth(\Omega t_l)(x^2_{v,i(l)}+x^2_{v',i'(l)})+\frac{\Omega}{\sinh(\Omega t_l)} x_{v,i(l)}\cdot x_{v',i'(l)}-\mu_0^2 t_l} \,.$$

For each line l of the graph joining positions $x_{v,i(l)}$ and $x_{v',i'(l)}$, we choose an orientation and we define the "short" variable $u_l = x_{v,i(l)} - x_{v',i'(l)}$ and the "long" variable $v_l = x_{v,i(l)} + x_{v',i'(l)}$.

With this notation, defining $\Omega t_l = \alpha_l$, the propagators in our graph can be written as

$$\int_0^\infty \prod_l \frac{\Omega d\alpha_l}{[2\pi \sinh(\alpha_l)]^2} e^{-\frac{\Omega}{4}\coth(\frac{\alpha_l}{2})u_l^2 - \frac{\Omega}{4}\tanh(\frac{\alpha_l}{2})v_l^2 - \frac{\mu_0^2}{\Omega}\alpha_l} \,. \tag{7.2}$$

As in matrix space we can slice each propagator according to the size of its α parameter and obtain the multiscale representation of each Feynman amplitude:

$$A_G = \sum_\mu A_{G,\mu} \quad , \quad A_{G,\mu} = \int \prod_{v,i=1,\ldots,4} dx_{v,i} \prod_l C_l^{i_\mu(l)}(u_l, v_l) \tag{7.3}$$

$$\prod_v \left[\delta(x_{v,1} - x_{v,2} + x_{v,3} - x_{v,4}) e^{\imath \sum_{i<j}(-1)^{i+j+1} x_{v,i} \theta^{-1} x_{v,j}} \right] ,$$

$$C^i(u,v) = \int_{M^{-2i}}^{M^{-2(i-1)}} \frac{\Omega d\alpha}{[2\pi \sinh(\alpha)]^2} e^{-\frac{\Omega}{4}\coth(\frac{\alpha}{2})u^2 - \frac{\Omega}{4}\tanh(\frac{\alpha}{2})v^2 - \frac{\mu_0^2}{\Omega}\alpha} , \tag{7.4}$$

where μ runs over scales attributions $\{i_\mu(l)\}$ for each line l of the graph, and the sliced propagator C^i in slice $i \in \mathbb{N}$ obeys the crude bound:

Lemma 7.1. *For some constants K (large) and c (small):*

$$C^i(u,v) \leqslant KM^{2i} e^{-c[M^i\|u\| + M^{-i}\|v\|]} \tag{7.5}$$

(which a posteriori justifies the terminology of "long" and "'short" variables).

The proof is elementary.

7.2. Routing, Filk moves

7.2.1. Oriented graphs. We pick a tree T of lines of the graph, hence connecting all vertices, pick with a root vertex and build an *orientation* of all the lines of the graph in an inductive way. Starting from an arbitrary orientation of a field at the root of the tree, we climb in the tree and at each vertex of the tree we impose cyclic order to alternate entering and exiting tree lines and loop half-lines, as in Figure 23a. Then we look at the loop lines. If every loop lines consist in the contraction of an entering and an exiting line, the graph is called orientable. Otherwise we call it non-orientable as in Figure 23b.

7.2.2. Position routing. There are n δ functions in an amplitude with n vertices, hence n linear equations for the $4n$ positions, one for each vertex. The *position routing* associated to the tree T solves this system by passing to another equivalent system of n linear equations, one for each branch of the tree. This is a triangular change of variables, of Jacobian 1. This equivalent system is obtained by summing up the arguments of the δ functions of the vertices in each branch. This change of variables is exactly the x-space analog of the resolution of momentum conservation called *momentum routing* in the standard physics literature of commutative field theory, except that one should now take care of the additional $\pm$ cyclic signs.

One can prove [47] that the rank of the system of δ functions in an amplitude with n vertices is

- $n - 1$ if the graph is orientable;
- n if the graph is non-orientable.

The position routing change of variables is summarized by the following lemma:

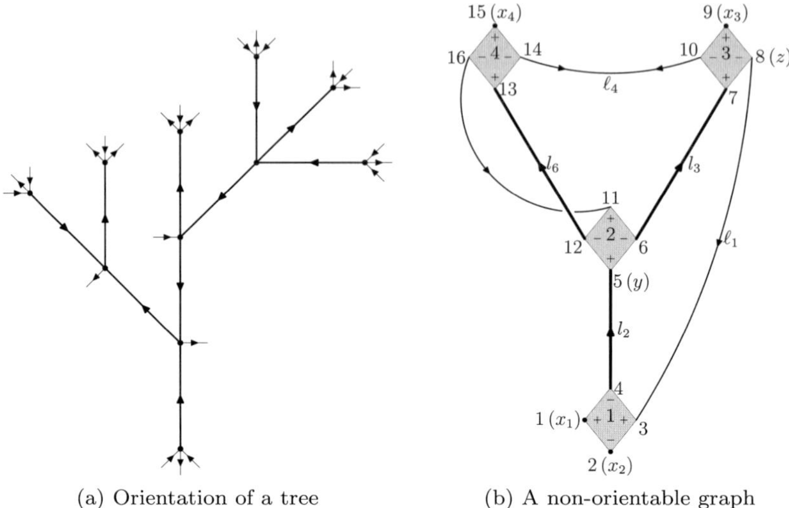

(a) Orientation of a tree   (b) A non-orientable graph

Figure 23: Orientation.

Lemma 7.2 (Position Routing). *We have, calling I_G the remaining integrand in (7.3):*

$$A_G = \int \left[\prod_v \left[\delta(x_{v,1} - x_{v,2} + x_{v,3} - x_{v,4}) \right] \right] I_G(\{x_{v,i}\}) \tag{7.6}$$

$$= \int \prod_b \delta \left(\sum_{l \in T_b \cup L_b} u_l + \sum_{l \in L_{b,+}} v_l - \sum_{l \in L_{b,-}} v_l + \sum_{f \in X_b} \epsilon(f) x_f \right) I_G(\{x_{v,i}\}),$$

where $\epsilon(f)$ is ± 1 depending on whether the field f enters or exits the branch.

We can now use the system of delta functions to eliminate variables. It is of course better to eliminate long variables as their integration costs a factor M^{4i} whereas the integration of a short variable brings M^{-4i}. Rough power counting, neglecting all oscillations of the vertices leads therefore, in the case of an orientable graph with N external fields, n internal vertices and $l = 2n - N/2$ internal lines at scale i to:

- a factor $M^{2i(2n-N/2)}$ coming from the M^{2i} factors for each line of scale i in (7.5);
- a factor $M^{-4i(2n-N/2)}$ for the $l = 2n - N/2$ short variables integrations;
- a factor $M^{4i(n-N/2+1)}$ for the long variables after eliminating $n - 1$ of them using the delta functions.

The total factor is therefore $M^{-(N-4)i}$, the ordinary scaling of ϕ_4^4, which means that only two- and four-point subgraphs ($N \leqslant 4$) diverge when i has to be summed.

In the non-orientable case, we can eliminate one additional long variable since the rank of the system of delta functions is larger by one unit! Therefore we get a power counting bound M^{-Ni}, which proves that only *orientable* graphs may diverge.

In fact we of course know that not all *orientable* two- and four-point subgraphs diverge but only the planar ones with a single external face. (It is easy to check that all such planar graphs are indeed orientable.)

Since only these planar subgraphs with a single external face can be renormalized by Moyal counterterms, we need to prove that orientable, non-planar graphs or orientable planar graphs with several external faces have in fact a better power counting than this crude estimate. This can be done only by exploiting their vertices oscillations. We explain now how to do this with minimal effort.

7.2.3. Filk moves and rosettes. Following Filk [87], we can contract all lines of a spanning tree T and reduce G to a single vertex with "tadpole loops" called a "rosette graph". This rosette is a cycle (which is the border of the former tree) bearing loop lines on it (see Figure 24): Remark that the rosette can also be

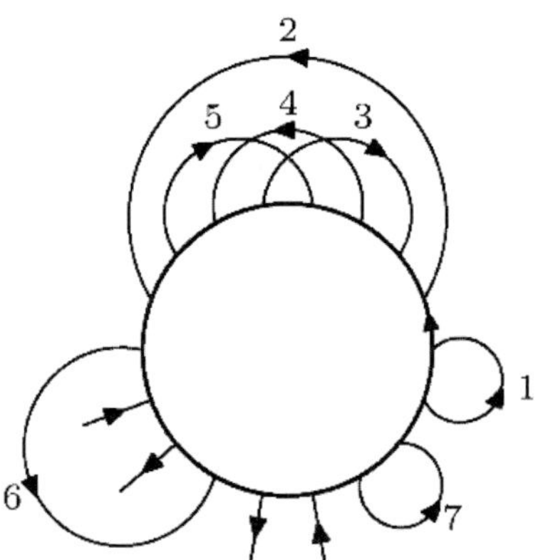

Figure 24: A rosette.

considered as a big vertex, with $r = 2n + 2$ fields, on which N are external fields with external variables x and $2n + 2 - N$ are loop fields for the corresponding $n + 1 - N/2$ loops. When the graph is orientable, the rosette is also orientable, which means that turning around the rosette the lines alternatively enter and exit. These lines correspond to the contraction of the fields on the border of the tree T before the Filk contraction, also called the "first Filk move".

7.2.4. Rosette factor. We start from the root and turn around the tree in the trigonometrical sense. We number separately all the fields as $1, \ldots, 2n + 2$ and all the tree lines as $1, \ldots, n - 1$ in the order they are met.

Lemma 7.3. *The rosette contribution after a complete first Filk reduction is exactly:*

$$\delta(v_1 - v_2 + \cdots - v_{2n+2} + \sum_{l \in T} u_l)e^{iVQV+iURU+iUSV} \tag{7.7}$$

where the v variables are the long or external variables of the rosette, counted with their signs, and the quadratic oscillations for these variables is

$$VQV = \sum_{0 \leqslant i < j \leqslant r} (-1)^{i+j+1} v_i \theta^{-1} v_j. \tag{7.8}$$

We have now to analyze in detail this quadratic oscillation of the remaining long loop variables since it is essential to improve power counting. We can neglect the secondary oscillations URU and USV which imply short variables.

The second Filk reduction [87] further simplifies the rosette factor by erasing the loops of the rosette which do not cross any other loops or arch over external fields. It can be shown that the loops which disappear in this operation correspond to those long variables who do not appear in the quadratic form Q.

Using the remaining *oscillating factors* one can prove that non-planar graphs with genus larger than one or with more than one external face *do not diverge*.

The basic mechanism to improve the power counting of a single non-planar subgraph is the following:

$$\int dw_1 dw_2 e^{-M^{-2i_1}w_1^2 - M^{-2i_2}w_2^2 - iw_1\theta^{-1}w_2 + w_1.E_1(x,u) + w_2 E_2(x,u)}$$

$$= \int dw_1' dw_2' e^{-M^{-2i_1}(w_1')^2 - M^{-2i_2}(w_2')^2 + iw_1'\theta^{-1}w_2' + (u,x)Q(u,x)}$$

$$= KM^{4i_1} \int dw_2' e^{-(M^{2i_1}+M^{-2i_2})(w_2')^2} = KM^{4i_1}M^{-4i_2} . \tag{7.9}$$

In these equations we used for simplicity M^{-2i} instead of the correct but more complicated factor $(\Omega/4)\tanh(\alpha/2)$ (of course this does not change the argument) and we performed a unitary linear change of variables $w_1' = w_1 + \ell_1(x,u)$, $w_2' = w_2 + \ell_2(x,u)$ to compute the oscillating w_1' integral. The gain in (7.9) is M^{-8i_2}, which is the difference between M^{-4i_2} and the normal factor M^{4i_2} that the w_2 integral would have cost if we had done it with the regular $e^{-M^{-2i_2}w_2^2}$ factor for long variables. To maximize this gain we can assume $i_1 \leqslant i_2$.

This basic argument must then be generalized to each non-planar subgraph in the multiscale analysis, which is possible.

Finally it remains to consider the case of subgraphs which are planar orientable but with more than one external face. In that case there are no crossing loops in the rosette but there must be at least one loop line arching over a nontrivial subset of external legs (see, e.g., line 6 in Figure 24). We have then a nontrivial integration over at least one external variable, called x, of at least one long loop variable called w. This "external" x variable without the oscillation improvement would be integrated with a test function of scale 1 (if it is a true external line of scale 1) or better (if it is a higher long loop variable)[23]. But we get now

$$\int dx\,dw\, e^{-M^{-2i}w^2 - iw\theta^{-1}x + w.E_1(x',u)} = KM^{4i} \int dx\, e^{-M^{+2i}x^2} = K' , \qquad (7.10)$$

so that a factor M^{4i} in the former bound becomes $\mathcal{O}(1)$, hence is improved by M^{-4i}.

In this way we can reduce the convergence of the multiscale analysis to the problem of renormalization of planar two- and four-point subgraphs with a single external face, which we treat in the next section.

Remark that the power counting obtained in this way is still not optimal. To get the same level of precision than with the matrix base requires, e.g., to display g independent improvements of the type (7.9) for a graph of genus g. This is doable but basically requires a reduction of the quadratic form Q for a single-faced rosette (also called "hyperrosette") into g standard symplectic blocks through the so-called "third Filk move" introduced in [68]. We return to this question in Section 8.2.

7.3. Renormalization

7.3.1. Four-point function. Consider the amplitude of a four-point graph G which in the multiscale expansion has all its internal scales higher than its four external scales.

The idea is that one should compare its amplitude to a similar amplitude with a "Moyal factor" $\exp\left(2i\theta^{-1}\left(x_1 \wedge x_2 + x_3 \wedge x_4\right)\right)\delta(\Delta)$ factorized in front, where $\Delta = x_1 - x_2 + x_3 - x_4$. But precisely because the graph is planar with a single external face we understand that the external positions x only couple to *short variables* U of the internal amplitudes through the global delta function and the oscillations. Hence we can break this coupling by a systematic Taylor expansion to first order. This separates a piece proportional to "Moyal factor", then absorbed into the effective coupling constant, and a remainder which has at least one additional small factor which gives him improved power counting.

This is done by expressing the amplitude for a graph with $N = 4$, $g = 0$ and $B = 1$ as:

$$A(G)(x_1, x_2, x_3, x_4) = \int \exp\left(2i\theta^{-1}\left(x_1 \wedge x_2 + x_3 \wedge x_4\right)\right) \prod_{\ell \in T_k^i} du_\ell \, C_\ell(u_\ell, U_\ell, V_\ell)$$

$$\times \left[\prod_{l \in G_k^i,\, l \notin T} du_l dv_l C_l(u_l, v_l) \right] e^{iURU + iUSV} \qquad (7.11)$$

$$\times \left\{ \delta(\Delta) + \int_0^1 dt \left[\mathfrak{U} \cdot \nabla \delta(\Delta + t\mathfrak{U}) + \delta(\Delta + t\mathfrak{U})[iXQU + \mathfrak{R}'(t)] \right] e^{itXQU + \mathfrak{R}(t)} \right\} .$$

---

where $C_\ell(u_\ell, U_\ell, V_\ell)$ is the propagator taken at $X_\ell = 0$, $\mathfrak{U} = \sum_\ell u_\ell$ and $\mathfrak{R}(t)$ is a correcting term involving $\tanh \alpha_\ell [X.X + X.(U+V)]$.

The first term is of the initial $\int Tr\phi \star \phi \star \phi \star \phi$ form. The rest no longer diverges, since the U and $\mathfrak{R}$ provide the necessary small factors.

7.3.2. Two-point function. Following the same strategy we have to Taylor-expand the coupling between external variables and U factors in two-point planar graphs with a single external face to *third order* and some nontrivial symmetrization of the terms according to the two external arguments to cancel some odd contributions. The corresponding factorized relevant and marginal contributions can be then shown to give rise only to

- a mass counterterm;
- a wave function counterterm;
- A harmonic potential counterterm,

and the remainder has convergent power counting. This concludes the construction of the effective expansion in this direct space multiscale analysis.

Again the BPHZ theorem itself for the renormalized expansion follows by developing the counterterms still hidden in the effective couplings and its finiteness follows from the standard classification of forests. See however the remarks at the end of Section 4.2.2.

Since the bound (7.5) works for any $\Omega \neq 0$, an additional bonus of the x-space method is that it proves renormalizability of the model for any Ω in $]0,1]^{24}$, while the matrix method proved it only for Ω in $]0.5, 1]$.

7.3.3. The Langmann-Szabo-Zarembo model. It is a four-dimensional theory of a bosonic complex field defined by the action

$$S = \int \frac{1}{2}\bar{\phi}(-D^\mu D_\mu + \Omega^2 x^2)\phi + \lambda\bar{\phi} \star \phi \star \bar{\phi} \star \phi \qquad (7.12)$$

where $D^\mu = \imath\partial_\mu + B_{\mu\nu}x^\nu$ is the covariant derivative in a magnetic field B.

The interaction $\bar{\phi} \star \phi \star \bar{\phi} \star \phi$ ensures that perturbation theory contains only orientable graphs. For $\Omega > 0$ the x-space propagator still decays as in the ordinary ϕ_4^4 case and the model has been shown renormalizable by an easy extension of the methods of the previous Section [47].

However at $\Omega = 0$, there is no longer any harmonic potential in addition to the covariant derivatives and the bounds are lost. We call models in this category *covariant*.

7.3.4. Covariant models. Consider the x-kernel of the operator

$$H^{-1} = \left(p^2 + \Omega^2\tilde{x}^2 - 2\imath B\left(x^0 p_1 - x^1 p_0\right)\right)^{-1}, \qquad (7.13)$$

---

[24]The case Ω in $[1, +\infty[$ is irrelevant since it can be rewritten by LS duality as an equivalent model with Ω in $]0, 1]$.

$$H^{-1}(x,y) = \frac{\widetilde{\Omega}}{8\pi} \int_0^\infty \frac{dt}{\sinh(2\widetilde{\Omega}t)} \exp\left(-\frac{\widetilde{\Omega}}{2}\frac{\cosh(2Bt)}{\sinh(2\widetilde{\Omega}t)}(x-y)^2\right.$$

$$-\frac{\widetilde{\Omega}}{2}\frac{\cosh(2\widetilde{\Omega}t)-\cosh(2Bt)}{\sinh(2\widetilde{\Omega}t)}(x^2+y^2)$$

$$\left.+2i\widetilde{\Omega}\frac{\sinh(2Bt)}{\sinh(2\widetilde{\Omega}t)}x\wedge y\right) \quad \text{with } \widetilde{\Omega} = \frac{2\Omega}{\theta}. \tag{7.14}$$

The Gross-Neveu model or the covariant Langmann-Szabo-Zarembo models correspond to the case $B = \widetilde{\Omega}$. In these models there is no longer any confining decay for the "long variables" but only an oscillation:

$$Q^{-1} = H^{-1} = \frac{\widetilde{\Omega}}{8\pi} \int_0^\infty \frac{dt}{\sinh(2\widetilde{\Omega}t)} \exp\left(-\frac{\widetilde{\Omega}}{2}\coth(2\widetilde{\Omega}t)(x-y)^2 + 2i\widetilde{\Omega}x\wedge y\right). \tag{7.15}$$

The construction of these covariant models is more difficult, since sufficiently many oscillations must be proven independent before power counting can be established. The prototype paper which solved this problem is [51], which we briefly summarize now.

The main technical difficulty of the covariant models is the absence of decreasing functions for the long v variables in the propagator replaced by an oscillation, see (7.15). Note that these decreasing functions are in principle created by integration over the u variables[25]:

$$\int du\, e^{-\frac{\widetilde{\Omega}}{2}\coth(2\widetilde{\Omega}t)u^2+iu\wedge v} = K\tanh(2\widetilde{\Omega}t)\,e^{-k\tanh(2\widetilde{\Omega}t)v^2}. \tag{7.16}$$

But to perform all these Gaussian integrations for a general graph is a difficult task (see [69]) and is in fact not necessary for a BPHZ theorem. We can instead exploit the vertices and propagators oscillations to get rational decreasing functions in some linear combinations of the long v variables. The difficulty is then to prove that all these linear combinations are independent and hence allow to integrate over all the v variables. To solve this problem we need the exact expression of the total oscillation in terms of the short and long variables. This consists in a generalization of the Filk's work [87]. This has been done in [51]. Once the oscillations are proven independent, one can just use the same arguments than in the Φ^4 case (see Section 7.2) to compute an upper bound for the power counting:

Lemma 7.4 (Power counting $\mathbf{GN}_\Theta^2$). *Let G be a connected orientable graph. For all $\Omega \in [0,1)$, there exists $K \in \mathbb{R}_+$ such that its amputated amplitude A_G integrated*

---

[25] In all the following we restrict ourselves to the dimension 2.

over test functions is bounded by

$$|A_G| \leqslant K^n M^{-\frac{1}{2}\omega(G)}$$
(7.17)

$$\text{with } \omega(G) = \begin{cases} N-4 & \text{if } (N=2 \text{ or } N \geqslant 6) \text{ and } g=0, \\ & \text{if } N=4,\, g=0 \text{ and } B=1, \\ & \text{if } G \text{ is critical}, \\ N & \text{if } N=4,\, g=0,\, B=2 \text{ and } G \text{ non-critical}, \\ N+4 & \text{if } g \geqslant 1. \end{cases}$$
(7.18)

As in the non-commutative Φ^4 case, only the planar graphs are divergent. But the behavior of the graphs with more than one broken face is different. Note that we already discussed such a feature in the matrix basis (see Section 6.4). In the multiscale framework, the Feynman diagrams are endowed with a scale attribution which gives each line a scale index. The only subgraphs we meet in this setting have all their internal scales higher than their external ones. Then a subgraph G of scale i is called *critical* if it has $N=4, g=0, B=2$ and that the two "external" points in the second broken face are only linked by a single line of scale $j < i$. The typical example is the graph of Figure 22a. In this case, the subgraph is logarithmically divergent whereas it is convergent in the Φ^4 model. Let us now show roughly how it happens in the case of Figure 22a but now in x-space.

The same arguments than in the Φ^4 model prove that the integrations over the internal points of the graph 22a lead to a logarithmic divergence which means that $A_{G^i} \simeq \mathcal{O}(1)$ in the multiscale framework. But remember that there is a remaining oscillation between a long variable of this graph and the external points in the second broken face of the form $v \wedge (x - y)$. But v is of order M^i which leads to a decreasing function implementing $x - y$ of order M^{-i}. If these points are true external ones, they are integrated over test functions of norm 1. Then thanks to the additional decreasing function for $x - y$ we gain a factor M^{-2i} which makes the graph convergent. But if x and y are linked by a single line of scale $j < i$ (as in Figure 22b), instead of test functions we have a propagator between x and y. This one behaves like (see (7.15)):

$$C^j(x,y) \simeq M^j\, e^{-M^{2j}(x-y)^2 + i x \wedge y}.$$
(7.19)

The integration over $x - y$ instead of giving M^{-2j} gives M^{-2i} thanks to the oscillation $v \wedge (x - y)$. Then we have gained a good factor $M^{-2(i-j)}$. But the oscillation in the propagator $x \wedge y$ now gives $x + y \simeq M^{2i}$ instead of M^{2j} and the integration over $x + y$ cancels the preceding gain. The critical component of Figure 22a is logarithmically divergent.

This kind of argument can be repeated and refined for more general graphs to prove that this problem appears only when the external points of the auxiliary broken faces are linked only by a *single* lower line [51]. This phenomenon can be seen as a mixing between scales. Indeed the power counting of a given subgraph now depends on the graphs at lower scales. This was not the case in the commutative realm. Fortunately this mixing does not prevent renormalization. Note that though the critical subgraphs are not renormalizable by a vertex-like counterterm, they are regularized by the renormalization of the two-point function at scale j. The proof of this point relies heavily on the fact that there is only one line of lower scale.

Let us conclude this section by mentioning the flows of the covariant models. One very interesting feature of the non-commutative Φ^4 model is the boundedness of its flows and even the vanishing of its beta function for a special value of its bare parameters [52, 57, 58]. Note that its commutative counterpart (the usual ϕ^4 model on $\mathbb{R}^4$) is asymptotically free in the infrared and has then an unbounded flow. It turns out that the flow of the covariant models are not regularized by the non-commutativity. The one-loop computation of the beta functions of the non-commutative Gross-Neveu model [95] shows that it is asymptotically free in the ultraviolet region as in the commutative case.

8. Parametric representation

8.1. Ordinary Symanzik polynomials

In ordinary commutative field theory, Symanzik's polynomials are obtained after integration over internal position variables. The amplitude of an amputated graph G with external momenta p is, up to a normalization, in space-time dimension D:

$$A_G(p) = \delta(\sum p) \int_0^\infty \frac{e^{-V_G(p,\alpha)/U_G(\alpha)}}{U_G(\alpha)^{D/2}} \prod_l (e^{-m^2 \alpha_l} d\alpha_l) \ . \tag{8.1}$$

The first and second Symanzik polynomials U_G and V_G are

$$U_G = \sum_T \prod_{l \notin T} \alpha_l \ , \tag{8.2a}$$

$$V_G = \sum_{T_2} \prod_{l \notin T_2} \alpha_l \Big(\sum_{i \in E(T_2)} p_i \Big)^2 \ , \tag{8.2b}$$

where the first sum is over spanning trees T of G and the second sum is over two trees T_2, i.e., forests separating the graph in exactly two connected components $E(T_2)$ and $F(T_2)$; the corresponding Euclidean invariant $(\sum_{i \in E(T_2)} p_i)^2$ is, by momentum conservation, also equal to $(\sum_{i \in F(T_2)} p_i)^2$.

There are many interesting features in the parametric representation:

- It is more compact than direct or momentum space for dimension $D > 2$, hence it is adapted to numerical computations.
- The dimension D appears now as a simple parameter. This allows us to make it non integer or even complex, at least in perturbation theory. This opens the road to the definition of dimensional regularization and renormalization,

which respect the symmetries of gauge theories. This technique was the key to the first proof of the renormalizability of non-abelian gauge theories [10].

- The form of the first and second Symanzik show an explicit *positivity* and *democracy* between trees (or two-trees): each of them appears with positive and equal coefficients.
- The locality of the counterterms is still visible (although less obvious than in direct space). It corresponds to the factorization of U_G into $U_S U_{G/S}$ plus smaller terms under scaling of all the parameters of a subgraph S, because the leading terms are the trees whose restriction to S are subtrees of S. One could remark that this factorization also plays a key role in the constructive RG analysis and multiscale bounds of the theory [9].

In the next two subsections we shall derive the analogs of the corresponding statements in NCVQFT. But before that let us give a brief proof of formulas (8.1). The proof of (8.2b) is similar.

Formula (8.1) is equivalent to the computation of the determinant, namely that of the quadratic form gathering the heat kernels of all the internal lines in position space, when we integrate over all vertices *save one*. The role of this saved vertex is crucial because otherwise the determinant of the quadratic form vanishes, i.e., the computation becomes infinite by translation invariance.

But the same determinants and problems already arose a century before Feynman graphs in the XIX century theory of electric circuits, where wires play the role of propagators and the conservation of currents at each node of the circuit play the role of conservation of momenta or translation invariance. In fact the parametric representation follows from the tree matrix theorem of Kirchhoff [96], which is a key result of combinatorial theory which in its simplest form may be stated as:

Theorem 8.1 (Tree Matrix Theorem). *Let A be an n by n matrix such that*

$$\sum_{i=1}^{n} A_{ij} = 0 \ \ \forall j \ . \tag{8.3}$$

Obviously $\det A = 0$. *But let A^{11} be the matrix A with line 1 and column 1 deleted. Then*

$$\det A^{11} = \sum_{T} \prod_{\ell \in T} A_{i_\ell, j_\ell}, \tag{8.4}$$

where the sum runs over all directed trees on $\{1, \ldots, n\}$, directed away from root 1.

This theorem is a particular case of a more general result that can compute any minor of a matrix as a graphical sum over forests and more [97].

To deduce (8.1) from that theorem one defines A_{ii} as the coordination of the graph at vertex i and A_{ij} as $-l(ij)$ where $l(ij)$ is the number of lines from vertex ic to vertex j. The line 1 and column 1 deleted correspond, e.g., to fix the first vertex 1 at the origin to break translation invariance.

We include now a proof of this theorem using Grassmann variables derived from [97], because this proof was essential for us to find the correct non-commutative generalization of the parametric representation. Recall that Grassmann variables anticommute,

$$\chi_i \chi_j + \chi_j \chi_i = 0, \tag{8.5}$$

hence in particular $\chi_i^2 = 0$, and that the Grassmann rules of integration are

$$\int d\chi = 0 \; ; \; \int \chi d\chi = 1. \tag{8.6}$$

Therefore we have:

Lemma 8.2. *Consider a set of $2n$ independent Grassmann variables*

$$\overline{\psi}_1, \dots, \overline{\psi}_n, \psi_1, \dots, \psi_n \tag{8.7}$$

and the integration measure

$$d\overline{\psi}d\psi = d\overline{\psi}_1, \dots, d\overline{\psi}_n, d\psi_1, \dots, d\psi_n \tag{8.8}$$

The bar is there for convenience, but it is not complex conjugation. Prove that for any matrix A,

$$\det A = \int d\overline{\psi}d\psi e^{-\overline{\psi}A\psi} . \tag{8.9}$$

More generally, if p is an integer $0 \leq p \leq m$, and $I = \{i_1, \dots, i_p\}$, $J = \{j_1, \dots, j_p\}$ are two ordered subsets with p elements $i_1 < \cdots < i_p$ and $j_1 < \cdots < j_p$, if also $A^{I,J}$ denotes the $(n-p) \times (n-p)$ matrix obtained by erasing the rows of A with index in I and the columns of A with index in J, then

$$\int d\overline{\psi}d\psi \, (\psi_J \overline{\psi}_I) e^{-\overline{\psi}A\psi} = (-1)^{\Sigma I + \Sigma J} \det(A^{I,J}) \tag{8.10}$$

where $(\psi_J \overline{\psi}_I) \overset{def}{=} \psi_{j_1} \overline{\psi}_{i_1} \psi_{j_2} \overline{\psi}_{i_2} \dots \psi_{j_p} \overline{\psi}_{i_p}$, $\Sigma I \overset{def}{=} i_1 + \cdots + i_p$ and likewise for ΣJ.

We return now to

Proof of Theorem 8.1. We use Grassmann variables to write the determinant of a matrix with one line and one row deleted as a Grassmann integral with two corresponding sources:

$$\det A^{11} = \int (d\overline{\psi}d\psi) \, (\psi_1 \overline{\psi}_1) e^{-\overline{\psi}A\psi}. \tag{8.11}$$

The trick is to use (8.3) to write

$$\overline{\psi}A\psi = \sum_{i,j=1}^{n} (\overline{\psi}_i - \overline{\psi}_j) A_{ij} \psi_j. \tag{8.12}$$

By Lemma 8.2:

$$\det A^{11} = \int d\overline{\psi} d\psi \ (\psi_1 \overline{\psi}_1) \exp\left(-\sum_{i,j=1}^{n} A_{ij}(\overline{\psi}_i - \overline{\psi}_j)\psi_j\right) \tag{8.13}$$

$$= \int d\overline{\psi} d\psi \ (\psi_1 \overline{\psi}_1) \left[\prod_{i,j=1}^{n} \left(1 - A_{ij}(\overline{\psi}_i - \overline{\psi}_j)\psi_j\right)\right] \tag{8.14}$$

by the Pauli exclusion principle. We now expand to get

$$\det A^{11} = \sum_{\mathcal{G}} \left(\prod_{\ell=(i,j)\in\mathcal{G}} (-A_{ij})\right) \Omega_{\mathcal{G}} \tag{8.15}$$

where $\mathcal{G}$ is *any* subset of $[n] \times [n]$, and we used the notation

$$\Omega_{\mathcal{G}} \stackrel{\text{def}}{=} \int d\overline{\psi} d\psi \ (\psi_1 \overline{\psi}_1) \left(\prod_{(i,j)\in\mathcal{G}} [(\overline{\psi}_i - \overline{\psi}_j)\psi_j]\right). \tag{8.16}$$

The theorem will now follow from the following

Lemma 8.3. $\Omega_{\mathcal{G}} = 0$ *unless the graph $\mathcal{G}$ is a tree directed away from 1 in which case $\Omega_{\mathcal{G}} = 1$.*

Proof. Trivially, if (i,i) belongs to $\mathcal{G}$, then the integrand of $\Omega_{\mathcal{G}}$ contains a factor $\overline{\psi}_i - \overline{\psi}_i = 0$ and therefore $\Omega_{\mathcal{G}}$ vanishes.

But the crucial observation is that if there is a loop in $\mathcal{G}$ then again $\Omega_{\mathcal{G}} = 0$. This is because then the integrand of $\Omega_{\mathcal{F},\mathcal{R}}$ contains the factor

$$\overline{\psi}_{\tau(k)} - \overline{\psi}_{\tau(1)} = (\overline{\psi}_{\tau(k)} - \overline{\psi}_{\tau(k-1)}) + \cdots + (\overline{\psi}_{\tau(2)} - \overline{\psi}_{\tau(1)}). \tag{8.17}$$

Now, upon inserting this telescoping expansion of the factor $\overline{\psi}_{\tau(k)} - \overline{\psi}_{\tau(1)}$ into the integrand of $\Omega_{\mathcal{F},\mathcal{R}}$, the latter breaks into a sum of $(k-1)$ products. For each of these products, there exists an $\alpha \in \mathbb{Z}/k\mathbb{Z}$ such that the factor $(\overline{\psi}_{\tau(\alpha)} - \overline{\psi}_{\tau(\alpha-1)})$ appears *twice*: once with the $+$ sign from the telescopic expansion of $(\overline{\psi}_{\tau(k)} - \overline{\psi}_{\tau(1)})$, and once more with a $+$ (resp. $-$) sign if $(\tau(\alpha), \tau(\alpha-1))$ (resp. $(\tau(\alpha-1), \tau(\alpha))$) belongs to $\mathcal{F}$. Again, the Pauli exclusion principle entails that $\Omega_{\mathcal{G}} = 0$.

Now every connected component of $\mathcal{G}$ must contain 1, otherwise there is no way to saturate the $d\psi_1$ integration.

This means that $\mathcal{G}$ has to be a directed tree on $\{1, \ldots, n\}$. It remains only to see now that $\mathcal{G}$ has to be directed away from 1, which is not too difficult.

Now Theorem 8.1 follows immediately.

8.2. Non-commutative hyperbolic polynomials, the non-covariant case

Since the Mehler kernel is still quadratic in position space it is possible to also integrate explicitly all positions to reduce Feynman amplitudes of, e.g., non-commutative $\Phi_4^{\star 4}$ purely to parametric formulas, but of course the analogs of Symanzik polynomials are now hyperbolic polynomials which encode the richer information about ribbon graphs. These polynomials were first computed in [68] in the case of the non-covariant vulcanized $\Phi_4^{\star 4}$ theory. The computation relies essentially on a Grassmann variable analysis of Pfaffians which generalizes the tree matrix theorem of the previous section.

Defining the antisymmetric matrix σ as

$$\sigma = \begin{pmatrix} \sigma_2 & 0 \\ 0 & \sigma_2 \end{pmatrix} \text{ with} \tag{8.18}$$

$$\sigma_2 = \begin{pmatrix} 0 & -i \\ i & 0 \end{pmatrix} \tag{8.19}$$

the δ-functions appearing in the vertex contribution can be rewritten as an integral over some new variables p_V. We refer to these variables as to *hypermomenta*. Note that one associates such a hypermomentum p_V to any vertex V *via* the relation

$$\delta(x_1^V - x_2^V + x_3^V - x_4^V) = \int \frac{dp_V'}{(2\pi)^4} e^{ip_V'(x_1^V - x_2^V + x_3^V - x_4^V)}$$

$$= \int \frac{dp_V}{(2\pi)^4} e^{p_V \sigma (x_1^V - x_2^V + x_3^V - x_4^V)} . \tag{8.20}$$

Consider a particular ribbon graph G. Specializing to dimension 4 and choosing a particular root vertex $\bar{V}$ of the graph, one can write the Feynman amplitude for G in the condensed way

$$\mathcal{A}_G = \int \prod_\ell [\frac{1 - t_\ell^2}{t_\ell}]^2 d\alpha_\ell \int dx\, dp\, e^{-\frac{\Omega}{2} X G X^t} \tag{8.21}$$

where $t_\ell = \tanh \frac{\alpha_\ell}{2}$, X summarizes all positions and hypermomenta and G is a certain quadratic form. If we call x_e and $p_{\bar{V}}$ the external variables we can decompose G according to an internal quadratic form Q, an external one M and a coupling part P so that

$$X = \begin{pmatrix} x_e & p_{\bar{V}} & u & v & p \end{pmatrix} \quad , \quad G = \begin{pmatrix} M & P \\ P^t & Q \end{pmatrix} , \tag{8.22}$$

Performing the gaussian integration over all internal variables one obtains:

$$\mathcal{A}_G = \int [\frac{1 - t^2}{t}]^2 d\alpha \frac{1}{\sqrt{\det Q}} e^{-\frac{\tilde{\Omega}}{2} \begin{pmatrix} x_e & \bar{p} \end{pmatrix} [M - P Q^{-1} P^t] \begin{pmatrix} x_e \\ \bar{p} \end{pmatrix}} . \tag{8.23}$$

This form allows to define the polynomials $HU_{G,\bar{v}}$ and $HV_{G,\bar{v}}$, analogs of the Symanzik polynomials U and V of the commutative case (see (8.1)). They are defined by

$$\mathcal{A}_{\bar{V}}(\{x_e\},\ p_{\bar{v}}) = K' \int_0^\infty \prod_l [d\alpha_l(1-t_l^2)^2] HU_{G,\bar{v}}(t)^{-2} e^{-\frac{HV_{G,\bar{v}}(t,x_e,p_{\bar{v}})}{HU_{G,\bar{v}}(t)}}. \tag{8.24}$$

They are polynomials in the set of variables t_ℓ $(\ell = 1,\ldots,L)$, the hyperbolic tangent of the half-angle of the parameters α_ℓ.

Using now (8.23) and (8.24) the polynomial $HU_{G,\bar{v}}$ writes

$$HU_{\bar{v}} = (\det Q)^{\frac{1}{4}} \prod_{\ell=1}^{L} t_\ell. \tag{8.25}$$

The main results ([68]) are:

- The polynomials $HU_{G,\bar{v}}$ and $HV_{G,\bar{v}}$ have a strong *positivity property*. Roughly speaking they are sums of monomials with positive integer coefficients. This positive integer property comes from the fact that each such coefficient is the square of a Pfaffian with integer entries,
- Leading terms can be identified in a given "Hepp sector", at least for *orientable graphs*. A Hepp sector is a complete ordering of the t parameters. These leading terms which can be shown strictly positive in $HU_{G,\bar{v}}$ correspond to super-trees which are the disjoint union of a tree in the direct graph and a tree in the dual graph. Hypertrees in a graph with n vertices and F faces have therefore $n + F - 2$ lines. (Any connected graph has hypertrees, and under reduction of the hypertree, the graph becomes a hyperrosette). Similarly one can identify "super-two-trees" $HV_{G,\bar{v}}$ which govern the leading behavior of $HV_{G,\bar{v}}$ in any Hepp sector.

From the second property, one can deduce the *exact power counting* of any orientable ribbon graph of the theory, just as in the matrix base.

Let us now borrow from [68] some examples of these hyperbolic polynomials. We put $s = (4\theta\Omega)^{-1}$. For the bubble graph of Figure 25:

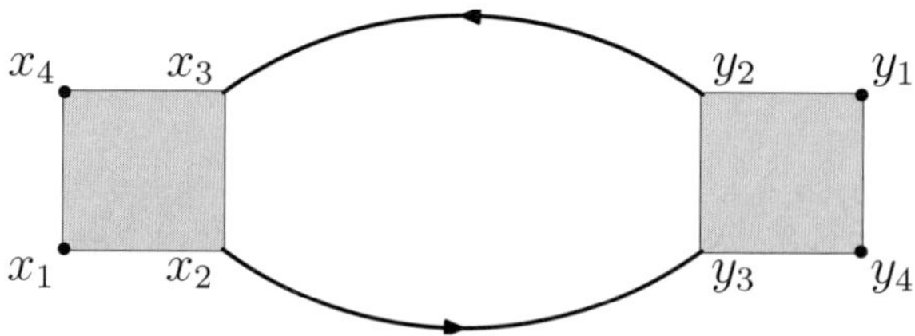

Figure 25: The bubble graph.

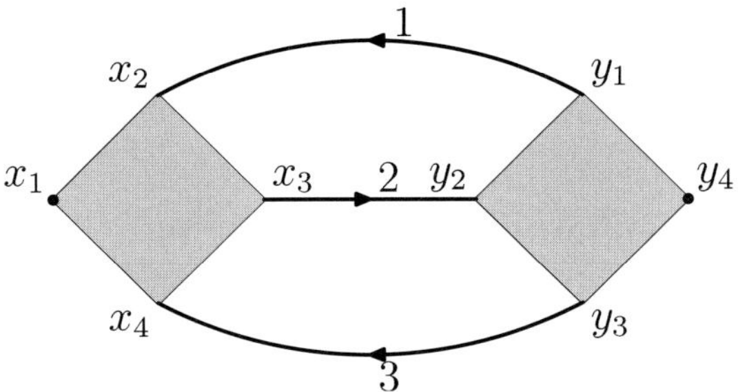

Figure 26: The sunshine graph.

$$HU_{G,v} = (1 + 4s^2)(t_1 + t_2 + t_1^2 t_2 + t_1 t_2^2)\,,$$

$$HV_{G,v} = t_2^2\Big[p_2 + 2s(x_4 - x_1)\Big]^2 + t_1 t_2\Big[2p_2^2 + (1 + 16s^4)(x_1 - x_4)^2\Big]$$

$$+ t_1^2\Big[p_2 + 2s(x_1 - x_4)\Big]^2. \tag{8.26}$$

For the sunshine graph Figure 26:

$$HU_{G,v} = \Big[t_1 t_2 + t_1 t_3 + t_2 t_3 + t_1^2 t_2 t_3 + t_1 t_2^2 t_3 + t_1 t_2 t_3^2\Big](1 + 8s^2 + 16s^4)$$

$$+ 16s^2(t_2^2 + t_1^2 t_3^2)\,. \tag{8.27}$$

For the non-planar sunshine graph (see Figure 27) we have:

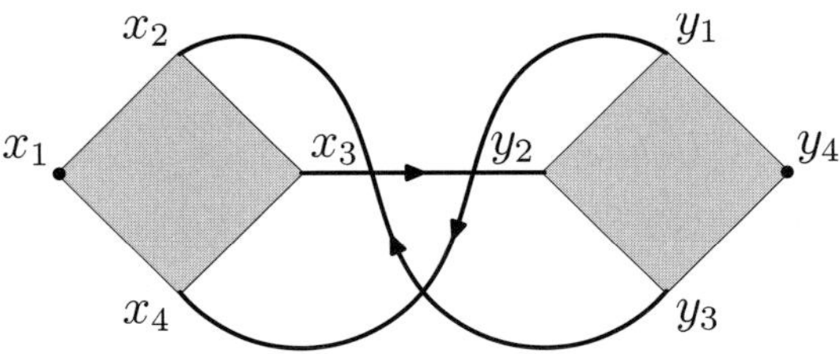

Figure 27: The non-planar sunshine graph.

$$HU_{G,v} = \Big[t_1 t_2 + t_1 t_3 + t_2 t_3 + t_1^2 t_2 t_3 + t_1 t_2^2 t_3 + t_1 t_2 t_3^2\Big](1 + 8s^2 + 16s^4)$$

$$+ 4s^2\Big[1 + t_1^2 + t_2^2 + t_1^2 t_2^2 + t_3^2 + t_1^2 t_3^2 + t_2^2 t_3^2 + t_1^2 t_2^2 t_3^2\Big]\,,$$

We note the improvement in the genus with respect to its planar counterparts.

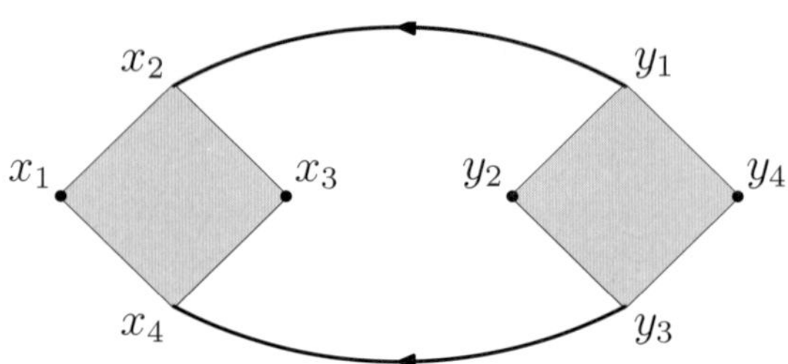

Figure 28: The broken bubble graph.

For the broken bubble graph (see Figure 28) we have:

$$HU_{G,v} =(1 + 4s^2)(t_1 + t_2 + t_1^2 t_2 + t_1 t_2^2)\,,$$

$$\begin{aligned}
HV_{G,v} =\,&t_2^2\left[4s^2(x_1 + y_2)^2 + (p_2 - 2s(x_3 + y_4))^2\right] + t_1^2\left[p_2 + 2s(x_3 - y_4)\right]^2 \\
&+ t_1 t_2\left[8s^2 y_2^2 + 2(p_2 - 2sy_4)^2 + (x_1 + x_3)^2 + 16s^4(x_1 - x_3)^2\right] \\
&+ t_1^2 t_2^2 4s^2(x_1 - y_2)^2\,.
\end{aligned}$$

Note that $HU_{G,v}$ is identical to the one of the bubble with only one broken face. The power counting improvement comes from the broken face and can be seen only in $HV_{G,v}$.

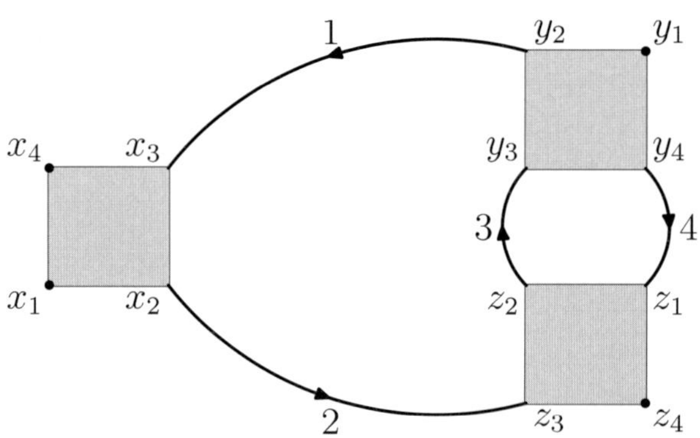

Figure 29: The half-eye graph.

Finally, for the half-eye graph (see Figure 29), we start by defining

$$A_{24} =t_1 t_3 + t_1 t_3 t_2^2 + t_1 t_3 t_4^2 + t_1 t_3 t_2^2 t_4^2\,. \tag{8.28}$$

The $HU_{G,v}$ polynomial with fixed hypermomentum corresponding to the vertex with two external legs is

$$
\begin{aligned}
HU_{G,v_1} =&(A_{24} + A_{14} + A_{23} + A_{13} + A_{12})(1 + 8s^2 + 16s^4) \\
&+ t_1 t_2 t_3 t_4 (8 + 16s^2 + 256s^4) + 4t_1 t_2 t_3^2 + 4t_1 t_2 t_4^2 \\
&+ 16s^2 (t_3^2 + t_2^2 t_4^2 + t_1^2 t_4^2 + t_1^2 t_2^2 t_3^2) \\
&+ 64s^4 (t_1 t_2 t_3^2 + t_1 t_2 t_4^2)\,,
\end{aligned}
\tag{8.29}
$$

whereas with another fixed hypermomentum we get

$$
\begin{aligned}
HU_{G,v_2} =&(A_{24} + A_{14} + A_{23} + A_{13} + A_{12})(1 + 8s^2 + 16s^4) \\
&+ t_1 t_2 t_3 t_4 (4 + 32s^2 + 64s^4) + 32s^2 t_1 t_2 t_3^2 + 32s^2 t_1 t_2 t_4^2 \\
&+ 16s^2 (t_3^2 + t_1^2 t_4^2 + t_2^2 t_4^2 + t_1^2 t_3^2 t_2^2)\,.
\end{aligned}
\tag{8.30}
$$

Note that the leading terms are identical and the choice of the root perturbs only the non-leading ones. Moreover note the presence of the t_3^2 term. Its presence can be understood by the fact that in the sector $t_1, t_2, t_4 > t_3$ the subgraph formed by the lines $1, 2, 4$ has two broken faces. This is the sign of a power counting improvement due to the additional broken face in that sector. To exploit it, we just have to integrate over the variables of line 3 in that sector, using the second polynomial $HV_{G',v}$ for the triangle subgraph G' made of lines $1, 2, 4$.

8.3. Non-commutative hyperbolic polynomials, the covariant case

In the *covariant case* the diagonal coefficients on the long variables disappear but there are new antisymmetric terms proportional to Ω due to the propagator oscillations.

It is possible to reproduce easily the positivity theorem of the previous non-covariant case, because we still have sums of squares of Pfaffians. But identifying the leading terms of the polynomials under a rescaling associating to a subgraph is more difficult. It is easy to see that for transcendental values of Ω, the desired leading terms cannot vanish because that would correspond to Ω being the root of a polynomial with integer coefficients. But power counting under a transcendentality condition is not very satisfying, especially because continuous RG flows also necessarily cross non transcendental points.

But thanks to a slightly more difficult analysis inspired by [93] and which involve a kind of new fourth Filk move, it is possible to prove that except again for some special cases of four-point graphs with two broken faces, the power counting goes through at $\Omega < 1$.

The corresponding analysis together with many examples are given in [69].

The covariant case at $\Omega = 1$, also called the *self-dual* covariant case is very interesting, because it may be the most relevant for the study of, e.g., the quantum Hall effect. Apparently it corresponds to a very degenerate non-renormalizable situation because even the four-point function has non logarithmic divergences as can be seen easily in the matrix basis, where the propagator is now either

$1/(2m + A)$ or $1/(2n + A)$ depending on the sign of the "magnetic field" Ω . But there is a huge gauge invariance and we feel that the Ward identities of Section 5 should allow renormalization of the theory even in that case.

Let us also recall that the parametric representation can be used to derive the dimensional regularization and renormalization of the theory. Dimensional *regularization* means that Feynman amplitudes are meromorphic in the space-time dimension D. This can be conveniently proved in the Complete Mellin representation [98, 99] in which the dependence in D occurs in explicit Γ functions. This representation is also the right tool to derive asymptotic expansions in powers and powers of logarithms for completely general scalings of subsets of external invariants. It has been recently extended to the non-commutative case in [100].

Dimensional *regularization* means that there is a recursive extraction of poles in D in which the divergent terms are counterterms of the form of the initial Lagrangian. In parametric space it means that a certain factorization occurs for the leading terms of the hyperbolic polynomials HU and HV under rescaling of the variables of the divergent subgraphs. This study will be done in [70] and may be useful for the future renormalization of non-commutative gauge theories.

9. Conclusion

Non-commutative QFT seemed initially to have non-renormalizable divergencies, due to UV/IR mixing. But following the Grosse-Wulkenhaar breakthrough, there has been recent rapid progress in our understanding of renormalizable QFT on Moyal spaces. We can already propose a preliminary classification of these models into different categories, according to the behavior of their propagators:

- ordinary models at $0 < \Omega < 1$ such as $\Phi_4^{\star 4}$ (which has non-orientable graphs) or $(\bar{\phi}\phi)^2$ models (which has none). Their propagator, roughly $(p^2 + \Omega^2 \tilde{x}^2 + A)^{-1}$ is LS covariant and has good decay both in matrix space (4.11–4.14) and direct space (7.2). They have non-logarithmic mass divergencies and definitely require "vulcanization", i.e., the Ω term.
- self-dual models at $\Omega = 1$ in which the propagator is LS invariant. Their propagator is even better. In the matrix base it is diagonal, e.g., of the form $G_{m,n} = (m+n+A)^{-1}$, where A is a constant. The supermodels seem generically ultraviolet fixed points of the ordinary models, at which nontrivial Ward identities force the vanishing of the beta function. The flow of Ω to the $\Omega = 1$ fixed point is very fast (exponentially fast in RG steps).
- covariant models such as orientable versions of LSZ or Gross-Neveu (and presumably orientable gauge theories of various kind: Yang-Mills, Chern-Simons...). They may have only logarithmic divergencies and apparently no perturbative UV/IR mixing. However the vulcanized version still appears the most generic framework for their treatment. The propagator is then roughly $(p^2 + \Omega^2 \tilde{x}^2 + 2\Omega \tilde{x} \wedge p)^{-1}$. In matrix space this propagator shows definitely a weaker decay (6.19) than for the ordinary models, because of the presence

of a nontrivial saddle point. In direct space the propagator no longer decays with respect to the long variables, but only oscillates. Nevertheless the main lesson is that in matrix space the weaker decay can still be used; and in x space the oscillations can never be completely killed by the vertices oscillations. Hence these models retain therefore essentially the power counting of the ordinary models, up to some nasty details concerning the four-point subgraphs with two external faces. Ultimately, thanks to a little conspiration in which the four-point subgraphs with two external faces are renormalized by the mass renormalization, the covariant models remain renormalizable. This is the main message of [51, 93].

- self-dual covariant models which are of the previous type but at $\Omega = 1$. Their propagator in the matrix base is diagonal and depends only on one index m (e.g., always the left side of the ribbon). It is of the form $G_{m,n} = (m+A)^{-1}$. In x space the propagator oscillates in a way that often exactly compensates the vertices oscillations. These models have definitely worse power counting than in the ordinary case, with, e.g., quadratically divergent four-point graphs (if sharp cut-offs are used). Nevertheless Ward identities can presumably still be used to show that they can still be renormalized. This probably requires a much larger conspiration to generalize the Ward identities of the supermodels.

Notice that the status of non-orientable covariant theories is not yet clarified.

Parametric representation can be derived in the non-commutative case. It implies hyperbolic generalizations of the Symanzik polynomials which condense the information about the rich topological structure of a ribbon graph. Using this representation, dimensional regularization and dimensional renormalization should extend to the non-commutative framework.

Remark that trees, which are the building blocks of the Symanzik polynomials, are also at the heart of (commutative) constructive theory, whose philosophy could be roughly summarized as "You shall use trees[26], but you shall *not* develop their loops or else you shall diverge". It is quite natural to conjecture that hypertrees, which are the natural non-commutative objects intrinsic to a ribbon graph, should play a key combinatoric role in the yet to develop non-commutative constructive field theory.

In conclusion we have barely started to scratch the world of renormalizable QFT on non-commutative spaces. The little we see through the narrow window now open is extremely tantalizing. There exists renormalizable NCQFTs, e.g., $\Phi^{\star 4}$ on $\mathbb{R}^4_\theta$, Gross-Neveu on $\mathbb{R}^2_\theta$ and they enjoy better properties than their commutative counterparts, since they have no Landau ghosts. The constructive program looks *easier* on non-commutative geometries than on commutative ones. Non-commutative non relativistic field theories with a chemical potential seem the right formalism for a study ab initio of various problems in presence of a magnetic field, and in particular of the quantum Hall effect. The correct scaling and RG

---

[26]These trees may be either true trees of the graphs in the fermionic case or trees associated to cluster or Mayer expansions in the bosonic case, but this distinction is not essential.

theory of this effect presumably requires to build a very singular theory (of the self-dual covariant type) because of the huge degeneracy of the Landau levels. To understand this theory and the gauge theories on non-commutative spaces seem the most obvious challenges ahead of us.

References

[1] Neil J. Cornish, David N. Spergel, Glenn D. Starkman, and Eiichiro Komatsu, *Constraining the Topology of the Universe*, Phys. Rev. Lett. **92** (2004), 201–302, astro-ph/0310233

[2] V. Rivasseau and F. Vignes-Tourneret, Non-Commutative Renormalization, in *Rigorous Quantum Field Theory*, a Festschrift for Jacques Bros, Birkhäuser Progress in Mathematics Vol. 251, 2007, `hep-th/0409312`.

[3] M. Peskin and Daniel V. Schroeder (Contributor), *An Introduction to Quantum Field Theory*, Perseus Publishing, (1995).

[4] C. Itzykson and J.-B. Zuber, Quantum Field Theory, McGraw Hill, 1980.

[5] P. Ramond, Field Theory, Addison-Wesley, 1994.

[6] J. Glimm and A. Jaffe, Quantum Physics. A functional integral point of view. McGraw and Hill, New York, 1981.

[7] Manfred Salmhofer, Renormalization: An Introduction, Texts and Monographs in Physics, Springer Verlag, 1999.

[8] Renormalization, Poincaré Seminar 2002, in Vacuum Energy Renormalization, Birkhäuser, Ed. by B. Duplantier and V. Rivasseau, Basel, 2003,

[9] V. Rivasseau, *From Perturbative to Constructive Renormalization*, Princeton University Press, 1991.

[10] G. 't Hooft and M. Veltman, Nucl. Phys. **B50** (1972), 318.

[11] D. Gross and F. Wilczek, *Ultraviolet Behavior of Non-Abelian Gauge Theories*, Phys. Rev. Lett. **30** (1973), 1343–1346.

[12] H.D. Politzer, *Reliable Perturbative Results for Strong Interactions?*, Phys. Rev. Lett. **30** (1973), 1346–1349.

[13] D. Gross and F. Wilczek, *Asymptotically Free Gauge Theories. I*, Phys. Rev. **D8** (1973), 3633–3652.

[14] K. Wilson, *Renormalization group and critical phenomena, II Phase space cell analysis of critical behavior*, Phys. Rev. B **4** (1974), 3184.

[15] J. Zinn-Justin, Quantum Field Theory and Critical Phenomena, Oxford University Press, 2002.

[16] G. Benfatto, G. Gallavotti, Renormalization Group, Princeton University Press, 1995.

[17] G. Gentile and V. Mastropietro, *Methods for the Analysis of the Lindstedt Series for KAM Tori and Renormalizability in Classical Mechanics. A Review with Some Applications*, Rev. Math. Phys. **8** (1996).

[18] W. Zimmermann, *Convergence of Bogoliubov's method for renormalization in momentum space*, Comm. Math. Phys. **15** (1969), 208.

[19] D. Kreimer, *On the Hopf algebra structure of perturbative quantum field theories*, Adv. Theor. Math. Phys. **2** (1998), 303–334, `q-alg/9707029`.

[20] A. Connes and D. Kreimer, *Renormalization in quantum field theory and the Riemann-Hilbert problem i: The Hopf algebra structure of graphs and the main theorem*, Commun. Math. Phys. **210** (2000), 249–273.

[21] A. Connes and D. Kreimer, *Renormalization in quantum field theory and the Riemann–Hilbert problem ii: The ß-function, diffeomorphisms and the renormalization group*, Commun. Math. Phys. **216** (2001), 215–241.

[22] G. Gallavotti, Perturbation Theory, in Mathematical Physics towards the XXI century, p. 275–294, ed. R. Sen, A. Gersten, Ben Gurion University Press, Ber Sheva, 1994.

[23] J. Feldman and E. Trubowitz, *Perturbation Theory for Many Fermions Systems*, Helv. Phys. Acta **63** (1990), 156.

[24] J. Feldman and E. Trubowitz, *The Flow of an Electron-Phonon System to the Superconducting State*, Helv. Phys. Acta **64** (1991), 213.

[25] G. Benfatto and G. Gallavotti, *Perturbation theory of the Fermi surface in a quantum liquid. A general quasi-particle formalism and one dimensional systems*, Journ. Stat. Physics **59** (1990), 541.

[26] J. Feldman, J. Magnen, V. Rivasseau and E. Trubowitz, *An Intrinsic 1/N Expansion for Many Fermion Systems*, Europhys. Letters **24** (1993), 437.

[27] A.B. Zamolodchikov, JETP Letters **43** (1986), 731.

[28] K. Wiese, *The functional renormalization group treatment of disordered systems, a review*, Ann. Henri Poincaré **4** (2003), 473.

[29] Michael Green, John H. Schwarz and Edward Witten, Superstring theory, Cambridge University Press (1987).

[30] A. Strominger and C. Vafa, *Microscopic Origin of the Bekenstein-Hawking Entropy*, Phys.Lett. **B379** (1996), 99–104, hep-th/9601029.

[31] E. d'Hoker and T. Phong, *Two-Loop Superstrings I, Main Formulas*, Phys. Lett. **B529** (2002), 241–255, hep-th/0110247.

[32] Lee Smolin, The Trouble With Physics: The Rise of String Theory, the Fall of a Science, and What Comes Next, Houghton-Mifflin, Sep. 2006.

[33] H.S. Snyders, *Quantized space-time*, Phys. Rev **71** (1947), 38.

[34] A. Connes, M.R. Douglas, and A. Schwarz, *Non-commutative geometry and matrix theory: Compactification on tori*, JHEP **02** (1998), 003, `hep-th/9711162`.

[35] N. Seiberg and E. Witten, *String theory and non-commutative geometry*, JHEP **09** (1999), 032, `hep-th/9908142`.

[36] M.R. Douglas and N.A. Nekrasov, *Non-commutative field theory*, Rev. Mod. Phys. **73** (2001), 977–1029, `hep-th/0106048`.

[37] S. Minwalla, M. Van Raamsdonk, and N. Seiberg, *Non-commutative perturbative dynamics*, JHEP **02** (2000), 020, `hep-th/9912072`.

[38] H. Grosse and R. Wulkenhaar, *Power-counting theorem for non-local matrix models and renormalization*, Commun. Math. Phys. **254** (2005), no. 1, 91–127, `hep-th/0305066`.

[39] H. Grosse and R. Wulkenhaar, *Renormalization of ϕ^4-theory on non-commutative $\mathbb{R}^2$ in the matrix base*, JHEP **12** (2003), 019, `hep-th/0307017`.

[40] H. Grosse and R. Wulkenhaar, *Renormalization of ϕ^4-theory on non-commutative $\mathbb{R}^4$ in the matrix base*, Commun. Math. Phys. **256** (2005), no. 2, 305–374, `hep-th/0401128`.

[41] E. Langmann and R.J. Szabo, *Duality in scalar field theory on non-commutative phase spaces*, Phys. Lett. **B533** (2002), 168–177, `hep-th/0202039`.

[42] V. Rivasseau, F. Vignes-Tourneret, and R. Wulkenhaar, *Renormalization of non-commutative ϕ^4-theory by multi-scale analysis*, Commun. Math. Phys. **262** (2006), 565–594, `hep-th/0501036`.

[43] E. Langmann, R.J. Szabo, and K. Zarembo, *Exact solution of quantum field theory on non-commutative phase spaces*, JHEP **01** (2004), 017, `hep-th/0308043`.

[44] E. Langmann, R.J. Szabo, and K. Zarembo, *Exact solution of non-commutative field theory in background magnetic fields*, Phys. Lett. **B569** (2003), 95–101, `hep-th/0303082`.

[45] E. Langmann, *Interacting fermions on non-commutative spaces: Exactly solvable quantum field theories in $2n + 1$ dimensions*, Nucl. Phys. **B654** (2003), 404–426, `hep-th/0205287`.

[46] J. Polchinski, *Renormalization and Effective Lagrangians*, Nucl. Phys. **B231** (1984), 269.

[47] R. Gurau, J. Magnen, V. Rivasseau and F. Vignes-Tourneret, *Renormalization of non-commutative ϕ_4^4 field theory in x space*, Commun. Math. Phys. **267** (2006), no. 2, 515–542, `hep-th/0512271`.

[48] I. Chepelev and R. Roiban, *Convergence theorem for non-commutative Feynman graphs and renormalization*, JHEP **03** (2001), 001, `hep-th/0008090`.

[49] I. Chepelev and R. Roiban, *Renormalization of quantum field theories on non-commutative $\mathbb{R}^d$. i: Scalars*, JHEP **05** (2000), 037, `hep-th/9911098`.

[50] R. Gurau, V. Rivasseau, and F. Vignes-Tourneret, *Propagators for non-commutative field theories*, Ann. H. Poincaré **7** (2006), 1601–1628, `hep-th/0512071`.

[51] F. Vignes-Tourneret, *Renormalization of the orientable non-commutative Gross-Neveu model*, Ann. H. Poincaré **8** (2007), `math-ph/0606069`.

[52] H. Grosse and R. Wulkenhaar, *The beta-function in duality-covariant non-commutative ϕ^4-theory*, Eur. Phys. J. **C35** (2004), 277–282, `hep-th/0402093`.

[53] K. Gawedzki and A. Kupiainen, *Gross-Neveu model through convergent perturbation expansions*, Comm. Math. Phys. **102** (1985), 1.

[54] J. Feldman, J. Magnen, V. Rivasseau and R. Sénéor, *A renormalizable field theory: the massive Gross-Neveu model in two dimensions*, Comm. Math. Phys. **103** (1986), 67.

[55] K. Gawedzki and A. Kupiainen, *Massless ϕ_4^4 theory: Rigorous control of a renormalizable asymptotically free model*, Comm. Math. Phys. **99** (1985), 197.

[56] J. Feldman, J. Magnen, V. Rivasseau and R. Sénéor, *Construction of infrared ϕ_4^4 by a phase space expansion*, Comm. Math. Phys. **109** (1987), 437.

[57] M. Disertori and V. Rivasseau, *Two and Three Loops Beta Function of Non-commutative Φ_4^4 Theory*, `hep-th/0610224`.

[58] M. Disertori, R. Gurau, J. Magnen and V. Rivasseau, *Vanishing of Beta Function of Non-commutative* Φ_4^4 *to all orders*, Submitted to Phys. Lett. B, **hep-th/0612251**.

[59] W. Metzner and C. Di Castro, *Conservation Laws and correlation functions in the Luttinger liquid*, Phys. Rev. **B 47** (1993), 16107.

[60] G. Benfatto and V. Mastropietro, *Ward Identities and Chiral Anomaly in the Luttinger Liquid*, Commun. Math. Phys. **258** (2005), 609–655.

[61] G. Benfatto and V. Mastropietro, *Ward Identities and Vanishing of the Beta Function for* $d = 1$ *Interacting Fermi Systems*, Journal of Statistical Physics, **115** (2004), 143–184.

[62] A. Sokal, *An improvement of Watson's theorem on Borel summability*, Journ. Math. Phys. **21** (1980), 261.

[63] A. Abdesselam, *A Complete Renormalization Group Trajectory Between Two Fixed Points*, **math-ph/0610018**.

[64] H. Grosse and H. Steinacker, *Renormalization of the non-commutative* ϕ^3 *model through the Kontsevich model*, Nucl. Phys. **B746** (2006), 202–226, **hep-th/0512203**.

[65] H. Grosse and H. Steinacker, *A nontrivial solvable non-commutative* ϕ^3 *model in* 4 *dimensions*, JHEP **0608** (2006), 008, **hep-th/0603052**.

[66] H. Grosse, H. Steinacker, *Exact renormalization of a non-commutative* ϕ^3 *model in* 6 *dimensions*, **hep-th/0607235**.

[67] Axel de Goursac, J.C. Wallet and R. Wulkenhaar, *Non-commutative Induced Gauge Theory*, **hep-th/0703075**.

[68] R. Gurau and V. Rivasseau, *Parametric representation of non-commutative field theory*, to appear in Commun. Math. Phys, **math-ph/0606030**.

[69] V. Rivasseau and A. Tanasa, *Parametric representation of "critical" non-commutative QFT models*, submitted to Commun. Math. Phys., **hep-th/0701034**.

[70] R. Gurau and A. Tanasa, work in preparation.

[71] L. Susskind, *The Quantum Hall Fluid and Non-Commutative Chern Simons Theory*, **hep-th/0101029**.

[72] A. Polychronakos, *Quantum Hall states as matrix Chern-Simons theory*, JHEP **0104** (2001), 011, **hep-th/0103013**.

[73] S. Hellerman and M. Van Raamsdonk, *Quantum Hall Physics = Non-commutative Field Theory*, JHEP **0110** (2001), 039, **hep-th/0103179**.

[74] B. Duplantier, Conformal Random Geometry in Les Houches, Session LXXXIII, 2005, Mathematical Statistical Physics, A. Bovier, F. Dunlop, F. den Hollander, A. van Enter and J. Dalibard, eds., pp. 101–217, Elsevier B.V. (2006), **math-ph/0608053**.

[75] G. 't Hooft, *A planar diagram theory for strong interactions*, Nuclear Physics **B 72** (1974), 461.

[76] V. Rivasseau, F. Vignes-Tourneret, Renormalization of non-commutative field theories, Luminy Lectures, **hep-th/0702068**

[77] V. Rivasseau, *An introduction to renormalization*, in Poincaré Seminar 2002, ed. by B. Duplantier and V. Rivasseau. Progress in Mathematical Physics 30, Birkhäuser (2003), ISBN 3-7643-0579-7.

[78] R. Feynman and A. Hibbs, Quantum Mechanics and Path Integrals, McGraw and Hill, New York 1965.

[79] C. Itzykson and J.M Drouffe, Statistical Field Theory, Volumes 1 and 2, Cambridge University Press 1991.

[80] Giorgio Parisi, Statistical Field Theory, Perseus Publishing 1998.

[81] J.P. Eckmann, J. Magnen and R. Sénéor, *Decay properties and Borel summability for the Schwinger functions in $P(\phi)_2$ theories*, Comm. Math. Phys. **39** (1975), 251.

[82] J. Magnen and R. Sénéor, *Phase space cell expansion and Borel summability for the Euclidean ϕ_3^4 theory*, Comm Math. Phys. **56** (1977), 237.

[83] C. Kopper Renormalization Theory based on Flow Equations, in *Rigorous Quantum Field Theory*, a Festschrift for Jacques Bros, Birkhäuser Progress in Mathematics Vol. 251, 2007, `hep-th/0508143`.

[84] M. Bergère and Y.M.P. Lam, *Bogoliubov-Parasiuk theorem in the α-parametric representation*, Journ. Math. Phys. **17** (1976), 1546.

[85] C. de Calan and V. Rivasseau, *Local existence of the Borel transform in Euclidean ϕ_4^4*, Comm. Math. Phys. **82** (1981), 69.

[86] J.M. Gracia-Bondía and J. C. Várilly, A*lgebras of distributions suitable for phase space quantum mechanics. I*, J. Math. Phys. **29** (1988), 869–879.

[87] T. Filk, *Divergencies in a field theory on quantum space*, Phys. Lett. **B376** (1996), 53–58.

[88] V. Gayral, *Heat-kernel approach to UV/IR mixing on isospectral deformation manifolds*, Annales Henri Poincaré **6** (2005), 991–1023, `hep-th/0412233`.

[89] B. Simon, *Functional integration and quantum physics*, vol. 86 of Pure and applied mathematics. Academic Press, New York, 1979.

[90] P.K. Mitter and P.H. Weisz, *Asymptotic scale invariance in a massive Thirring model with $U(n)$ symmetry*, Phys. Rev. **D8** (1973), 4410–4429.

[91] D.J. Gross and A. Neveu, *Dynamical symmetry breaking in asymptotically free field theories*, Phys. Rev. **D10** (1974), 3235.

[92] C. Kopper, J. Magnen, and V. Rivasseau, *Mass generation in the large N Gross-Neveu model*, Commun. Math. Phys. **169** (1995), 121–180.

[93] F. Vignes-Tourneret, *Renormalization des théories de champs non-commutatives*. Physique théorique, Université Paris 11, september, 2006, `math-ph/0612014`.

[94] M. Disertori and V.Rivasseau, *Continuous constructive fermionic renormalization*, Annales Henri Poincaré **1** (2000), 1, `hep-th/9802145`.

[95] A. Lakhoua, F. Vignes-Tourneret and J.C. Wallet, *One-loop Beta Functions for the Orientable Non-commutative Gross-Neveu Model*, submitted to JHEP, `hep-th/0701170`.

[96] G. Kirchhoff, *Über die Auflösung der Gleichungen, auf welche man bei der Untersuchung der linearen Verteilung galvanischer Ströme geführt wird*, Ann. Physik Chemie **72**(1847), 497–508.

[97] A. Abdesselam, *Grassmann-Berezin Calculus and Theorems of the Matrix-Tree Type*, `math.CO/0306396`.

[98] C. de Calan and A.P.C. Malbouisson, *Complete Mellin representation and asymptotic behaviours of Feynman amplitudes*, Annales de l'Institut Henri Poincaré, physique théorique **32** (1980), 91–107.

[99] C. de Calan, F. David and V. Rivasseau, *Renormalization in the complete Mellin representation of Feynman amplitudes*, Commun. Math. Phys. **78** (1981), 531–544.

[100] R. Gurau, A.P.C. Malbouisson, V. Rivasseau and A. Tanasa, *Non-commutative Complete Mellin representation for Feynman Amplitudes*, math-ph/0705.3437v1.

Vincent Rivasseau
Laboratoire de Physique Théorique
Bât. 210
Université Paris XI
F-91405 Orsay Cedex
France
e-mail: rivass@th.u-psud.fr

Quantum Spaces, 109–159
© 2007 Birkhäuser Verlag Basel

Non-commutative Fluids

Alexios P. Polychronakos

Abstract. We review the connection between non-commutative gauge theory, matrix models and fluid mechanical systems. The non-commutative Chern-Simons description of the quantum Hall effect and bosonization of collective fermion states are used as specific examples.

1. Introduction

The idea that space may be a derived or emergent concept is a relatively old theme in theoretical physics. In the context of quantum mechanics, observables are operators and it is only their spectrum and mutual relations (commutators) that define their physical content. Space, to the extent that it is observable, should be not different. The properties attributed to space from everyday experience – and postulated in newtonian mechanics and special relativity – could be either exact or approximate, emerging in some particular or partial classical limit. Other structures, extending or deforming the concepts of classical geometry, and reducing to it under appropriate conditions, are conceivable.

This possibility has had an early emergence in speculations by Heisenberg himself. It made reappearances in various guises and contexts [1]. One of the most strikingly prescient of later developments in non-commutative gauge theory was the work of Eguchi and Kawai in large-N single-plaquette lattice gauge theory [2]. It was, however, after the seminal and celebrated work of Alain Connes that non-commutative geometry achieved the mathematical rigor and conceptual richness that made it a major component of modern theoretical physics. The concept made further inroads when it emerged as a property of space-time solutions derived from string theory [4, 5] and, by now, it claims a huge body of research literature.

One of the reasons that makes the idea of non-commutative spaces attractive is the common language and connections that it provides between apparently disparate topics. Indeed, as will be reviewed in this write-up, non-commutative physics unifies such *a priori* different objects as gauge fields, membranes, fluids, matrix models and many-body systems. (Some of the above connections can be established independently, but the full continuum emerges only in the non-commutative setting.)

Unification of description usually brings unification of concepts. This raises the stakes and elevates noncommutativity into a possibly fundamental property of nature. We could ask, for instance, whether the eventual bringing together of gravity, quantum mechanics and thermodynamics will arise out of some underlying fully non-commutative structure that shapes into space-time, quantum mechanics and statistical ensembles in some appropriate limit. Whether this is indeed true is, of course, unclear and leaves room for wild speculation.

At this point, we should refrain from fantasizing any further and take a more pragmatic point of view. The obvious question is: does noncommutativity buy us any advantage for physics as we presently know it? It will be the purpose of this exposé (as, I imagine, of the other contributions to this book) to demonstrate that this, indeed, is the case.

2. Review of non-commutative spaces

The concepts of non-commutative geometry will be covered by other contributions to this book and there is probably little use in repeating them here. Moreover, there are many excellent and complete review articles, of which [6–8] are only a small sample.

Nevertheless, a brief summary will be presented here, for two main reasons. Firstly, it will make this write-up essentially self-contained and will minimize the need to refer to other sources for a coherent reading; and secondly, the level and tone of the presentation will be adapted to our needs, and hopefully will serve as a low-key alternative to more rigorous and complete treatments.

2.1. The operator formulation

The simplest starting point for the definition of non-commutative spaces is through the definition of non-commutative coordinates. This is the approach that is most closely related to physics, making the allusions to quantum mechanics most explicit, and is therefore also the most common one in physics texts. In this, the non-commutative spaces are defined in terms of their coordinates x^μ, which are abstracted into (linear) operators. Such coordinates can be added and multiplied (associatively), forming a full operator algebra, but are not (necessarily) commutative. Instead, they obey the commutation relations

$$[x^\mu, x^\nu] = i\theta^{\mu\nu} \ , \ \mu, \nu = 1, \ldots, d. \tag{1}$$

The antisymmetric two-tensor $\theta^{\mu\nu}$ could be itself an operator, but is usually taken to commute with all x^μ (for 'flat' non-commutative spaces) and is, thus, a set of ordinary, constant c-numbers. Its inverse, when it exists,

$$\omega_{\mu\nu} = (\theta^{-1})_{\mu\nu}, \tag{2}$$

defines a constant two-form ω characterizing the noncommutativity of the space.

Clearly the form of θ can be changed by redefining the coordinates of the space. Linear redefinitions of the x^μ, in particular, would leave $\theta^{\mu\nu}$ a c-number (nonlinear redefinitions will be examined later). We can take advantage of this to give a simple form to $\theta^{\mu\nu}$. Specifically, by an orthogonal transformation of the x^μ we can bring $\theta^{\mu\nu}$ to a Darboux form consisting of two-dimensional blocks proportional to $i\sigma_2$ plus a set of zero eigenvalues. This would decompose the space into a direct sum of mutually commuting two-dimensional non-commutative subspaces, plus possibly a number of commuting coordinates (odd-dimensional spaces necessarily have at least one commuting coordinate). In general, there will be $2n$ properly noncommuting coordinates x^α ($\alpha = 1, \ldots, 2n$) and $q = d - 2n$ commuting ones Y^i ($i = 1, \ldots, q$). In that case ω will be defined as the inverse of the projection $\bar\theta$ of θ on the fully noncommuting subspace:

$$\omega_{\alpha\beta} = (\bar\theta^{-1})_{\alpha\beta} \ , \ \omega_{ij} = 0. \tag{3}$$

The actual non-commutative space can be thought of as a representation of the above operator algebra (1), acting on a set of states. For real spaces the operators x^μ will be considered hermitian, their eigenvalues corresponding to possible values of the corresponding coordinate. Not all coordinates can be diagonalized simultaneously, so the notion of 'points' (sets of values for all coordinates x^μ) is absent. The analogy with quantum mechanical coordinate and momentum is clear, with each 'Darboux' pair of non-commutative coordinates being the analog of a canonical quantum pair. Nevertheless, a full set of geometric notions survives, in particular relating to fields on the space, as will become clear.

The representation of x^μ can be reducible or irreducible. For the commuting components Y^i any useful representation must necessarily be reducible, else the corresponding directions would effectively be absent (consisting of a single point). States are labeled by the values of these coordinates y^i, taken to be continuous. The rest of the space, consisting of canonical Heisenberg pairs, admits the tensor product of Heisenberg-Fock Hilbert spaces (one for each two-dimensional noncommuting subspace $k = 1, \ldots, n$) as its unique irreducible representation. In general, we can have a reducible representation consisting of the direct sum of N such irreducible components for each set of values y^i, labeled by an extra index $a = 1, \ldots, N$ (we shall take N not to depend on y^i). A complete basis for the states, then, can be

$$|n_1, \ldots, n_n; y^1, \ldots, y^q; a\rangle \tag{4}$$

where n_k is the Fock (oscillator) excitation number of the kth two-dimensional subspace.

Due to the reducibility of the above representation, the operators x^μ do not constitute a complete set. To make the set complete, additional operators need be introduced. To deal with the reducibility due to the values y^i, we consider translation (derivative) operators ∂_μ. These are defined through their action on x^μ, generating constant shifts:

$$[\partial_\mu, x^\nu] = \delta_\mu^\nu. \tag{5}$$

On the fully non-commutative subspace these are inner automorphisms generated by

$$\partial_\alpha = -i\omega_{\alpha\beta}x^\beta. \tag{6}$$

For the commutative coordinates, however, extra operators have to be appended, shifting the Casimirs Y^i and thus acting on the coordinates y^i as usual derivatives.

To deal with the reducibility due to the components $a = 1, \ldots, N$, we need to introduce yet another set of operators in the full representation space mixing the above N components. Such a set are the hermitian $U(N)$ operators G^r, $r = 1, \ldots, N^2$ that commute with the x^μ, ∂_μ and mix the components a. (We could, of course, choose these operators to be the $SU(N)$ subset, eliminating the trivial identity operator.) The set of operators $x^\alpha, \partial_i, G^r$ is now complete.

Within the above setting, we can define field theories on a non-commutative space. Fields are the analogs of functions of coordinates x^μ; that is, arbitrary operators in the universal enveloping algebra of the x^μ. In general, the above fields are *not* arbitrary operators on the full representation space, since they commute with ∂_i and G^r. In particular, they act 'pointwise' on the commutative coordinates Y^i are, therefore, ordinary functions of the y^i.

We can, of course, define fields depending also on the remaining operators. Fields involving operators G^r are useful, as they act as $N \times N$ matrices on components a. They are the analogs of matrix-valued fields and will be useful in constructing gauge theories. We could further define operators that depend on the commutative derivatives ∂_i. These have no commutative analog, and will not be considered here. Notice, however, that on fully non-commutative spaces (even-dimensional spaces without commutative components), the matrix-valued fields $f^{ab}(x^\mu)$ constitute the *full* set of operators acting on the representation space.

The fundamental notions completing the discussion of non-commutative field theory are the definitions of derivatives and space integral. Derivatives of a function f are defined as commutators with the corresponding operator:

$$\partial_\mu \cdot f = [\partial_\mu, f]. \tag{7}$$

That is, through the adjoint action of the operator ∂_μ on fields (we use the dot to denote this action). For the commutative derivatives ∂_i this is the ordinary partial derivative ∂/∂_{y^i}. For the non-commutative coordinates, however, such action is generated by the x^α themselves, as $\partial_\alpha = -i\omega_{\alpha\beta}x^\beta$. So the notion of coordinates and derivatives on purely non-commutative spaces fuses, the distinction made only upon specifying the action of the operators x^α on fields (left- or right- multiplication, or adjoin action).

The integral over space is defined as the trace in the representation space, normalized as:

$$\int d^d x = \int d^q y \, \mathrm{tr}' \, \sqrt{\det(2\pi\theta)} \, \mathrm{tr} \equiv \mathrm{Tr} \tag{8}$$

where tr is the trace over the Fock spaces and tr$'$ is the trace over the degeneracy index $a = 1, \ldots, N$. This corresponds to a space integral and a trace over the matrix indices a. The extra determinant factor ensures the recovery of the proper commutative limit (think of semiclassical quantization, or the transition from quantum to classical statistical mechanical partition functions.)

All manipulations within ordinary field theory can be transposed here, with a non-commutative twist. For instance, the fact that the integral of a total derivative vanishes (under proper boundary conditions), translates to the statement that the trace of a commutator vanishes, and its violation by fields with nontrivial behavior at infinity is mirrored in the nonvanishing trace of the commutator of unbounded, non-trace class operators, such as the non-commutative coordinates themselves.

2.2. Weyl maps, Wigner functions and $*$-products

The product of non-commutative fields is simply the product of the corresponding operators, which is clearly associative but not commutative. It is also not 'point-wise', as the notion of points does not even exist. Nevertheless, in the limit $\theta^{\mu\nu} \to 0$ we recover the usual (commutative) geometry and algebra of functions. Points are recovered as any set of states whose spread Δx^μ in each coordinate x^μ goes to zero in the commutative limit. Such a useful set is, e.g., the set of coherent states in each non-commutative (Darboux) pair of coordinates with average values x^μ.

Observations like that can form the basis of a complete mapping between non-commutative fields and commutative functions $f(x)$, leading to the notion of the 'symbol' of $f(x)$ and the star-product. Specifically, by expressing fields as functions of the fundamental operators x^μ and ordering the various x^μ in the expressions for the fields in a prescribed way, using their known commutators, establishes a one-to-one correspondence between functions of operators and ordinary functions. This is reminiscent of, and in fact equivalent to, the Wigner function mapping of a quantum mechanical operator onto the classical phase space (see [9] for a simple review).

The ordering that is most usually adopted is the fully symmetric Weyl ordering, in which monomials in the x^μ are fully symmetrized. It is simplest to work with the Fourier transforms of functions, since exponentials of linear combinations of x^μ are automatically Weyl ordered. So a classical function $f(x)$, with Fourier transform $\tilde{f}(k)$, is mapped to the operator (non-commutative field) f as:

$$f = \int dk \, e^{ik_\mu x^\mu} \tilde{f}(k) \tag{9}$$

(the integral over k is of the appropriate dimensionality). Conversely, the 'symbol' (commutative function) corresponding to an operator f can be expressed as:

$$\tilde{f}(k) = \sqrt{\det(\theta/2\pi)} \, \mathrm{tr} \, f e^{-ik_\mu x^\mu} \tag{10}$$

where the above trace is taken over an irreducible representation of the non-commutative coordinates. This reproduces scalar functions. For matrix-valued non-commutative fields f, acting nontrivially on a direct sum of N copies of the irreducible representation, the above expression generalizes to

$$\tilde{f}^{ab}(k) = \sqrt{\det(\theta/2\pi)} \sum_n \langle n, a| f e^{-ik_\mu x^\mu} |n, b\rangle \tag{11}$$

where $|n, a\rangle$ are a complete set of states for the ath copy of the irreducible representation, reproducing a matrix function of commutative variables. Hermitian operators f map to hermitian matrix functions $f^{ab}(x)$ or, in the case $N = 1$, real functions.

On can show that, under the above mapping, derivatives and integrals of non-commutative fields map to the standard commutative ones for their symbol. The product of operators, however, maps to a new function, called the star-product of the corresponding functions [10]:

$$f \leftrightarrow f(x) \ , \ g \leftrightarrow g(x) \implies fg \leftrightarrow (f * g)(x). \tag{12}$$

The star product can be written explicitly in terms of the Fourier transforms of functions as

$$(f * g)(k) = \int dk \, \tilde{f}(q) \, \tilde{g}(k - q) \, e^{\frac{i}{2}\theta^{\mu\nu} k_\mu k_\nu} . \tag{13}$$

This is the standard convolution of Fourier transforms, but with an extra phase factor. The resulting $*$-product is associative but non-commutative and also non-local in the coordinates x^μ. The commutator of two non-commutative fields maps to the so-called star, or Moyal, brackets of their symbols.

The above mapping has the advantage that it circumnavigates the conceptual problems of non-commutative geometry by working with familiar objects such as ordinary functions and their integral and derivatives, trading the effects of non-commutativity for a nonlocal, non-commutative function product. It can, however, obscure the beauty and conceptual unification that arises from noncommutativity and make some issues or calculations unwieldy. In what follows, we shall stick with the operator formulation as exposed above. Translation into the $*$-product language can always be done at any desired stage.

3. Non-commutative gauge theory

Gauge theory on non-commutative spaces becomes particularly attractive [11–13]. Gauge fields A_μ are hermitian operators acting on the representation space. Since they do not depend on ∂_i they cannot shift the values of y^i, while they act nontrivially on the fully noncommuting subspace. They have effectively become big matrices acting on the full Fock space with elements depending on the commuting coordinates. Derivatives of these fields are defined through the adjoint action of ∂_μ,

$$\partial_\mu \cdot A_\nu = [\partial_\mu, A_\nu]. \tag{14}$$

Using the above formalism, gauge field theory can be built in a way analogous to the commuting case. Gauge transformations are unitary transformations in the full representation space. Restricting A_μ to depend on the coordinates only, as above, produces the so-called $U(1)$ gauge theory. $U(N)$ gauge theory can be obtained by relaxing this restriction and allowing A_μ to also be a function of the G^r and thus act on the index a.

3.1. Background-independent formulation

The basic moral of the previous section is that non-commutative gauge theory can be written in a universal way [14–16]. In the operator formulation no special distinction needs be done between $U(1)$ and $U(N)$ theories, nor need gauge and space-time degrees of freedom be treated distinctly. The fundamental operators of the theory are

$$D_\mu = -i\partial_\mu + A_\mu \tag{15}$$

corresponding to covariant derivatives. Gauge transformations are simply unitary conjugations of the covariant derivative operators by a unitary field U. That is, the D_μ transform covariantly:

$$D_\mu \to U^{-1} D_\mu U. \tag{16}$$

This reproduces the (non-commutative version of the) standard gauge transformation of A_μ:

$$A_\mu \to -iU^{-1}\partial_\mu \cdot U + U^{-1} A_\mu U. \tag{17}$$

For the fully non-commutative components, covariant derivative operators assume the form

$$D_\alpha = \omega_{\alpha\beta} x^\beta + A_\alpha = \omega_{\alpha\beta}(x^\beta + \theta^{\beta\gamma} A_\gamma) = \omega_{\alpha\beta} X^\beta. \tag{18}$$

The above rewriting is important in various ways. It stresses the fact that, on fully non-commutative spaces, the separation of D_α into x^α (coordinate) and A_α (gauge) is largely arbitrary and artificial: both are operators acting on the Hilbert space on an equal footing, the distinction between 'derivative' and 'coordinate' having been eliminated. This separation is also gauge dependent, since a unitary transformation will mix the two parts. In effect, gauge transformations mix spatial and gauge degrees of freedom! Further, it is not consistent any more to consider strictly $SU(N)$ gauge fields. Even if A_μ is originally traceless in the N-dimensional index a, gauge transformations U cannot meaningfully be restricted to $SU(N)$: the notion of partial trace of an operator with respect to one component of a direct product space makes sense, but the notion of partial determinant does not. A gauge transformation will always generate a $U(1)$ part for A_μ, making $U(N)$ gauge theory the only theory that arises naturally.

The above rewriting also introduces the 'covariant coordinate' field X^α that combines the ordinary coordinate and gauge fields in a covariant way and is dual to the covariant derivative. Non-commutative gauge theory can be constructed entirely in terms of the X^α. These, in turn, can be thought of as 'deformed' coordinates, the deformation being generated by (the dual of) gauge fields, which alludes to stretching membranes and fluids. All this is relevant in the upcoming story.

Any lagrangian built entirely out of D_μ will lead to a gauge invariant action, since the trace will remain invariant under any unitary transformation. The standard Maxwell-Yang-Mills action is built by defining the field strength

$$F_{\mu\nu} = \partial_\mu \cdot A_\nu - \partial_\nu \cdot A_\mu + i[A_\mu, A_\nu] = i[D_\mu, D_\nu] - \omega_{\mu\nu} \tag{19}$$

and writing the standard action

$$S_{LYM} = \frac{1}{4g^2} \mathrm{Tr} F_{\mu\nu} F^{\mu\nu} = -\frac{1}{4g^2} \mathrm{Tr}([D_\mu, D_\nu] + i\omega_{\mu\nu})^2 \tag{20}$$

where Tr also includes integration over commutative components y^i. In the above we used some c-number metric tensor $g^{\mu\nu}$ to raise the indices of F. Note that the operators $\partial_\alpha\cdot$, understood to act in the adjoint on fields, commute, while the operators $\partial_\alpha = -i\omega_{\alpha\beta} X^\beta$ have a nonzero commutator equal to

$$[\partial_\alpha, \partial_\beta] = i\omega_{\alpha\beta}. \tag{21}$$

This explains the extra ω-term appearing in the definition of F in terms of covariant derivative commutators.

One can, however, just as well work with the action

$$\hat{S}_{LYM} = \frac{1}{4g^2} \mathrm{Tr} \hat{F}_{\mu\nu} \hat{F}^{\mu\nu} = -\frac{1}{4g^2} \mathrm{Tr}[D_\mu, D_\nu][D^\mu, D^\nu]. \tag{22}$$

Indeed, $\hat{S}$ differs from S by a term proportional to $\mathrm{Tr}\omega^2$, which is an irrelevant (infinite) constant, as well as a term proportional to $\omega^{\mu\nu} \mathrm{Tr}[D_\mu, D_\nu]$, which, being the trace of a commutator (a 'total derivative'), does not contribute to the equations of motion. The two actions lead to the same classical theory. Note that $\theta^{\mu\nu}$ or $\omega_{\mu\nu}$ do not appear in the action. These quantities arise only in the commutator of non-commutative coordinates. Since the x^μ do not explicitly appear in the action either (being just a gauge-dependent part of D_μ), all reference to the specific non-commutative space has been eliminated! This is the 'background independent' formulation of non-commutative gauge theory that stresses its universality.

3.2. Superselection of the non-commutative vacuum

How does, then, a particular non-commutative space arise in this theory? The equations of motion for the operators D_μ are

$$[D^\mu, [D_\mu, D_\nu]] = 0. \tag{23}$$

The general operator solution of this equation is not fully known. Apart from the trivial solution $D_\mu = 0$, it admits as solution all operators with c-number commutators, satisfying

$$[D_\mu, D_\nu] = -i\omega_{\mu\nu} \tag{24}$$

for some ω. This is the classical 'non-commutative vacuum', where $D_\mu = -i\partial_\mu$, and expanding D_μ around this vacuum leads to a specific non-commutative gauge theory.

Quantum mechanically, $\omega_{\mu\nu}$ are superselection parameters and the above vacuum is stable. To see this, assume that the time direction is commutative and consider the collective mode

$$D_\alpha = -i\lambda_{\alpha\beta}\partial_\beta \tag{25}$$

with $\lambda_{\alpha\beta}$ parameters depending only on time. This mode would change the non-commutative vacuum while leaving the gauge field part of D_α unexcited. ω gets modified into

$$\omega'_{\mu\nu} = \lambda_{\mu\alpha}\omega_{\alpha\beta}\lambda_{\beta\nu}. \tag{26}$$

The action implies a quartic potential for this mode, with a strength proportional to Tr1, and a kinetic term proportional to $\mathrm{Tr}\partial_\alpha\partial_\beta$. (There is also a gauge constraint which does not alter the qualitative dynamical behavior of λ.) Both potential and kinetic terms are infinite, and to regularize them we should truncate each Fock space trace up to some highest state Λ, corresponding to a finite volume regularization (the area of each non-commutative two-dimensional subspace has effectively become Λ). One can check that the potential term would grow as Λ^n while the kinetic term would grow as Λ^{n+1}. Thus the kinetic term dominates; the above collective degrees of freedom acquire an infinite mass and will remain "frozen" to whatever initial value they are placed, in spite of the nontrivial potential. (This is analogous to the θ-angle of the vacuum of four-dimensional non-abelian gauge theories: the vacuum energy depends on θ which is still superselected.) Quantum mechanically there is no interference between different values of λ and we can fix them to some c-number value, thus fixing the noncommutativity of space [17]. This phenomenon is similar to symmetry breaking, but with the important difference that the potential is not flat along changes of the "broken" vacuum, and consequently there are no Goldstone bosons.

In conclusion, we can start with the action (22) as the definition of our theory, where D_μ are arbitrary operators (matrices) in some space. Gauge theory is then defined as a perturbation around a (stable) classical vacuum. Particular choices of this vacuum will lead to standard non-commutative gauge theory, with $\theta^{\mu\nu}$ and N appearing as vacuum parameters. Living in any specific space and gauge group amounts to landscaping!

3.3. Non-commutative Chern-Simons action

A particularly useful and important type of action in gauge theory is the Chern-Simons term [18]. This is a topological action, best written in terms of differential forms. In the commutative case, we define the one- and two-forms

$$\mathrm{A} = iA_\mu dx^\mu \ , \ \ \mathrm{F} = d\mathrm{A} + \mathrm{A}^2 = \frac{i}{2}\Big(\partial_\mu A_\nu - \partial_\nu A_\mu + i[A_\mu, A_\nu]\Big)dx^\mu dx^\nu. \tag{27}$$

The Chern-Simons action S_{2n+1} is the integral of the $2n+1$-form C_{2n+1} satisfying

$$d\mathrm{C}_{2n+1} = \mathrm{tr}\mathrm{F}^{n+1}. \tag{28}$$

By virtue of (28) and the gauge invariance of $\mathrm{tr}\mathrm{F}^n$ it follows that S_{2n+1} is gauge invariant up to total derivatives, since, if δ stands for an infinitesimal gauge transformation,

$$d\delta\mathrm{C}_{2n+1} = \delta d\mathrm{C}_{2n+1} = \delta\mathrm{tr}\mathrm{F}^n = 0 \ , \ \text{so} \ \delta C_{2n+1} = d\Omega_{2n}. \tag{29}$$

The integrated action is therefore invariant under infinitesimal gauge transformations. Large gauge transformations may lead to an additive change in the action and they usually imply a quantization of its coefficient [18, 19]. As a result, the equations of motion derived from this action are gauge covariant and read

$$\frac{\delta S_{2n+1}}{\delta A} = \frac{\delta}{\delta A} \int C_{2n+1} = (n+1)F^n. \tag{30}$$

The above can be considered as the defining relation for C_{2n+1}.

We can define corresponding non-commutative Chern-Simons actions [20–28]. To this end, we shall adopt the differential form language [17] and define the usual basis of one-forms dx^μ as a set of formal anticommuting parameters with the property

$$dx^\mu dx^\nu = -dx^\nu dx^\mu \ , \ \ dx^{\mu_1} \cdots dx^{\mu_d} = \epsilon^{\mu_1 \cdots \mu_d}. \tag{31}$$

Topological actions do not involve the metric tensor and can be written as integrals of d-forms. The only dynamical objects available in non-commutative gauge theory are D_μ and thus the only form that we can write is

$$D = idx^\mu D_\mu = d + A \tag{32}$$

where we defined the exterior derivative and gauge field one-forms

$$d = dx^\mu \partial_\mu \ , \ \ A = idx^\mu A_\mu \tag{33}$$

(note that both D and A as defined above are antihermitian). The action of the exterior derivative d on an operator p-form H, d·H, yields the $p+1$-form $dx^\mu[\partial_\mu, H]$ and is given by

$$d \cdot H = dH - (-)^p Hd. \tag{34}$$

In particular, on the gauge field one-form A it acts as

$$d \cdot A = dA + Ad. \tag{35}$$

Correspondingly, the covariant exterior derivative of H is

$$D \cdot H = DH - (-)^p HD. \tag{36}$$

As a result of the noncommutativity of the operators ∂_μ, the exterior derivative operator is not nilpotent but rather satisfies

$$d^2 = \omega \ , \ \ \omega = \frac{i}{2} dx^\mu dx^\nu \omega_{\mu\nu}. \tag{37}$$

We stress, however, that d· is still nilpotent since ω commutes with all operator forms:

$$d \cdot d \cdot H = [d, [d, H]_\mp]_\pm = \pm[\omega, H] = 0. \tag{38}$$

The two-form $\hat{F} = \frac{i}{2} dx^\mu dx^\nu \hat{F}_{\mu\nu}$ is simply

$$\hat{F} = D^2 = \frac{1}{2} D \cdot D = \omega + dA + Ad + A^2 = \omega + F \tag{39}$$

where $F = \frac{i}{2} dx^\mu dx^\nu F_{\mu\nu}$ is the conventionally defined field strength two-form.

The most general d-form that we can write involves arbitrary combinations of D and ω. If, however, we adopt the view that ω should arise as a superselection (vacuum) parameter and not as a term in the action, the unique form that we can write is D^d and the unique action

$$\hat{S}_d = \frac{d+1}{2d}\,\mathrm{Tr}\,\mathrm{D}^d = \mathrm{Tr}\,\mathrm{C}_d. \tag{40}$$

This is the Chern-Simons action. The coefficient was chosen to conform with the commutative definition, as will be discussed shortly. In even dimensions $\hat{S}_d$ reduces to the trace of a commutator $\mathrm{Tr}[\mathrm{D},\mathrm{D}^{d-1}]$, a total derivative that does not affect the equations of motion and corresponds to a topological term. In odd dimensions it becomes a nontrivial action.

$\hat{S}_d$ is by construction gauge invariant. To see that it also satisfies the defining property of a Chern-Simons form (30) is almost immediate: $\delta/\delta\mathrm{A} = \delta/\delta\mathrm{D}$ and thus, for $d = 2n + 1$:

$$\frac{\delta}{\delta\mathrm{A}}\,\mathrm{Tr}\,\mathrm{D}^{2n+1} = (2n+1)\,\mathrm{D}^{2n} = (2n+1)\,\hat{\mathrm{F}}^n. \tag{41}$$

So, with the chosen normalization in (40) we have the defining condition (30) with $\hat{\mathrm{F}}$ in the place of F. What is less obvious is that $\hat{S}_D$ can be written entirely in terms of F and A and that, for commutative spaces, it reduces to the standard Chern-Simons action. To achieve that, one must expand C_D in terms of d and A, make use of the cyclicity of trace and the condition $\mathrm{d}^2 = \omega$ and reduce the expressions into ones containing $\mathrm{dA} + \mathrm{Ad}$ rather than isolated d s. The condition

$$\mathrm{Tr}\,\omega^n\mathrm{d} = 0 \tag{42}$$

which is a result of the fact that ∂_μ is off-diagonal for both commuting and non-commuting dimensions, can also be used to get rid of overall constants. This is a rather involved procedure for which we have no algorithmic approach. (Specific cases will be worked out later.) Note, further, that the use of the cyclicity of trace implies that we dismiss total derivative terms (traces of commutators). Such terms do not affect the equations of motion. For $d = 1$ the result is simply

$$\hat{S}_1 = \mathrm{Tr}\mathrm{A} \tag{43}$$

which is the 'abelian' one-dimensional Chern-Simons term. For $d = 3$ we obtain

$$\hat{S}_3 = \mathrm{Tr}(\mathrm{AF} - \frac{1}{3}\mathrm{A}^3) + 2\mathrm{Tr}(\omega\mathrm{A}) \tag{44}$$

where we used the fact that $\mathrm{Tr}[\mathrm{A}(\mathrm{dA} + \mathrm{Ad})] = 2\mathrm{Tr}(\mathrm{A}^2\mathrm{d})$. The first term is the non-commutative version of the standard three-dimensional Chern-Simons term, while the second is a lower-dimensional Chern-Simons term involving explicitly ω.

We can get the general expression for $\hat{S}_d$ by referring to the defining relation. This reads

$$\frac{\delta}{\delta\mathrm{A}}\,\hat{S}_{2n+1} = (n+1)\hat{\mathrm{F}}^n = (n+1)(\mathrm{F}+\omega)^n = (n+1)\sum_{k=0}^{n}\binom{n}{k}\omega^{n-k}\mathrm{F}^k \tag{45}$$

and by expressing F^k as the A-derivative of the standard Chern-Simons action S_{2k+1} we get

$$\frac{\delta}{\delta A} \left\{ \hat{S}_{2n+1} - \sum_{k=0}^{n} \binom{n+1}{k+1} \omega^{n-k} S_{2k+1} \right\} = 0. \tag{46}$$

So the expression in brackets must be a constant, easily seen to be zero by setting $A = 0$. We therefore have

$$\hat{S}_{2n+1} = \sum_{k=0}^{n} \binom{n+1}{k+1} \mathrm{Tr}\, \omega^{n-k} C_{2k+1}. \tag{47}$$

We observe that we get the $2n+1$-dimensional Chern-Simons action plus all lower-dimensional actions with tensors ω inserted to complete the dimensions. Each term is separately gauge invariant and we could have chosen to omit them, or include them with different coefficient. It is the specific combination above, however, that has the property that it can be reformulated in a way that does not involve ω explicitly. The standard Chern-Simons action can also be written in terms of D alone by inverting (47):

$$S_{2n+1} = (n+1)\mathrm{Tr} \int_0^1 D(t^2 D^2 - \omega)^n dt = \mathrm{Tr} \sum_{k=0}^{n} \binom{n+1}{k+1} \frac{k+1}{2k+1} (-\omega)^{n-k} D^{2k+1}. \tag{48}$$

For example, the simplest nontrivial non-commutative action in 2+1 dimensions reads

$$S_3 = \mathrm{Tr} \left(\frac{2}{3} D^3 - 2\omega D \right). \tag{49}$$

The above can be written more explicitly in terms of the two spatial covariant derivatives $D_{1,2}$, which are operators acting on the non-commutative space, and the temporal covariant derivative $D_0 = dt(\partial_t + iA_0)$, which contains a proper derivative operator in the commutative direction $x^0 = t$ and a non-commutative gauge field A_0:

$$S_3 = \int dt\, 2\pi\theta\, \mathrm{Tr} \left\{ \epsilon^{ij} (\dot{D}_i + i[A_0, D_i])D_j + \frac{2}{\theta} A_0 \right\}. \tag{50}$$

Note that the overall coefficient of the last, linear term is independent of θ.

We also point out a peculiar property of the Chern-Simons form $\hat{C}_{2n+1}$. Its covariant derivative yields $\hat{F}^{n+1}$:

$$D \cdot \hat{C}_{2n+1} = D\hat{C}_{2n+1} + \hat{C}_{2n+1}D = \frac{2n+2}{2n+1} \hat{F}^{n+1}. \tag{51}$$

A similar relation holds between C_d (understood as the form appearing inside the trace in the right-hand side of (48)) and F. Clearly the standard Chern-Simons form does not share this property. Our C_d differs from the standard one by commutators that cannot all be written as ordinary derivatives (such as, $e.g.$, $[d, dA]$). These unconventional terms turn C_d into a covariant quantity that satisfies (51).

3.4. Level quantization for the non-commutative Chern-Simons action

We conclude our consideration of the non-commutative Chern-Simons action by considering the quantization requirements for its coefficient [27, 28].

In the commutative case, a quantization condition for the coefficient of non-abelian Chern-Simons actions ('level quantization') is required for global gauge invariance. This has its roots in the topology of the group of gauge transformations in the given manifold. E.g., for the three-dimensional term, the fact that $\pi_3[SU(N)] = Z$ for any $N > 1$ implies the existence of topologically nontrivial gauge transformations and corresponding level quantization.

For the non-commutative actions we have not studied the topology of the gauge group. This would appear to be a hard question for a 'fuzzy' non-commutative space, but in fact is well defined and easy to answer: gauge transformations are simply unitary transformations on the full representation space on which X^μ or D_μ act. This space is infinite-dimensional, so we are dealing with (some version of) $U(\infty)$. Two observations, however, elucidate the answer. First, for odd-dimensional non-commutative spaces there is always one (and in general only one) commutative dimension t, conventionally called time and compactified to a circle; and second, if we require gauge transformations to act trivially at infinity, we are essentially restricting the corresponding unitary operators to have finite support on the representation space and be bounded. So the relevant gauge transformations are essentially $U(N)$ matrices of the form $U(t)$, where N is the 'support' of U, that is, the dimension of the subspace of the Hilbert space on which U acts nontrivially. The relevant topology is $S^1 \to U(N)$ and is nontrivial due to the $U(1)$ factor in $U(N)$:

$$\pi_1[U(N)] = \pi_1[U(1)] = Z. \tag{52}$$

This is true for *any* non-commutative gauge theory, abelian or non-abelian. A 'winding number one' transformation would be a matrix of the form

$$U(t) = e^{i\frac{2\pi}{N}t}\tilde{U}(t) \ , \ t \in [0, 1] \tag{53}$$

with $\tilde{U}$ an $SU(N)$ matrix satisfying $\tilde{U}(0) = 1$ and $\tilde{U}(1) = \exp(-i\frac{2\pi}{N})$, a Z_N matrix. This satisfies $U(0) = U(1) = 1$ but cannot be smoothly deformed to $U(t) = 1$.

What is the change, if any, of the non-commutative Chern-Simons action under the above transformation? We may look at the explicit form (50) of S_3 to decide it. The first, cubic term is completely gauge invariant. Indeed, under a gauge transformation the quantity inside the trace and integral transforms covariantly

$$\epsilon^{ij}(\dot{D}_i + i[A_0, D_i])D_j \to U(t)^{-1}\left[\epsilon^{ij}(\dot{D}_i + i[A_0, D_i])D_j\right]U(t) \tag{54}$$

and upon tracing it remains invariant. The term A_0, however, transforms as

$$A_0 \to U(t)^{-1}A_0 U(t) - iU(t)^{-1}\dot{U}(t). \tag{55}$$

The last term gives a nontrivial contribution to the action equal to

$$\Delta S_3 = -i4\pi \int_0^1 dt\, \mathrm{tr} U(t)^{-1}\dot{U}(t). \tag{56}$$

122   A.P. Polychronakos

The $SU(N)$ part $\tilde{U}$ of $U(t)$ does not contribute to the above, since $\tilde{U}^{-1}\dot{\tilde{U}}$ is traceless. The $U(1)$ factor, however, contributes a part equal to

$$\Delta S_3 = -i4\pi \int_0^1 dt\, i\frac{2\pi}{N}\mathrm{tr}1 = 8\pi^2. \tag{57}$$

The coefficient of the action λ should be such that the overall change of the action be quantum mechanically invisible, that is, a multiple of 2π. We get

$$\lambda\, 8\pi^2 = 2\pi n \text{ or } \lambda = \frac{n}{4\pi} \tag{58}$$

with n an integer.

The above quantization condition is independent of θ and conforms with the level quantization of the commutative non-abelian Chern-Simons theory. It also holds for the *abelian* (or, rather, $U(1)$) theory, for which there is no quantization in the commutative case. In the commutative limit the corresponding topologically nontrivial gauge transformations become singular and decouple from the theory, thus eliminating the need for quantization. This result will be relevant in the upcoming considerations of the quantum Hall effect.

4. Connection with fluid mechanics

At this point we take a break from non-commutative gauge theory to bring into the picture fluid mechanics and review its two main formulations, Euler and Lagrange. As will become apparent, the two subjects are intimately related. Already we saw that non-commutative gauge theory can be formulated in terms of covariant deformed coordinate operators X^μ. These parallel the spatial coordinates of particle fluids, with the undeformed background coordinates x^μ playing the role of body-fixed labels of the particles.

4.1. Lagrange and Euler descriptions of fluids

We start with a summary review of the two main formulations of fluid mechanics, the particle-fixed (Lagrange) and space-fixed (Euler) descriptions. For more extensive reviews see [29, 30].

A fluid can be viewed as a dense collection of (identical) particles moving in some d-dimensional space, evolving in time t. The Lagrange description uses the coordinates of the particles comprising the fluid: $X^i(x,t)$. These are labeled by a set of parameters x^i, which are the coordinates of some fiducial reference configuration and are called particle-fixed or comoving coordinates. They serve, effectively, as particle 'labels'. Summation over particles amounts to integration over the comoving coordinates x times the density of particles in the fiducial configuration $\rho_0(x)$, which is usually taken to be homogeneous.

In the Euler description the fluid is described by the space-time-dependent density $\rho(r,t)$ and velocity fields $v^i(r,t)$ at each point of space with coordinates r^i. The two formulations are related by considering the particles at space coordinates r^i, that is, $X^i = r^i$, and expressing the density and velocity field in terms of the

Lagrange variables. We assume sufficient regularity so that (single-valued) inverse functions $\chi^i(r,t)$ exist:

$$X^i(t,x)\Big|_{x=\chi(t,r)} = r^i. \tag{59}$$

$X^i(x,t)$ provides a mapping of the fiducial particle position x^i to position at time t, while $\chi^i(r,t)$ is the inverse mapping. The Euler density then is defined by

$$\rho(r,t) = \rho_0 \int dx\, \delta\big(X(x,t) - r\big) . \tag{60}$$

(The integral and the δ-function carry the dimensionality of the relevant space.) This evaluates as

$$\frac{1}{\rho(r,t)} = \frac{1}{\rho_0} \det \frac{\partial X^i(x,t)}{\partial x^j}\Big|_{x=\chi(r,t)} \tag{61}$$

which is simply the change of volume element from fiducial to real space. The Euler velocity is

$$v^i(r,t) = \dot{X}^i(x,t)\Big|_{x=\chi(r,t)} \tag{62}$$

where overdot denotes differentiation with respect to the explicit time dependence. (Evaluating an expression at $x = \chi(r,t)$ is equivalent to eliminating x in favor of X, which is then renamed r.)

The number of particles in the fluid is conserved. This is a trivial (kinematical) condition in the Lagrange formulation, where comoving coordinates directly relate to particles. In the Euler formulations this manifests through conservation of the particle current $j^i = \rho v^i$, given in terms of Lagrange variables by

$$j^i(r,t) = \rho_0 \int dx\, \dot{X}^i(r,t)\delta\big(X(x,t) - r\big). \tag{63}$$

As a consequence of the above definition it obeys the continuity equation

$$\dot{\rho} + \partial_i j^i = 0 . \tag{64}$$

The kinetic part of the lagrangian K for the Lagrange variables is simply the single-particle lagrangian for each particle in terms of the particle coordinates, $K_{\rm sp}(X)$, summed over all particles,

$$K = \rho_0 \int dx K_{\rm sp}\big(X(x,t)\big). \tag{65}$$

The exact form of $K_{\rm sp}$ depends on whether the particles are relativistic or non-relativistic, the presence of magnetic fields etc. As an example, the kinetic term for a non-relativistic plasma in an external magnetic field generated by an electromagnetic vector potential $\mathcal{A}_i$ is

$$K = \rho_0 \int dx \left[\frac{1}{2} m\, g_{ij}(X)\, \dot{X}^i \dot{X}^j + q\mathcal{A}_i(X,t)\, \dot{X}^i \right] \tag{66}$$

with m and q the mass and charge of each fluid particle and g_{ij} the metric of space.

124   A.P. Polychronakos

Single-particle (external) potentials can be written in a similar way, while many-body and near-neighbor (density dependent) potentials will be more involved.

4.2. Reparametrization symmetry and its non-commutative avatar

The Lagrange description has an obvious underlying symmetry. Comoving coordinates are essentially arbitrary particle labels. All fluid quantities are invariant under particle relabeling, that is, under reparametrizations of the variables x^i, provided that the density of the fiducial configuration ρ_0 remains invariant. Such transformations are volume-preserving diffeomorphisms of the variables x^i.

For the minimal nontrivial case of two spatial dimensions, this symmetry corresponds to area-preserving diffeomorphisms. They can be thought of as canonical transformations on a two-dimensional phase space and are parametrized by a function of the two spatial variables, the generator of canonical transformation. Infinitesimal transformations are written

$$\delta x^i = \epsilon^{ij} \frac{\partial f}{\partial x^j} \tag{67}$$

with $f(x)$ the generating function. Obviously δx^i satisfies the area-preserving condition

$$\det \frac{\partial (x^i + \delta x^i)}{\partial x^j} = 1 \text{ or } \frac{\partial \delta x^i}{\partial x^i} = 0. \tag{68}$$

The same condition can be written in an even more suggestive way. Define a canonical structure for the two-dimensional space in terms of the Poisson brackets

$$\{x^1, x^2\} = \theta \text{ or } \{x^i, x^j\} = \theta \epsilon^{ij} = \theta^{ij} \tag{69}$$

for some constant θ. Rescaling f by a factor θ^{-1}, we can re-write δx^i as

$$\delta x^i = \theta^{ij} \partial_j f = \{x^i, f\}. \tag{70}$$

Similarly, the transformation of the fundamental (Lagrange) fluid variables under the above redefinition is

$$\delta X^i = \partial_j X^i \delta x^j = \theta^{jk} \partial_j X^i \partial_k f = \{X^i, f\}. \tag{71}$$

The above look like the classical analog (or precursor) of the gauge transformations of the covariant non-commutative gauge coordinates X^i of the previous sections. This is not accidental: the area-preserving transformations for the fluid correspond to relabeling the parameters x and do not generate a physically distinct fluid configurations. They represent simply a redundancy in the description of the fluid in terms of Lagrange coordinates; that is, a gauge symmetry. Physical fluid quantities, such as the Euler variables, or the fluid lagrangian, are expressed as integrals of quantities transforming 'covariantly' under the above transformation; that is, transforming by the Poisson bracket of the quantity with the generator of the transformation f, as in (71). They are, therefore, invariant under such transformations; that is, gauge invariant.

The analogy with non-commutative gauge theory becomes manifest by writing the Lagrange particle coordinates in terms of their deviation from the fiducial coordinates [31–34]:

$$X^i(x,t) = x^i + a^i(x,t) = x^i + \theta^{ij} A_j(x,t). \tag{72}$$

The deviation a^i, and its dual A_i do not transform covariantly any more; rather

$$\delta A_i = \partial_i f + \{A_i, f\}. \tag{73}$$

The similarity with the gauge transformation of a gauge field is obvious. The duals of the X^i,

$$D_i = \omega_{ij} X^j = \omega_{ij} x^j + A_i, \tag{74}$$

obviously correspond to covariant derivatives (although at this stage they are just rewritings of the comoving particle coordinates). The analog of the field strength is

$$\hat{F}_{ij} = \{D_i, D_j\} = \omega_{ij} + \partial_i A_j - \partial_j A_i + \{A_i, A_j\}. \tag{75}$$

This is related to the fluid density, which in the Poisson bracket formulation reads

$$\frac{\rho_0}{\rho} = \det \frac{\partial X^k(x,t)}{\partial x^l} = \frac{1}{\theta}\{X^1, X^2\}. \tag{76}$$

The field strength calculates as

$$\hat{F}_{ij} = \omega^{ij}\{X^1, X^2\} = \frac{\rho_0}{\rho}\epsilon_{ij}. \tag{77}$$

The field strength essentially becomes the (inverse) fluid density!

Similar considerations generalize to higher dimensions, with one twist: canonical transformations, the classical version of non-commutative gauge transformations, are only a symplectic subgroup of full volume-preserving diffeomorphisms. Higher-dimensional non-commutative gauge theory is analogous to a special version of fluid mechanics that enjoys a somewhat limited particle relabeling invariance. For the purposes of describing the quantum Hall effect, an essentially two-dimensional situation, this is inconsequential.

4.3. Gauging the symmetry

In the above discussion the role of time was not considered. The particle relabeling (x-space reparametrization) considered above were time-independent. Time-dependent transformations are not, a priori, invariances of the fluid since they introduce extra, nonphysical terms in the particle velocities $\dot{X}^i(x,t)$. To promote this transformation into a full space-time gauge symmetry we must gauge time derivatives by introducing a temporal gauge field A_0:

$$D_0 X^i = \dot{x}^i + \{A_0, X^i\}. \tag{78}$$

Under the transformation (71) with a time-dependent function f the above derivative will transform covariantly,

$$\delta D_0 X^i = \{D_0 X^i, f\}, \tag{79}$$

provided that the gauge field A_0 transforms as

$$\delta A_0 = \dot{f} + \{A_0, f\}. \tag{80}$$

This gauging, however, has dynamical consequences. We can gauge fix the theory by choosing the temporal gauge, putting $A_0 = 0$. The action becomes identical to the ungauged action, with the exception that now we have to satisfy the Gauss law for the gauge-fixed symmetry, that is, the equation of motion for the reduced field A_0. The exact form of the constraint depends on the kinetic term of the lagrangian for the fluid:

$$G = \left\{ X^i, \frac{\partial K}{\partial \dot{X}^i} \right\} = 0. \tag{81}$$

As an example, for the plasma of (66) the Gauss law reads

$$G = \{\dot{X}^i, m g_{ij}(X)\, \dot{X}^j + q\mathcal{A}_i(X)\} = 0. \tag{82}$$

Interesting two-dimensional special cases are ($g_{ij} = \delta_{ij}$, $q = 0$), when

$$G = \{\dot{X}^i, X^i\} = 0 \tag{83}$$

and the 'lowest Landau level' case of massless particles in a constant magnetic field ($m = 0$, $\mathcal{A}_i = (B/2)\epsilon_{ij} X^j$), when

$$G = \{X^1, X^2\} = 0. \tag{84}$$

We conclude by mentioning that the fluid structure we described in this section can also be interpreted as membrane dynamics. Indeed, a membrane is, in principle, a sheet of fluid in a higher-dimensional space. A two-dimensional membrane in two space dimensions is space-filling, and thus indistinguishable from a fluid, the density expressing the way in which the membrane shrinks or expands locally. The full correspondence of membranes, non-commutative (matrix) theory and fluids, relativistic and non-relativistic, has been examined elsewhere [35]. We shall not expand on it here.

4.4. Non-commutative fluids and the Seiberg-Witten map

In the previous section we alluded to the connection between non-commutative gauge theory and fluid mechanics. It is time to make the connection explicit [34]. We shall work specifically in two (flat) spatial dimensions, as the most straightforward case and relevant to the quantum Hall effect.

The transition from (classical) fluids to non-commutative fluids is achieved the same way as the transition from classical to quantum mechanics. We promote the canonical Poisson brackets introduced in the previous section to (operator) commutators. All Poisson brackets that appear become commutators:

$$\{\,,\,\} \;\to\; -i[\,,\,]. \tag{85}$$

So the comoving parameters satisfy

$$[x^i, x^j] = i\theta^{ij}. \tag{86}$$

They have become a non-commutative plane. This means that the particle labels cannot have 'sharp' values and pinpointing the particles of the fluid is no more possible. In effect, we have a 'fuzzification' of the underlying fluid particles and a corresponding 'fuzzy' fluid.

The remaining structure smoothly goes over to non-commutative gauge theory, as already alluded. We assume that the non-commutative coordinates x^1, x^2 act on a single irreducible representation of their Heisenberg algebra; this effectively assigns a single particle state for each 'point' of space (each state in the representation). Inclusion of multiple copies of the irreducible representations would correspond to multiple particle states per 'point' of space and would endow the particles with internal degrees of freedom.

Integration over the comoving parameters becomes $2\pi\theta$ times trace over the representation space. Summation over particles, then, becomes

$$\sum_{\text{particles}} = \rho_0 \int dx \ \to \ 2\pi\theta\rho_0 \text{Tr}. \tag{87}$$

The parameter θ, or its inverse ω, was introduced arbitrarily and plays no role in the fluid description. This is similar to the background-independent formulation of non-commutative gauge theory in terms of covariant derivatives or coordinates. Presently, we relate θ to the inverse density of the fiducial configuration ρ_0^{-1}

$$2\pi\theta = \frac{1}{\rho_0} \tag{88}$$

in which case the factor in the preceding equation disappears. Particle summation becomes a simple trace, so particles are identified with states in the representation space. This relation between fiducial density and noncommutativity parameter will always be assumed to hold from now on.

The Lagrange coordinates of particles X^i and the gauge field A_0 are functions of the underlying 'fuzzy' (non-commutative) particle labels, and thus become non-commutative fields. Area-preserving reparametrizations, which are canonical transformations in the classical case, become unitary transformations in the non-commutative case (think, again, of quantum mechanics). Operators X^i transform by unitary conjugations; infinitesimally,

$$\delta X^i = i[f, X^i]. \tag{89}$$

The deviations of X^i from the fiducial coordinates x^i, on the other hand, as defined in (72), and the temporal gauge field pick up extra terms and transform as proper gauge fields:

$$\delta A_\mu = \partial_\mu f - i[A_\mu, f]. \tag{90}$$

The remaining question is the form of the (gauge invariant) lagrangian that corresponds to the non-commutative fluid. This depends on the specific fluid dynamics and will be dealt with in the next section. Before we go there, we would like to examine further the properties of the non-commutative fluid that derives from the present construction. Just because the underlying particles become fuzzy does not necessarily mean that the emerging fluid cannot be described in traditional terms. Indeed, fluids are dense distributions of particles and we are not supposed to be able to distinguish individual particles in any case. The Euler description, which talks about collective fluid properties like density and velocity, remains valid in the non-commutative case as we shall see.

The non-commutative version of equation (76) for the density becomes (with $2\pi\theta\rho_0 = 1$)

$$[X^1, X^2] = \frac{i}{2\pi\rho}. \tag{91}$$

This relation would suggest that the density, too, becomes a non-commutative field. The difficulty with this expression is that it gives the density as a function of the underlying comoving coordinates, which we know are non-commutative.

A better expression is (60), which gives the density as a function of a point in space r. This formula directly transcribes into

$$\rho(r,t) = \mathrm{Tr}\delta(X - r) \tag{92}$$

in the non-commutative case. r is still an ordinary space variable, and the trace eliminates the operator nature of the expression in the right-hand side, rendering a classical function of r and t. The only difficulty is in the definition of the delta function for the non-commutative argument $X^i - r^i$: the various X^i (two in our case) are operators and do not commute, so there are ordering issues in defining any function of the two. In fact, the operator $\delta(X - r)$ may not even be hermitian unless properly ordered, which would produce a complex density.

In dealing with such problems, a procedure similar to the definition of the 'symbol' of a non-commutative field is followed: a standard ordering of all monomials involving various X^is is prescribed. The Weyl (totally symmetrized) ordering is usually adopted. Under this ordering, the delta function above is defined as

$$\delta(X - r) = \int dk\, e^{ik_i(r^i - X^i)} \tag{93}$$

where k_i are classical (c-number) Fourier integration parameters. The above operator has also the advantage of being hermitian. The spatial Fourier transform of the density with respect to r is simply

$$\rho(k,t) = \mathrm{Tr}\, e^{-ik_i X^i}. \tag{94}$$

In a similar vein, we use the classical expression for the particle current

$$j^i(r,t) = \rho_0 \int dx\, \dot{X}^i \delta(X - r) \tag{95}$$

to write the corresponding expression for the non-commutative fluid as

$$j^i(k,t) = \mathrm{Tr}\, D_0 X^i\, e^{-ik_j X^j}. \tag{96}$$

In the above, we used the covariant time derivative in order to make the expression explicitly gauge invariant. The corresponding current is real, as the trace ensures that the change of ordering between $D_0 X$ and the exponential is immaterial.

The crucial observation is that the above density and current still satisfy the continuity equation, which in Fourier space becomes

$$\dot{\rho} + ik_i j^i = 0. \tag{97}$$

The proof is straightforward and relies on the following two facts, true due to the cyclicity of trace:

$$\frac{d}{dt}\mathrm{Tr}\, e^{-ik_i X^i} = -i\mathrm{Tr}\, k_j \dot{X}^j e^{-ik_i X^i} \tag{98}$$

and

$$\mathrm{Tr}\,[A_0, k_j X^j]e^{-ik_i X^i} = 0. \tag{99}$$

The non-commutative fluid, therefore, has an Euler description in terms of a traditional conserved particle density and current.

The above observation is the basis for a mapping between commutative and non-commutative gauge theories, which first arose in the context of string theory and is known as the Seiberg-Witten map [5]. The key element is that, in 2+1 dimensions, a conserved current can be written in terms of its dual two-form, which then satisfies the Bianchi identity. Specifically, define

$$J_{\mu\nu} = \epsilon_{\mu\nu\lambda}j^\lambda \tag{100}$$

where $j^0 = \rho$. Then, due to the continuity equation $\partial_\mu j^\mu = 0$, $J_{\mu\nu}$ satisfies

$$\partial_\mu J_{\nu\lambda} + \text{cyclic perms.} = 0 \text{ or } \mathrm{d}J = 0. \tag{101}$$

This means that J can be considered as an abelian field strength, which allows us to define an abelian commutative gauge field $\tilde{A}_\mu$. The reference configuration of the fluid, in which particles are in their fiducial positions $X^i = x^i$ and corresponds to vanishing non-commutative gauge field, gives $j_0^\mu = (\rho_0, 0, 0)$ or $J_0 = \rho_0 \mathrm{d}x^1 \mathrm{d}x^2$. If we want to have this configuration correspond to vanishing abelian gauge field $\tilde{F}_{\mu\nu}$, we have to define

$$\tilde{F} = J - J_0 \tag{102}$$

or, more explicitly

$$\tilde{F}_{0i} = \epsilon_{ik}j^k \ , \ \ \tilde{F}_{ij} = \epsilon_{ij}(\rho - \rho_0). \tag{103}$$

Substituting the explicit expressions (94,96) for ρ and j^i, and expressing X^i in them in terms of non-commutative fields, gives an explicit mapping between the non-commutative fields A_μ and the commutative fields $\tilde{A}_\mu$.

Similar considerations extend to higher dimensions but, again, we shall not dwell on them here [34, 36, 37]. The moral lesson of the above is that the Lagrange formulation of fuzzy fluids is inherently non-commutative, while the Euler formulation is commutative. The Seiberg-Witten map between them becomes the transition from the particle-fixed Lagrange to the space-fixed Euler formulation.

5. The non-commutative description of quantum Hall states

We reach, now, one of the main topics of this presentation. Is the above useful to anything? Can we use it to describe or solve any physical system or does it remain an interesting peculiarity?

To find an appropriate application, we must look for systems with 'fuzzy' particles. This is not hard: quantum mechanical particles on their phase spaces are fuzzy, due to Heisenberg uncertainty. This can be carried through, and eventually leads to the description of one-dimensional fermions in terms of matrix models.

A more interesting situation arises in lowest Landau level physics, in which particles become fuzzy on the *coordinate* space. Spatial coordinates become non-commuting when restricted to the lowest Landau level [38,39], already introducing a non-commutative element (although quite distinct from the one introduced in the sequel). This is also the setting for the description of quantum Hall states and will be the topic of the present section.

5.1. Non-commutative Chern-Simons description of the quantum Hall fluid

The system to be described consists of a large number $N \to \infty$ of electrons on the plane in the lowest Landau level of an external constant magnetic field B (we take the electron charge $e = 1$). Upon proper dynamical conditions, they form quantum Hall states (for a review of the quantum Hall effect see [40]). According to the observations of the previous section, we can parametrize their coordinates as a fuzzy fluid in terms of two non-commutative Lagrange coordinates (infinite hermitian 'matrices') X^i, $i = 1, 2$, that is, by two operators on an infinite Hilbert space. The density of these electrons is not fixed at this point, but will eventually relate to the noncommutativity parameter as $\rho_0 = 1/2\pi\theta$.

The action is the non-commutative fluid analog of the gauge action of massless particles in an external constant magnetic field. In the symmetric gauge for the magnetic field, this would read

$$ S = \int dt \, \frac{B}{2} \operatorname{Tr} \left\{ \epsilon_{ij} D_0 X^i X^j \right\} = \int dt \, \frac{B}{2} \operatorname{Tr} \left\{ \epsilon_{ij} (\dot{X}^i + i[A_0, X^i]) X^j \right\}. \quad (104) $$

The above expression was made gauge invariant by gauging the time derivative and introducing a non-commutative temporal gauge field A_0. As explained in previous sections, however, this introduces a Gauss law constraint, which in the present case reads

$$ [X^1, X^2] = 0. \quad (105) $$

This is undesirable in many ways. The would-be non-commutative coordinates become commutative, eliminating the fuzziness of the description. More seriously, the density of the fluid classically becomes singular, as can be seen from the expression (91) for the inverse fluid density. (It can also be deduced from the commutative expression (94), although in a slightly more convoluted way.)

Taking care of the above difficulty also gives the opportunity to introduce an important piece of physics for the system: fractional quantum Hall states (Laughlin states, in their simplest form) are incompressible and have a constant spatial density ρ_0. The *filling fraction* ν of the state is defined as the fraction of the Landau level density $\rho_{LL} = B/2\pi$ that ρ_0 represents:

$$\nu = \frac{\rho_0}{\rho_{LL}} = \frac{2\pi\rho}{B} = \frac{1}{\theta B} \tag{106}$$

where the non-commutative parameter θ is related to the desired fluid density in the standard way, spelled out again as

$$\rho_0 = \frac{1}{2\pi\theta}. \tag{107}$$

We can introduce this constant density ρ_0 in the system by modifying the Gauss law constraint by an appropriate constant, achieved by adding a term linear in A_0. The resulting action reads

$$S = \int dt \, \frac{B}{2} \, \mathrm{Tr} \left\{ \epsilon_{ij} (\dot{X}^i + i[A_0, X^i]) X^j + 2\theta A_0 \right\}. \tag{108}$$

This was first proposed by Susskind [33], motivated by the earlier, classical mapping of the quantum Hall fluid to a gauge action [31] and related string theory work [41]. The equation of motion for A_0, now, imposes the Gauss law constraint

$$[X^1, X^2] = i\theta \tag{109}$$

essentially identifying X^1, X^2 with a non-commutative plane.

Interestingly, the above action is exactly the non-commutative CS action in 2+1 dimensions! A simple comparison of expression (50) and (108) above reveals that they are the same, upon identifying $\theta D_i = \epsilon_{ij} X^j$. The coefficient of the CS term λ relates to B and the filling fraction as

$$\lambda = \frac{B\theta}{4\pi} = \frac{1}{4\pi\nu}. \tag{110}$$

This establishes the connection of the non-commutative Chern-Simons action with the quantum Hall effect.

As before, gauge transformations are conjugations of X^i or D_i by arbitrary time-dependent unitary operators. In the quantum Hall fluid context they take the meaning of reshuffling the electrons. Equivalently, the X^i can be considered as coordinates of a two-dimensional fuzzy membrane, $2\pi\theta$ playing the role of an area quantum and gauge transformations realizing area preserving diffeomorphisms. The canonical conjugate of X^1 is $P_2 = BX^2$, and the generator of gauge transformations is

$$G = -iB[X^1, X^2] = B\theta = \frac{1}{\nu} \tag{111}$$

by virtue of (109). Since gauge transformations are interpreted as reshuffling of particles, the above has the interpretation of endowing the particles with quantum statistics of order $1/\nu$.

5.2. Quasiparticle and quasihole classical states

The classical equation (109) has a unique solution, modulo gauge (unitary) trans-formations, namely the unique irreducible representation of the Heisenberg al-gebra. Representation states can be conveniently written in a Fock basis $|n\rangle$, $n = 0, 1, \ldots$, for the ladder operators $X^1 \pm iX^2$, $|0\rangle$ representing a state of mini-mal spread at the origin. The classical theory has this representation as its unique state, the vacuum.

Deviations from the vacuum (109) can be achieved by introducing sources in the action [33]. A localized source at the origin has a density of the form $\rho = \rho_0 - q\delta^2(x)$ in the continuous (commutative) case, representing a point source of particle number $-q$, that is, a hole of charge q for $q > 0$. The non-commutative analog of such a density is

$$[X^1, X^2] = i\theta(1 + q|0\rangle\langle 0|). \tag{112}$$

In the membrane picture the right-hand side of (112) corresponds to area and implies that the area quantum at the origin has been increased to $2\pi\theta(1 + q)$, therefore piercing a hole of area $A = 2\pi\theta q$ and creating a particle deficit $q = \rho_0 A$. We shall call this a quasihole state. For $q > 0$ we find the quasihole solution of (112) as

$$X^1 + iX^2 = \sqrt{2\theta} \sum_{n=1}^{\infty} \sqrt{n+q}\,|n-1\rangle\langle n|. \tag{113}$$

Such solutions are called non-commutative gauge solitons [14, 15, 53–55].

The case of quasiparticles, $q < 0$ is more interesting. Clearly the area quantum cannot be diminished below zero, and equations (112) and (113) cannot hold for $-q > 1$. The correct equation is, instead,

$$[X^1, X^2] = i\theta \left(1 - \sum_{n=0}^{k-1} |n\rangle\langle n| - \epsilon|k\rangle\langle k|\right) \tag{114}$$

where k and ϵ are the integer and fractional part of the quasiparticle charge $-q$. The solution of (114) is

$$X^1 + iX^2 = \sum_{n=0}^{k-1} z_n|n\rangle\langle n| + \sqrt{2\theta} \sum_{n=k+1}^{\infty} \sqrt{n-k-\epsilon}\,|n-1\rangle\langle n|. \tag{115}$$

(For $k = 0$ the first sum in (114,115) drops.) In the membrane picture, k quanta of the membrane have 'peeled' and occupy positions $z_n = x_n + iy_n$ on the plane, while the rest of the membrane has a deficit of area at the origin equal to $2\pi\theta\epsilon$, leading to a charge surplus ϵ. Clearly the quanta are electrons that sit on top of the continuous charge distribution. If we want all charge density to be concentrated at the origin, we must choose all $z_n = 0$. The above quasiparticle states for integer q are the non-commutative solitons and flux tubes that are also solutions of non-commutative gauge theory, while the quasihole states are not solutions of the non-commutative gauge theory action and have no direct analog.

Laughlin theory predicts that quasihole excitations in the quantum Hall state have their charge $-q$ quantized in integer units of ν, $q = m\nu$, with m a positive integer. We see that the above discussion gives no hint of this quantization, while we see at least some indication of electron quantization in (114, 115). Quasihole quantization will emerge in the quantum theory, as we shall see shortly, and is equivalent to a quantization condition of the non-commutative Chern-Simons term.

5.3. Finite number of electrons: the Chern-Simons matrix model

Describing an infinitely plane filled with electrons is not the most interesting situation. We wish to describe quantum Hall states of finite extent consisting of N electrons. Obviously the coordinates X^i of the non-commutative fluid description would have to be represented by finite $N \times N$ matrices. The action (108), however, and the equation (109) to which it leads, are inconsistent for finite matrices, and a modified action must be written which still captures the physical features of the quantum Hall system. Such an action exists, and leads to a matrix model truncation of the non-commutative Chern-Simons action involving a 'boundary field' [42]. It is

$$S = \int dt \frac{B}{2} \mathrm{Tr} \left\{ \epsilon_{ij} (\dot{X}^i + i[A_0, X^i]) X^j + 2\theta A_0 - \omega(X^i)^2 \right\} + \Psi^\dagger (i\dot{\Psi} - A_0 \Psi). \quad (116)$$

It has the same form as the planar CS action, but with two extra terms. The first, and most crucial, involves Ψ, a complex N-vector that transforms in the fundamental of the gauge group $U(N)$:

$$X^i \to U X^i U^{-1} , \quad \Psi \to U\Psi. \quad (117)$$

Its action is a covariant kinetic term similar to a complex scalar fermion. We shall, however, quantize it as a boson; this is perfectly consistent, since there is no spatial kinetic term that would lead to a negative Dirac sea and the usual inconsistencies of first-order bosonic actions.

The term proportional to ω (not to be confused with θ^{-1}) serves as a spatial regulator: since we will be describing a finite number of electrons, there is nothing to keep them localized anywhere in the plane. We added a confining harmonic potential which serves as a 'box' to keep the particles near the origin.

We can again impose the A_0 equation of motion as a Gauss constraint and then put $A_0 = 0$. In our case it reads

$$G \equiv -iB[X^1, X^2] + \Psi\Psi^\dagger - B\theta = 0. \quad (118)$$

Taking the trace of the above equation gives

$$\Psi^\dagger \Psi = N B\theta. \quad (119)$$

The equation of motion for Ψ in the $A_0 = 0$ gauge is $\dot{\Psi} = 0$. So we can take it to be

$$\Psi = \sqrt{N B\theta} \, |v\rangle \quad (120)$$

where $|v\rangle$ is a constant vector of unit length. Then (118) reads

$$[X^1, X^2] = i\theta \left(1 - N|v\rangle\langle v|\right). \quad (121)$$

This is similar to (109) for the infinite plane case, with an extra projection operator. Using the residual gauge freedom under time-independent unitary transformations, we can rotate $|v\rangle$ to the form $|v\rangle = (0, \ldots 0, 1)$. The above commutator then takes the form $i\theta\,\mathrm{diag}\,(1, \ldots, 1, 1 - N)$ which is the 'minimal' deformation of the planar result (109) that has a vanishing trace.

In the fluid (or membrane) picture, Ψ is like a boundary term. Its role is to absorb the 'anomaly' of the commutator $[X^1, X^2]$, much like the case of a boundary field theory required to absorb the anomaly of a bulk (commutative) Chern-Simons field theory.

The equations of motion for X^i read

$$\dot{X}^i + \omega\epsilon_{ij}X^j = 0. \tag{122}$$

This is just a matrix harmonic oscillator. It is solved by

$$X^1 + iX^2 = e^{i\omega t}A \tag{123}$$

where A is any $N \times N$ matrix satisfying the constraint

$$[A, A^\dagger] = 2\theta(1 - N|v\rangle\langle v|). \tag{124}$$

The classical states of this theory are given by the set of matrices $A = X^1 + iX^2$ satisfying (124) or (121). We can easily find them by choosing a basis in which one of the Xs is diagonal, say, X^1. Then the commutator $[X^1, X^2]$ is purely off-diagonal and the components of the vector $|v\rangle$ must satisfy $|v_n|^2 = 1/N$. We can use the residual $U(1)^N$ gauge freedom to choose the phases of v_n so that $v_n = 1/\sqrt{N}$. So we get

$$(X^1)_{mn} = x_n\delta_{mn} \ , \quad (X^2)_{mn} = y_n\delta_{mn} + \frac{i\theta}{x_m - x_n}(1 - \delta_{mn}). \tag{125}$$

The solution is parametrized by the N eigenvalues of X^1, x_n, and the N diagonal elements of X^2, y_n.

5.4. Quantum Hall 'droplet' vacuum

Not all solutions found above correspond to quantum Hall fluids. In fact, choosing all x_n and y_n much bigger than $\sqrt{\theta}$ and not too close to each other, both X^1 and X^2 become almost diagonal; they represent N electrons scattered in positions (x_n, y_n) on the plane and performing rotational motion around the origin with angular velocity ω. This is the familiar motion of charged particles in a magnetic field along lines of equal potential when their proper kinetic term is negligible. Quantum Hall states will form when particles coalesce near the origin, that is, for states of low energy.

To find the ground state, we must minimize the potential

$$V = \frac{B\omega}{2}\mathrm{Tr}[(X^1)^2 + (X^2)^2] = \frac{B\omega}{2}\mathrm{Tr}(A^\dagger A) \tag{126}$$

while imposing the constraint (121) or (124). This can be implemented with a matrix Lagrange multiplier Λ (essentially, solving the equations of motion including

$A_0 \equiv \Lambda$ and putting the time derivatives to zero). We obtain

$$A = [\Lambda, A] \ , \ \text{or} \ X^i = i\epsilon_{ij}[\Lambda, X^j]. \tag{127}$$

This is reminiscent of canonical commutation relations for a quantum harmonic oscillator, with Λ playing the role of the Hamiltonian. We are led to the solution

$$A = \sqrt{2\theta} \sum_{n=0}^{N-1} \sqrt{n}|n-1\rangle\langle n| \ , \ \Lambda = \sum_{n=0}^{N-1} n|n\rangle\langle n| \ , \ |v\rangle = |N-1\rangle. \tag{128}$$

This is essentially a quantum harmonic oscillator and Hamiltonian projected to the lowest N energy eigenstates. It is easy to check that the above satisfies both (124) and (127). Its physical interpretation is clear: it represents a circular quantum Hall 'droplet' of radius $\sqrt{2N\theta}$. Indeed, the radius-squared matrix coordinate R^2 is

$$R^2 \ = \ (X^1)^2 + (X^2)^2 = A^\dagger A + \frac{1}{2}[A, A^\dagger] \tag{129}$$

$$= \ \sum_{n=0}^{N-2} \theta(2n+1)|n\rangle\langle n| + \theta(N-1)|N-1\rangle\langle N-1|. \tag{130}$$

The highest eigenvalue of R^2 is $(2N-1)\theta$. The particle density of this droplet is $\rho_0 = N/(\pi R^2) \sim 1/(2\pi\theta)$ as in the infinite plane case.

The matrices X^i are known and can be explicitly diagonalized in this case. Their eigenvalues are given by the zeros of the Nth Hermite polynomial (times $\sqrt{2\theta}$). In the large-N limit the distribution of these zeros obeys the famous Wigner semi-circle law, with radius $\sqrt{N}$. Since these eigenvalues are interpreted as electron coordinates, this confirms once more the fact that the electrons are evenly distributed on a disk of radius $\sqrt{2N\theta}$.

5.5. Excited states of the model

Excitations of the classical ground state can now be considered. Any perturbation of (128) in the form of (125) is, of course, some excited state. We shall concentrate, however, on two special types of excitations.

The first is obtained by performing on $A, A^\dagger$ all transformations generated by the infinitesimal transformation

$$A' = A + \sum_{n=0}^{N-1} \epsilon_n (A^\dagger)^n \tag{131}$$

with ϵ_n infinitesimal complex parameters. The sum is truncated to $N-1$ since $A^\dagger$ is an $N \times N$ matrix and only its first N powers are independent. It is obvious that (124) remains invariant under the above transformation and therefore also under the finite transformations generated by repeated application of (131).

If $A, A^\dagger$ were true oscillator operators, these would be canonical (unitary) transformations, that is, gauge transformations that would leave the physical state invariant. For the finite $A, A^\dagger$ in (128), however, these are *not* unitary transformations and generate a new state. To understand what is that new state, examine what happens to the 'border' of the circular quantum Hall droplet under this transformation. This is defined by $A^\dagger A \sim 2N\theta$ (for large N). To find the new boundary parametrize $A \sim \sqrt{2N\theta}e^{i\phi}$, with ϕ the polar angle on the plane and calculate $(A^\dagger A)'$. The new boundary in polar coordinates is

$$R'(\phi) = \sqrt{2N\theta} + \sum_{n=-N}^{N} c_n e^{in\phi} \qquad (132)$$

where the coefficients c_n are

$$c_n = c_{-n}^* = \frac{R^n}{2}\epsilon_{n-1} \; (n > 0), \; c_0 = 0. \qquad (133)$$

This is an arbitrary area-preserving deformation of the boundary of the droplet, truncated to the lowest N Fourier modes. The above states are, therefore, arbitrary area-preserving boundary excitations of the droplet [56–58], appropriately truncated to reflect the finite non-commutative nature of the system (the fact that there are only N electrons).

Note that on the plane there is an infinity of area-preserving diffeomorphisms that produce a specific deformation of a given curve. From the droplet point of view, however, these are all gauge equivalent since they deform the outside of the droplet (which is empty) or the inside of it (which is full and thus invariant). The finite theory that we examine has actually broken this infinite gauge freedom, since most of these canonical transformations of $a, a^\dagger$ do not preserve the Gauss constraint (124) when applied on $A, A^\dagger$. The transformations (131) pick a representative in this class which respects the constraint.

The second class of excitations are the analogs of quasihole and quasiparticle states. States with a quasihole of charge $-q$ at the origin can be written quite explicitly in the form

$$A = \sqrt{2\theta} \left(\sqrt{q}|N-1\rangle\langle 0| + \sum_{n=1}^{N-1} \sqrt{n+q}|n-1\rangle\langle n| \right) , \; q > 0. \qquad (134)$$

It can be verified that the eigenvalues of $A^\dagger A$ are

$$(A^\dagger A)_n = 2\theta(n+q) , \; n = 0, 1, \ldots, N-1 \qquad (135)$$

so it represents a circular droplet with a circular hole of area $2\pi\theta q$ at the origin, that is, with a charge deficit q. The droplet radius has appropriately swelled, since the total number of particles is always N.

Note that (134) stills respects the Gauss constraint (124) (with $|v\rangle = |N-1\rangle$) *without* the explicit introduction of any source. So, unlike the infinite plane case, this model contains states representing quasiholes without the need to introduce external sources. What happens is that the hole and the boundary of the droplet together cancel the anomaly of the commutator, the outer boundary part absorbing an amount $N + q$ and the inner (hole) boundary producing an amount q. This possibility did not exist in the infinite plane, where the boundary at infinity was invisible, and an explicit source was needed to nucleate the hole.

Quasiparticle states are a different matter. In fact, there are *no* quasiparticle states with the extra particle number localized anywhere within the droplet. Such states do not belong to the $\nu = 1/B\theta$ Laughlin state. There are quasiparticle states with an *integer* particle number $-q = m$, and the extra m electrons occupying positions *outside* the droplet. The explicit form of these states is not so easy to write. At any rate, it is interesting that the matrix model 'sees' the quantization of the particle number and the inaccessibility of the interior of the quantum Hall state in a natural way.

Having said all that, we are now making the point that *all types of states defined above are the same.* Quasihole and quasiparticle states are non-perturbative boundary excitations of the droplet, while perturbative boundary excitations can be viewed as marginal particle states.

To clarify this point, note that the transformation (131) or (132) defining infinitesimal boundary excitations has $2N$ real parameters. The general state of the system, as presented in (125) also depends on $2N$ parameters (the x_n and y_n). The configuration space is connected, so all states can be reached continuously from the ground state. Therefore, all states can be generated by exponentiating (131). This is again a feature of the finite-N model: there is no sharp distinction between 'perturbative' (boundary) and 'soliton' (quasiparticle) states, each being a particular limit of the other.

5.6. Equivalence to the Calogero model

The model examined above should feel very familiar to Calogero model aficionados. Indeed, it is equivalent to the harmonic rational Calogero model [43–45], whose connection to fractional statistics [46] and anions [47–49] has been established in different contexts. This is an integrable system of N nonrelativistic particles on the line interacting with mutual inverse-square potential and an external harmonic potential, with Hamiltonian

$$H = \sum_{n=1}^{N} \left(\frac{\omega}{2B}p_n^2 + \frac{B\omega}{2}x_n^2 \right) + \sum_{n\neq m} \frac{\nu^{-2}}{(x_n - x_m)^2}. \tag{136}$$

In terms of the parameters of the model, the mass of the particles is B/ω and the coupling constant of the two-body inverse-square potential is ν^{-2}. We refer the reader to [50–52] for details on the Calogero model and its connection with the matrix model. Here we simply state the relevant results and give their connection to quantum Hall quantities.

The positions of the Calogero particles x_n are the eigenvalues of X^1, while the momenta p_n are the diagonal elements of X^2, specifically $p_n = By_n$. The motion of the x_n generated by the Hamiltonian (136) is compatible with the evolution of the eigenvalues of X^1 as it evolves in time according to (123). So the Calogero model gives a one-dimensional perspective of the quantum Hall state by monitoring some effective electron coordinates along X^1 (the eigenvalues of X^1).

The Hamiltonian of the Calogero model (136) is equal to the matrix model potential $V = \frac{1}{2}B\omega\mathrm{Tr}(X^i)^2$. Therefore, energy states map between the two models. The ground state is obtained by putting the particles at their static equilibrium positions. Because of their repulsion, they will form a lattice of points lying at the roots of the Nth Hermite polynomial and reproducing the semi-circle Wigner distribution mentioned before.

Boundary excitations of the quantum Hall droplet correspond to small vibrations around the equilibrium position, that is, sound waves on the lattice. Quasiholes are large-amplitude (nonlinear) oscillations of the particles at a localized region of the lattice. For a quasihole of charge q at the center, on the average q particles near $x = 0$ participate in the oscillation.

Finally, quasiparticles are excitations where one of the particles is isolated outside the ground state distribution (a 'soliton') [59]. As it moves, it 'hits' the distribution on one side and causes a solitary wave of net charge 1 to propagate through the distribution. As the wave reaches the other end of the distribution another particle emerges and gets emitted there, continuing its motion outside the distribution. So a quasiparticle is more or less identified with a Calogero particle, although its role, at different times, is assumed by different Calogero particles, or even by soliton waves within the ground state distribution.

Overall, we have a 'holographic' description of the two-dimensional quantum Hall states in terms of the one-dimensional Calogero particle picture. Properties of the system can be translated back-and-forth between the two descriptions. Further connections at the quantum level will be described in subsequent sections.

6. The quantum matrix Chern-Simons model

The properties of the model analyzed in the previous section are classical. The 'states' and 'oscillators' that we encountered were due to the non-commutative nature of the coordinates and were referring to the classical matrix model.

The full physical content of the model, and its complete equivalence to quantum Hall (Laughlin) states, is revealed only upon quantization. In fact, some of the most interesting features of the states, such as filling fraction and quasihole charge quantization, manifest only in the quantum domain. This will be the subject of the present section.

6.1. Quantization of the filling fraction

The quantization of the Chern-Simons matrix model has been treated in [51]. We shall repeat here the basic arguments establishing their relevance to the quantum Hall system.

We shall use double brackets for quantum commutators and double kets for quantum states, to distinguish them from matrix commutators and N-vectors.

Quantum mechanically the matrix elements of X^i become operators. Since the lagrangian is first-order in time derivatives, X^1_{mn} and X^2_{kl} are canonically conjugate:

$$[[X^1_{mn}, X^2_{kl}]] = \frac{i}{B}\delta_{ml}\delta_{kn} \tag{137}$$

or, in terms of $A = X^1 + iX^2$

$$[[A_{mn}, A^\dagger_{kl}]] = \frac{1}{B}\delta_{mk}\delta_{nl}. \tag{138}$$

The Hamiltonian, ordered as $\frac{1}{2}B\omega \mathrm{Tr}A^\dagger A$, is

$$H = \sum_{mn} \frac{1}{2}B\omega A^\dagger_{mn} A_{mn}. \tag{139}$$

This is just N^2 harmonic oscillators. Further, the components of the vector Ψ_n correspond to N harmonic oscillators. Quantized as bosons, their canonical commutator is

$$[[\Psi_m, \Psi^\dagger_n]] = \delta_{mn}. \tag{140}$$

So the system is a priori just $N(N+1)$ uncoupled oscillators. What couples the oscillators and reduces the system to effectively $2N$ phase space variables (the planar coordinates of the electrons) is the Gauss law constraint (118). In writing it, we in principle encounter operator ordering ambiguities. These are easily fixed, however, by noting that the operator G is the quantum generator of unitary rotations of both X^i and Ψ. Therefore, it must satisfy the commutation relations of the $U(N)$ algebra. The X-part is an orbital realization of $SU(N)$ on the manifold of $N \times N$ hermitian matrices. Specifically, expand $X^{1,2}$ and $A, A^\dagger$ in the complete basis of matrices $\{1, T^a\}$ where T^a are the $N^2 - 1$ normalized fundamental $SU(N)$ generators:

$$X^1 = x_0 + \sum_{a=1}^{N^2-1} x_a T^a \ , \quad \sqrt{B}A = a_o + \sum_{a=1}^{N^2-1} a_a T^a, \tag{141}$$

x_a, a_a are scalar operators. Then, by (137,138) the corresponding components of BX^2 are the conjugate operators $-i\partial/\partial x_a$, while $a_a, a^\dagger_a$ are harmonic oscillator operators. We can write the components of the matrix commutator $G_X = -iB[X^1, X^2]$ in G in the following ordering:

$$G^a_X = -if^{abc}x_b\frac{\partial}{\partial x_a} \tag{142}$$

$$= -i(A^\dagger_{mk}A_{nk} - A^\dagger_{nk}A_{mk}) \tag{143}$$

$$= -ia^\dagger_b f^{abc}a_c \tag{144}$$

where f^{abc} are the structure constants of $SU(N)$. Similarly, expressing $G_\Psi = \Psi\Psi^\dagger$ in the $SU(N)$ basis of matrices, we write its components in the ordering

$$G^a_\Psi = \Psi^\dagger_m T^a_{mn}.\Psi_n \tag{145}$$

The operators above, with the specific normal ordering, indeed satisfy the $SU(N)$ algebra. The expression of G^a_X in terms of x_a is like an angular momentum. The expression of G^a_Ψ in terms of the oscillators Ψ_i and of G^a_X in terms of the oscillators a_a is the well-known Jordan-Wigner realization of the $SU(N)$ algebra in the Fock space of bosonic oscillators. Specifically, let $R^a_{\alpha\beta}$ be the matrix elements of the generators of $SU(N)$ in any representation of dimension d_R, and $a_\alpha, a^\dagger_\alpha$ a set of d_R mutually commuting oscillators. Then the operators

$$G^a = a^\dagger_\alpha R^a_{\alpha\beta} a_\beta \tag{146}$$

satisfy the $SU(N)$ algebra. The Fock space of the oscillators contains all the symmetric tensor products of R-representations of $SU(N)$; the total number operator of the oscillators identifies the number of R components in the specific symmetric product. The expressions for G^a_Ψ and G^a_X are specific cases of the above construction for R^a the fundamental (T^a) or the adjoin $(-if^a)$ representation respectively.

So, the traceless part of the Gauss law (118) becomes

$$(G^a_X + G^a_\Psi)|\text{phys}\rangle\rangle = 0 \tag{147}$$

where $|\text{phys}\rangle\rangle$ denotes the physical quantum states of the model. The trace part, on the other hand, expresses the fact that the total $U(1)$ charge of the model must vanish. It reads

$$(\Psi^\dagger_n \Psi_n - NB\theta)|\text{phys}\rangle\rangle = 0. \tag{148}$$

We are now set to derive the first nontrivial quantum mechanical implication: the inverse-filling fraction is quantized to integer values. To see this, first notice that the first term in (148) is nothing but the total number operator for the oscillators Ψ_n and is obviously an integer. So we immediately conclude that $NB\theta$ must be quantized to an integer.

However, this is not the whole story. Let us look again at the $SU(N)$ Gauss law (147). It tells us that physical states must be in a singlet representation of G^a. The orbital part G^a_X, however, realizes only representations arising out of products of the adjoin, and therefore it contains only irreps whose total number of boxes in their Young tableau is an integer multiple of N. Alternatively, the $U(1)$ and Z_N part of U is invisible in the transformation $X^i \to UX^iU^{-1}$ and thus the Z_N charge of the operator realizing this transformation on states must vanish. (For instance, for $N = 2$, G^a is the usual orbital angular momentum in 3 dimensions which cannot be half-integer.)

Since physical states are invariant under the sum of G_X and G_Ψ, the representations of G_Ψ and G_X must be conjugate to each other so that their product contain the singlet. Therefore, the irreps of G_Ψ must also have a number of boxes which is a multiple of N. The oscillator realization (148) contains all the symmetric irreps of $SU(N)$, whose Young tableau consists of a single row. The number of boxes equals the total number operator of the oscillators $\Psi^\dagger_n \Psi_n$. So we conclude that $NB\theta$ must be an integer multiple of N [51], that is,

$$B\theta = \frac{1}{\nu} = k \ , \ k = \text{integer} \, . \tag{149}$$

The above effect has a purely group theoretic origin. The same effect, however, can be recovered using topological considerations, by demanding invariance of the quantum action $\exp(iS)$ under gauge $U(N)$ transformations with a nontrivial winding in the temporal direction [51]. This is clearly the finite-N counterpart of the level quantization for the non-commutative Chern-Simons term as exposed in a previous section, namely $4\pi\lambda =$ integer. By (110) this is equivalent to (149).

By reducing the model to the dynamics of the eigenvalues of X^1 we recover a quantum Calogero model with Hamiltonian

$$H = \sum_{n=1}^{N} \left(\frac{\omega}{2B} p_n^2 + \frac{B\omega}{2} x_n^2 \right) + \sum_{n \neq m} \frac{k(k+1)}{(x_n - x_m)^2}. \tag{150}$$

Note the shift of the coupling constant from k^2 to $k(k+1)$ compared to the classical case. This is a quantum reordering effect which results in the shift of ν^{-1} from k to $k+1 \equiv n$. The above model is, in fact, perfectly well defined even for fractional values of ν^{-1}, while the matrix model that generated it requires quantization. This is due to the fact that, by embedding the particle system in the matrix model, we have augmented its particle permutation symmetry S_N to general $U(N)$ transformations; while the smaller symmetry S_N is always well defined, the larger $U(N)$ symmetry becomes anomalous unless ν^{-1} is quantized.

6.2. Quantum states

We can now examine the quantum excitations of this theory. The quantum states of the model are simply states in the Fock space of a collection of oscillators. The total energy is the energy carried by the N^2 oscillators A_{mn} or a_a. We must also impose the constraint (147) and (148) on the Fock states. Overall, this becomes a combinatorics group theory problem which is in principle doable, although quite tedious.

Fortunately, we do not need to go through it here. The quantization of this model is known and achieves its most intuitive description in terms of the states of the corresponding Calogero model. We explain how.

Let us work in the X^1 representation, X^2 being its canonical momentum. Writing $X^1 = U\Lambda_1 U^{-1}$ with $\Lambda = \text{diag}\{x_i\}$ being its eigenvalues, we can view the state of the system as a wavefunction of U and x_n. The gauge generator G_X^a appearing in the Gauss law (147) is actually the conjugate momentum to the variables U. Due to the Gauss law, the angular degrees of freedom U are constrained to be in a specific angular momentum state, determined by the representation of $SU(N)$ carried by the Ψ_n. From the discussion of the previous section, we understand that this is the completely symmetric representation with $nN = N/\nu$ boxes in the Young tableau. So the dynamics of U are completely fixed, and it suffices to consider the states of the eigenvalues. These are described by the states of the quantum Calogero model. The Hamiltonian of the Calogero model corresponds to the matrix potential $V = \frac{1}{2} B\omega \text{Tr}(X^i)^2$, which contains all the relevant information for the system.

Calogero energy eigenstates are expressed in terms of N positive, integer 'quasi-occupation numbers' n_j (quasinumbers, for short), with the property

$$n_j - n_{j-1} \geq n = \frac{1}{\nu} \ , \ j = 1, \ldots, N. \tag{151}$$

In terms of the n_j the spectrum becomes identical to the spectrum of N independent harmonic oscillators

$$E = \sum_{j=1}^{N} E_j = \sum_{j=1}^{N} \omega \left(n_j + \frac{1}{2} \right). \tag{152}$$

The constraint (151) means that the n_j cannot be packed closer than $n = \nu^{-1}$, so they have a 'statistical repulsion' of order n. For filling fraction $\nu = 1$ these are ordinary fermions, while for $\nu^{-1} = n > 1$ they behave as particles with an enhanced exclusion principle.

The scattering phase shift between Calogero particles is $\exp(i\pi/\nu)$. So, in terms of the phase that their wavefunction picks upon exchanging them, they look like fermions for odd n and bosons for even n [46]. Since the underlying particles (electrons) must be fermions, we should pick n odd.

The energy 'eigenvalues' E_j are the quantum analogs of the eigenvalues of the matrix $\frac{1}{2}B\omega(X^i)^2$. The radial positions R_j are determined by

$$\frac{1}{2}B\omega R_j^2 = E_j \ \rightarrow \ R_j^2 = \frac{2n_j + 1}{B}. \tag{153}$$

So the quasinumbers $2n_j+1$ determine the radial positions of electrons. The ground state values are the smallest non-negative integers satisfying (151)

$$n_{j,gs} = n(j - 1) \ , \ j = 1, \ldots, N. \tag{154}$$

They form a 'Fermi sea' but with a density of states dilated by a factor ν compared to standard fermions. This state reproduces the circular quantum Hall droplet. Its radius maps to the Fermi level, $R \sim \sqrt{(2n_{N,gs} + 1)/B} \sim \sqrt{2N\theta}$.

Quasiparticle and quasihole states are identified in a way analogous to particles and holes of a Fermi sea. A quasiparticle state is obtained by peeling a 'particle' from the surface of the sea (quasinumber $n_{N,gs}$) and putting it to a higher value $n'_N > n(N-1)$. This corresponds to an electron in a rotationally invariant state at radial position $R' \sim \sqrt{2(n'_N + 1)/B}$. Successive particles can be excited this way. The particle number is obviously quantized to an integer (the number of excited quasinumbers) and we can only place them outside the quantum Hall droplet.

Quasiholes are somewhat subtler: they correspond to the minimal excitations of the ground state *inside* the quantum Hall droplet. This can be achieved by leaving all quasinumber n_j for $j \leq k$ unchanged, and increasing all n_j, $j > k$ by one,

$$n_j \ = \ n(j - 1) \ j \leq k \tag{155}$$
$$= \ n(j - 1) + 1 \ k < j \leq N. \tag{156}$$

This increases the gap between n_k and n_{k+1} to $n + 1$ and creates a minimal 'hole'.

This hole has a particle number $-q = -1/n = -\nu$. To see it, consider removing a particle altogether from quasinumber n_k. This would create a gap of $2n$ between n_{k-1} and n_{k+1}. The extra gap n can be considered as arising out of the formation of n holes (increasing n_j for $j \geq k$ n times). Thus the absence of a particle corresponds to n holes. We therefore obtain the important result that the quasihole charge is naturally quantized to units of

$$q_h = \nu = \frac{1}{n} \tag{157}$$

in accordance with Laughlin theory.

We conclude by stressing once more that there is no fundamental distinction between particles and holes for finite N. A particle can be considered as a non-perturbative excitation of many holes near the Fermi level, while a hole can be viewed as a coherent state of many particles of minimal excitation.

6.3. Final remarks on the matrix model

The quantization of the inverse filling fraction and, importantly, the quasihole charge quantization emerged as quantum mechanical consequences of this model. The quantizations of the two parameters had a rather different origin. We can summarize here the basic meaning of each:

Quantization of the inverse filling fraction is basically angular momentum quantization. The matrix commutator of $[X_1, X_2]$ is an orbital angular momentum in the compact space of the angular parameters of the matrices, and it must be quantized. Alternatively (and equivalently), it can be understood as a topological quantization condition due to a global gauge anomaly of the model.

Quantization of the quasihole charge, on the other hand, is nothing but harmonic oscillator quantization. Quasiholes are simply individual quanta of the oscillators A_{mn}. The square of the radial coordinate $R^2 = (X^1)^2 + (X^2)^2$ is basically a harmonic oscillator. $\sqrt{B}X_1$ and $\sqrt{B}X_2$ are canonically conjugate, so the quanta of R^2 are $2/B$. Each quantum increases R^2 by $2/B$ and so it increases the area by $2\pi/B$. This creates a charge deficit q equal to the area times the ground state density $q = (2\pi/B) \cdot (1/2\pi\theta) = 1/\theta B = \nu$. So the fundamental quasihole charge is ν.

An important effect, which can be both interesting and frustrating, is the quantum shift in the effective value of the inverse filling fraction from k to $n = k+1$. This is the root of the famous fermionization of the eigenvalues of the matrix model in the singlet sector ($k = 0$). Its presence complicates some efforts to reproduce layered quantum Hall states, as it frustrates the obvious charge density counting.

There are many questions on the above model that we left untouched, some of them already addressed and some still open [60–68]. Their list includes the description of Hall states with spin, the treatment of cylindrical, spherical or toroidal space topologies, the description of states with nontrivial filling fraction, the exact mapping between quantities of physical interest in the two descriptions, the inclusion of electron interactions etc. The interested reader is directed to the numerous papers in the literature dealing with these issues. In the concluding section we prefer to present an alternative non-commutative fluid description for quantum many-body states.

7. The non-commutative Euler picture and bosonization

In the previous sections we reviewed the non-commutative picture of the Lagrange formulation of fluids and its use in the quantum Hall effect. The Euler formulation, on the other hand, was peculiar in that it allowed for a fully commutative description, leading to the Seiberg-Witten map.

This, however, is not the only possibility. Indeed, we saw that there were *two* potential descriptions for the density of the fluid, one inherently commutative (94) and one inherently non-commutative (91). Although the commutative one was adopted, one could just as well work with the non-commutative one, expecting to recover the standard Euler description only at the commutative limit. As it turns out, this is a very natural description of fluids consisting of fermions. Since the non-commutative density is an inherently bosonic field, it affords a description of fermionic systems in terms of bosonic field variables, naturally leading to bosonization.

7.1. Density description of fermionic many-body systems

The starting point will be a system of N non-interacting fermions in $D = 1$ spatial dimensions. The restriction of the dimensionality of space at this point is completely unnecessary and inconsequential, and is imposed only for conceptual and notational simplification and easier comparison with previous sections. In fact, much of the formalism will not even make specific reference to the dimensionality of space.

We shall choose our fermions to be noninteracting and carrying no internal degrees of freedom such as spin, color etc. (there is no conflict with the spin-statistics theorem in this first-quantized, many-body description). Again, this is solely for convenience and to allow us to focus on the main conceptual issue of their fluid description rather than other dynamical questions. The only remaining physical quantity is the single-particle Hamiltonian defining their dynamics, denoted $H_{\mathrm{sp}}(x, p)$. Here x, p are single-particle coordinate and momentum operators, together forming a 'non-commutative plane', with the role of θ played by $\hbar$ itself:

$$[x, p]_{\mathrm{sp}} = i\hbar \tag{158}$$

The subscript sp will be appended to single-particle operators or relations (except x and p) to distinguish them from upcoming field theory quantities.

Single-particle states are elements of the irreducible representation of the above Heisenberg commutator. A basis would be the eigenstates $|n\rangle$ of H_{sp} corresponding to eigenvalues E_n (assumed nondegenerate for simplicity). The states of the N-body system, on the other hand, are fully antisymmetrized elements of the N-body Hilbert space consisting of N copies of the above space. They can be expressed in a Fock description in terms of the occupation number basis

$N_n = 0, 1$ for each single particle level. The ground state, in particular, is the state $|1, \ldots, 1, 0, \ldots\rangle$ with the N lowest levels occupied by fermions.

An alternative description, however, working with a *single* copy of the above space is possible, in terms of a single-particle density-like operator [69, 70]. Specifically, define the (hermitian) single-particle operator ρ whose eigenvalues correspond to the occupation numbers $N_i = 1$ for a set of N specific filled single-particle states and $N_i = 0$ for all other states:

$$\rho = \sum_{n=1}^{N} |\psi_n\rangle\langle\psi_n|. \tag{159}$$

Clearly ρ is a good description of the N-body fermion system whenever the fermions occupy N single-particle states. The ground state ρ_0, in particular, is such a state and would correspond to

$$\rho_0 = \sum_{n=1}^{N} |n\rangle\langle n|. \tag{160}$$

Due to the Schrödinger evolution of the single-particle states $|n\rangle$, the operator ρ satisfies the evolution equation

$$i\hbar\dot{\rho} = [H_{\mathrm{sp}}, \rho]_{\mathrm{sp}}. \tag{161}$$

This description has several drawbacks. It is obviously limited from the fact that it can describe only 'factorizable' states, that is, basis states in some appropriate Fock space, but not their linear combinations ('entangled' states). This is serious, as it violates the quantum mechanical superposition principle, and makes it clear that this cannot be a full quantum description of the system. Further, the operator ρ must be a projection operator with exactly N eigenvalues equal to one and the rest of them vanishing, which means that it must satisfy the algebraic constraint

$$\rho^2 = \rho \ , \ \mathrm{Tr}\rho = N. \tag{162}$$

So ρ is similar to the density matrix, except for its trace.

In spite of the above, we shall see that this is a valid starting point for a full description of the many-body quantum system in a second-quantized picture. To give ρ proper dynamics, we must write an action that leads to the above equations (evolution plus constraints) in a canonical setting. The simplest way to achieve this is by 'solving' the constraint in terms of a unitary field U as

$$\rho = U^{-1}\rho_0 U \tag{163}$$

with ρ_0 the ground state. Any ρ can be expressed as above, U being a unitary operator mapping the first N energy eigenstates to the actual single-particle states entering the definition of ρ. An appropriate action for U is

$$S = \int dt \mathrm{Tr}\left(i\hbar\rho_0\dot{U}U^{-1} - U^{-1}\rho_0 U H_{\mathrm{sp}}\right). \tag{164}$$

It is easy to check that it leads to (161) for (163). Note that the first term in the action is a first-order kinetic term, defining a canonical one-form. The matrix elements of U, therefore, encode both coordinates and momenta and constitute the full phase space variable of the system. The Poisson brackets of U and, consequently, ρ can be derived by inverting the above canonical one-form. The result is that the matrix elements ρ_{mn} of ρ have Poisson brackets

$$\{\rho_{m_1 n_1}, \rho_{m_2 n_2}\} = \frac{1}{i\hbar}(\rho_{m_1 n_2}\delta_{m_2 n_1} - \rho_{m_2 n_1}\delta_{m_1 n_2}). \tag{165}$$

The second term in the action is the Hamiltonian $H = \mathrm{Tr}(\rho H_{\mathrm{sp}})$ and represents the sum of the energy expectation values of the N fermions.

7.2. The correspondence to a non-commutative fluid

It should be clear that the above description essentially defines a non-commutative fluid. Indeed, the operators U and ρ act on the Heisenberg Hilbert space and can be expressed in terms of the fundamental operators x, p. As such, they are non-commutative fields. The constraint for ρ is the non-commutative version of the relation $f^2 = f$ defining the characteristic function of a domain. We can, therefore, visualize ρ as a 'droplet' of a non-commutative fluid that fills a 'domain' of the non-commutative plain with a droplet 'height' equal to 1. The actual density of the fluid is fixed by the integration formula on the non-commutative plane, assigning an area of $2\pi\hbar$ to each state on the Hilbert space. So the value of the density inside the droplet becomes $1/2\pi\hbar$.

A similar picture is obtained by considering the classical 'symbol' of the above operator, using the Weyl-ordering mapping. The corresponding commutative function represents a droplet with a fuzzy boundary (the field drops smoothly from 1 to 0, and can even become negative at some points), but the bulk of the droplet and its exterior are at constant density (0 or 1).

As one should expect, this is the value of the density of states on phase space according to the semiclassical quantization condition assigning one quantum state per phase space area $h = 2\pi\hbar$. The above description is the quantum, fuzzy, non-commutative analog of the classical phase space density. According to the Liouville theorem, a collection of particles with some density on the phase space evolves in an area-preserving way, so a droplet of constant density evolves into a droplet of different shape but the same constant density [71].

The ground state ρ_0 corresponds to a droplet filling a 'lake' in phase space in which the classical value of the single particle energy satisfies

$$H_{\mathrm{sp}}(x, p) \leq E_F. \tag{166}$$

This ensures the minimal energy for the full state. The boundary of the droplet is at the line defined by the points $H_{\mathrm{sp}} = E_F$, the highest energy of any single particle. This is the Fermi energy.

The unitary transformation U maps to a 'star-unitary' commutative function satisfying $U * U^* = 1$. One could think that in the commutative (classical) limit it becomes a phase, $U = \exp[i\phi(x,p)]$. This, however, is not necessarily so. U

enters into the definition of ρ only through the adjoint action $\rho = U^* * \rho_0 * U$. If U became a phase in the commutative limit, it would give $\rho = \rho_0$ (upon mapping star products to ordinary products), creating no variation. The trick is that $U(x,p)$ can contain terms of order $\hbar^{-1}$: since the star-products in the definition of ρ in terms of U reproduce ρ_0 plus terms of order $\hbar$, the overall result will be of order $\hbar^0$ and remain finite in the classical limit. So $U(x,p)$ may not map to a finite function in this limit; its action on ρ_0, however, is finite and defines a canonical transformation, changing the shape of the droplet. Overall, we have a correspondence with a fuzzy, incompressible phase space fluid in the density (Euler) description.

7.3. Quantization and the full many-body correspondence

What makes this description viable and useful is that it reproduces the *full Hilbert space* of the N fermions upon quantization.

The easiest way to see this is to notice that the action (164) is of the Kirillov-Kostant-Souriau form for the group of unitary transformations on the Hilbert space. For concreteness, we may introduce a cutoff and truncate the Hilbert space to the K first energy levels $K \gg N$. Then the above becomes the KKS action for the group $U(K)$. Its properties and quantization are fully known, and we summarize the basic points.

Both $\rho = U^{-1}\rho_0 U$ and the action (164) are invariant under time-dependent transformations

$$U(t) \to V(t)U(t) \ , \ [\rho_0, V(t)] = 0 \tag{167}$$

for any unitary operator ($K \times K$ unitary matrix) commuting with ρ_0. This means that the corresponding 'diagonal' degrees of freedom of U are redundant and correspond to a gauge invariance of the description in terms of U. This introduces a Gauss law as well as a 'global gauge anomaly' for the action that requires a quantization condition, akin to the magnetic monopole quantization or level quantization for the Chern-Simons term. The end result is:

- The eigenvalues of the constant matrix ρ_0 must be integers for a consistent quantization.

On the other hand, the classical Poisson brackets for ρ (165) become, upon quantization,

$$[[\rho_{m_1 n_1}, \rho_{m_2 n_2}]] = \rho_{m_1 n_2}\delta_{m_2 n_1} - \rho_{m_2 n_1}\delta_{m_1 n_2} \tag{168}$$

where we used, again, double brackets for quantum commutators to distinguish from matrix (single-particle) commutators. The above is nothing but the $U(K)$ algebra in a 'cartesian' basis (notice how $\hbar$ has disappeared). The quantum Hilbert space, therefore, will form representations of $U(K)$. The Gauss law, however, imposes constraints on what these can be. The end result is:

- The quantum states form an irreducible representation of $U(K)$ determined by a Young tableau with the number of boxes in each row corresponding to the eigenvalues of ρ_0.

In our case, the eigenvalues are N 1s and $K - N$ 0s, already properly quantized. So the Young tableau corresponds to a single column of N boxes; that is, the N-fold fully antisymmetric representation of $U(K)$.

This is exactly the Hilbert space of N fermions on K single-particle states! The dimensionality of this representation is

$$D = \frac{K!}{N!(K - N)!} \tag{169}$$

matching the number quantum states of N fermions in K levels. The matrix elements of the operator ρ_{mn} in the above representation can be realized in a Jordan-Wigner construction involving K fermionic oscillators Ψ_n, as

$$\rho_{mn} = \Psi_n^\dagger \Psi_m \tag{170}$$

satisfying the constraint

$$\sum_{n=1}^{K} \Psi_n^\dagger \Psi_n = N. \tag{171}$$

This Ψ is essentially the second-quantized fermion field, the above relation being the constraint to the N-particle sector. The quantized Hamiltonian operator for ρ in this realization becomes

$$H = \mathrm{Tr}(\rho H_{\mathrm{sp}}) = \sum_{m,n} \Psi_m^\dagger (H_{\mathrm{sp}})_{mn} \Psi_n \tag{172}$$

and thus also corresponds to the second-quantized many-body Hamiltonian. Overall, this becomes a *complete* description of the many-body fermion system in terms of a quantized non-commutative density field ρ or, equivalently, the unitary non-commutative field U.

It is worth pointing out that in the limit $K \to \infty$ the algebra (168) becomes infinite-dimensional and reproduces the so-called "W_∞ algebra". This algebra has a host of representations, one of which corresponds to the Hilbert space of N fermions. In the finite K case the conditions $\rho^2 = \rho$ and $\mathrm{tr}\rho = N$ fixed the Hilbert space. Similar conditions, corresponding to the appropriate choice of a 'vacuum' (highest-weight) state, fix the desired representation of the W_∞ density algebra. The commutative limit of this algebra, on the other hand, corresponds to the standard Poisson brackets of phase space density functions, as implied by the underlying canonical structure of x and p.

7.4. Higher-dimensional non-commutative bosonization

The above also constitutes an exact bosonization of the fermion system. Indeed, the fields ρ or U are bosonic, so they afford a description of fermions without use of Grassmann variables. The price to pay is the increase of dimensionality (two phase space rather than one space dimensions) and the non-commutative nature of the classical ρ-dynamics, even before quantization.

The correspondence to traditional bosonization can be achieved through the Seiberg-Witten map on the field U. We shall not enter into any detail here, but the upshot of the story is that the action (164) maps to the (commutative) action of a one-dimensional chiral boson under this map. The corresponding space derivative of the field is an abelian 'current' that maps to the boundary of the classical fluid droplet, which parametrizes the full shape of the fluid. Overall this recovers standard abelian bosonization results [72] in the non-commutative hydrodynamic setting. Generalizations to particles carrying internal degrees of freedom are possible and lead to the Wess-Zumino-Witten action for non-abelian bosonization [73].

Most intriguingly, much of the above discussion can be exported to higher dimensions. The formalism extends naturally to higher dimensions, the matrix ρ now acting on the space of states of a single particle in D spatial dimensions. The crucial difference, however, is that the Seiberg-Witten map of the higher-dimensional action yields a nontrivial action in $2D$ (phase space) dimensions that, unlike the $D = 1$ case, does not reduce to a D-dimensional chiral boson action.

We can obtain a more economical description by performing the Seiberg-Witten map only on a two-dimensional non-commutative subspace, leaving the rest of the $2D$-dimensional space untouched. This transformation works similarly to the $D = 1$ case, leading to a description in terms of a field in one residual (commutative) dimension as well as the remaining $2D - 2$ non-commutative ones. This constitutes a 'minimal' bosonization in the non-commutative field theory setting [75]. (For other approaches on higher-dimensional bosonization see [74].)

The form of the above theory can be motivated by starting with the fully classical, commutative picture of our density droplet in phase space of constant density $\rho_0 = 1/(2\pi\hbar)^D$, whose shape is fully determined in terms of its boundary. A convenient way to parametrize the boundary is in terms of the value of one of the phase space coordinates, say p_D, on the boundary as a function of the $2D - 1$ remaining ones. We write

$$p_D\big|_{\text{boundary}} \equiv R(x_1, p_1; \ldots x_D) \tag{173}$$

R will be the boundary field of the theory. For notational convenience, we rename the variable conjugate to the eliminated variable p_D (that is, x_D) σ and write ϕ^α ($\alpha = 1, \ldots, 2D - 2$) for the remaining $2D - 2$ phase space dimensions (x_n, p_n) ($n = 1, \ldots, D - 1$).

The dynamics of the classical system are determined by the canonical Poisson brackets of the field $R(\sigma, \phi)$. These can be derived through a Hamiltonian reduction of the full density Poisson brackets on the phase space [71] and we simply quote the result. We use $\theta^{\alpha\beta} = \{\phi^\alpha, \phi^\beta\}_{\text{sp}}$ for the standard (Darboux) single-particle Poisson brackets of ϕ (that is, $\theta^{\alpha\beta} = \epsilon^{\alpha\beta}$ if α and β correspond to x_n and p_n, otherwise zero), as well as the shorthand $R_{1,2} = R(\sigma_{1,2}, \phi_{1,2})$, with 1 and 2 labeling the two points in the $2D - 1$-dimensional space (σ, ϕ) at which we shall calculate the brackets. The field theory Poisson brackets for R_1 and R_2 read, in an obvious notation:

$$\{R_1, R_2\} = \frac{1}{\rho_0} \left[-\delta'(\sigma_1 - \sigma_2)\,\delta(\phi_1 - \phi_2) - \delta(\sigma_1 - \sigma_2)\,\{R(\sigma_1, \phi_1), \delta(\phi_1 - \phi_2)\}_{\text{sp1}} \right].$$
$$\tag{174}$$

Similarly, the Hamiltonian for the field R is the integral of the single-particle Hamiltonian over the bulk of the droplet and reads

$$H = \rho_o \int dp_D \, d\sigma \, d^{2d}\phi H_{\mathrm{sp}}(\sigma, \phi) \, \vartheta(R - p_D) \tag{175}$$

where $\vartheta(x) = \frac{1}{2}[1 + \mathrm{sgn(x)}]$ is the step function. (174) and (175) define a bosonic field theory (in a Hamiltonian setting) that describes the droplet classically.

The correct quantum version of the theory *cannot* simply be obtained by turning the above Poisson brackets into quantum commutators. We have already encountered a similar situation in the previous subsection: the commutative, classical Poisson algebra of the density operator ρ is deformed into the W_∞ algebra (or its finite $U(K)$ truncation) in the quantum case.

This observation will guide us in motivating the correct quantum commutators for the boundary field. We observe that the first, R-independent term of the above Poisson brackets reproduces a current algebra in the σ-direction, exactly as in one-dimensional bosonization. The second, homogeneous term, on the other hand, has the form of a density algebra in the residual $2D - 2$ phase space dimensions. In this sense, the field R is partly current and partly density. Taking our clues from standard bosonization and the story of the previous subsections, we propose that the current algebra part remains undeformed upon quantization, while the density part gets deformed to the corresponding non-commutative structure. A simple way to do that and still use the same (commutative) phase space notation is in the $*$-product language. Specifically, we turn the single-particle Poisson brackets to non-commutative Moyal brackets $\{.,.\}_*$ on the $2d$-dimensional phase space manifold ϕ^α. The full deformed field theory Poisson brackets, now, read:

$$\{R_1, R_2\} = \frac{1}{\rho_0}\left[-\delta'(\sigma_1 - \sigma_2)\delta(\phi_1 - \phi_2) - \delta(\sigma_1 - \sigma_2)\{R_1, \delta(\phi_1 - \phi_2)\}_{*1}\right]. \tag{176}$$

The Moyal brackets between two functions of ϕ are expressed in terms of the non-commutative Groenewald-Moyal star-product on the phase space ϕ [10]:

$$\{F(\phi), G(\phi)\}_* = \frac{1}{i\hbar}\left[F(\phi) * G(\phi) - G(\phi) * F(\phi)\right] \tag{177}$$

with $\hbar$ itself being the noncommutativity parameter. Correspondingly, the Hamiltonian H is given by expression (175) but with $*$-products replacing the usual products between its terms.

The transition to the matrix ('operator') notation can be done in the standard way, as exposed in the introductory sections, by choosing any basis of states ψ_a in the single-particle Hilbert space. This would map the field $R(\sigma, \phi)$ to dynamical matrix elements $R^{ab}(\sigma)$. The only extra piece that we need is the matrix representation of the delta function $\delta(\phi_1 - \phi_2)$, with defining property

$$\int d^{2d}\phi_1 F(\phi_1) \, \delta(\phi_1 - \phi_2) = F(\phi_2). \tag{178}$$

Since $\delta(\phi_1 - \phi_2)$ is a function of two variables, its matrix transform in each of them will produce a symbol with four indices $\delta^{a_1 b_1; a_2 b_2}$. The above defining relation in the matrix representation becomes

$$(2\pi\hbar)^{(D-1)} F^{a_1 b_1} \delta^{b_1 a_1; a_2 b_2} = F^{a_2 b_2} \tag{179}$$

which implies

$$\delta^{a_1 b_1; a_2 b_2} = \frac{1}{(2\pi\hbar)^{(D-1)}} \delta^{a_1 b_2} \delta^{a_2 b_1}. \tag{180}$$

With the above, and using $\rho_o = 1/(2\pi\hbar)^D$, the canonical Poisson brackets of the matrix R^{ab} become

$$\{R_1^{ab}, R_2^{cd}\} = -2\pi\hbar\delta'(\sigma_1 - \sigma_2)\delta^{ad}\delta^{cb} + 2\pi i\delta(\sigma_1 - \sigma_2)(R_1^{ad}\delta^{cb} - R_1^{cb}\delta^{ad}). \tag{181}$$

Not surprisingly, we recover a structure for the second term similar to the one for ρ of the previous subsection as expressed in (168).

We are now ready to perform the quantization of the theory. The fields $R^{ab}(\sigma)$ become operators whose quantum commutator is given by the above Poisson brackets times $i\hbar$. Defining, further, the Fourier modes

$$R_k^{ab} = \int \frac{d\sigma}{2\pi\hbar} R^{ab}(\sigma) e^{-ik\sigma} \tag{182}$$

the quantum commutators become

$$[[R_k^{ab}, R_{k'}^{cd}]] = k\delta(k + k')\delta^{ad}\delta^{cb} - R_{k+k'}^{ad}\delta^{cb} + R_{k+k'}^{cb}\delta^{ad}. \tag{183}$$

The zero mode $R_0^{aa} \equiv N$ is a Casimir and represents the total fermion number. For a compact dimension σ, normalized to a circle of length 2π, the Fourier modes become discrete.

The above is also recognized as a chiral current algebra for the matrix field R_k^{ab} on the unitary group of transformations of the first-quantized states ψ_a. To make this explicit, consider again the finite-dimensional truncation of the Hilbert space into K single-particle states; that is, $a, b, c, d = 1, \ldots, K$ (this would automatically be the case for a compact phase space $\{\phi^\alpha\}$). As remarked before, the homogeneous part of the above commutator is the $U(K)$ algebra in a 'cartesian' parametrization. To bring it into the habitual form, define the hermitian $K \times K$ fundamental generators of $U(K)$, T^A, $A = 0, \ldots, K^2 - 1$, normalized as $\text{tr}(T^A T^B) = \frac{1}{2}\delta^{AB}$, which fix the $U(M)$ structure constants $[T^A, T^B] = if^{ABC}T^C$ (with $f^{0AB} = 0$). Using the T^A as a basis we express the quantum commutators (183) in terms of the $R^A = \text{tr}(T^A R)$ as

$$[[R_k^A, R_{k'}^B]] = \frac{1}{2}k\delta(k + k')\delta^{AB} + if^{ABC}R_{k+k'}^C. \tag{184}$$

This is the so-called Kac-Moody algebra for the group $U(K)$.

The coefficient k_{KM} of the central extension of the Kac-Moody algebra (the first, affine term) must be quantized to an integer to have unitary representations. Interestingly, this coefficient in the above commutators emerges quantized to the value $k_{KM} = 1$. This is crucial for bosonization [73]. The $k_{KM} = 1$ algebra has a unique irreducible unitary representation over each 'vacuum'; that is, over highest weight states annihilated by all $R^A(k)$ for $k > 0$ and transforming under a fully antisymmetric $SU(K)$ representations under $T^A(0)$. These Fock-like representations correspond exactly to the perturbative Hilbert space of excitations of the many-body fermionic system over the full set of possible Fermi sea ground states. The $U(1)$ charge R_0^0, which is a Casimir, corresponds to the total fermion number; diagonal operators R_k^H, for $k < 0$ and H in the Cartan subgroup of $U(K)$ generate 'radial' excitations in the Fermi sea along each direction in the residual phase space variables; while off-diagonal operators R_k^T, for $k < 0$ and T off the Cartan subgroup, generate transitions of fermions between different points of the Fermi sea.

In the above, we have suddenly introduced the word 'perturbative' in the mapping between states of the field R and many-body fermion states. We had started with a full, non-perturbative description of the system before we reduced it to boundary variables. Where did perturbative come from?

This is a standard feature of bosonization, true also in the one-dimensional case. The boundary of the droplet could in principle 'hit' upon itself, breaking the droplet into disconnected components. The field R in such cases would develop 'shock waves' and lose single-valuedness. Quantum mechanically, the above situation corresponds to locally depleting the Fermi sea. This is an essentially non-perturbative phenomenon, whose account would require the introduction of branches for the field R after the formation of shock waves and corresponding boundary conditions between the branches. Quantum mechanically it would require nontrivial truncations and identifications of states in the Hilbert space of the quantum field R. In the absence of that, the bosonic theory gives an exact description of the Fermi system up to the point that the Dirac sea would be depleted. This is adequate for many-body applications.

Finally, the Hamiltonian of the bosonic theory becomes

$$H = \int \frac{dp_D \, d\sigma}{2\pi\hbar} \mathrm{tr} H_{\mathrm{sp}}(\sigma, p_D, \hat{\phi})\vartheta(R - p_D) \tag{185}$$

where p_D remains a scalar integration parameter while $\hat{\phi}$ become (classical) matrices and R is an operator matrix field as before. Clearly there are issues of ordering in the above expression, matrix (non-commutative) as well as quantum, just as in standard $1 + 1$-dimensional bosonization.

To demonstrate the applicability of this theory we shall work out explicitly the simplest nontrivial example of higher-dimensional bosonization: a system of N noninteracting two-dimensional fermions in a harmonic oscillator potential. The single-particle Hamiltonian is

$$H_{\mathrm{sp}} = \frac{1}{2}(p_1^2 + x_1^2 + p_2^2 + x_2^2). \tag{186}$$

For simplicity we chose the oscillator to be isotropic and of unit frequency. The single-body spectrum is the direct sum of two simple harmonic oscillator spectra, $E_{mn} = \hbar(m + n + 1)$, $m, n = 0, 1, \dots$. Calling $m + n + 1 = K$, the energy levels are $E_K = \hbar K$ with degeneracy K.

The N-body ground state consists of fermions filling states E_K up to a Fermi level $E_F = \hbar K_F$. In general, this state is degenerate, since the last energy level of degeneracy K_F is not fully occupied. Specifically, for a number of fermions N satisfying

$$N = \frac{K_F(K_F - 1)}{2} + M \ , \ 0 \leq M \leq K_F \tag{187}$$

the Fermi sea consists of a fully filled bulk (the first term above) and M fermions on the K_F-degenerate level at the surface. The degeneracy of this many-body state is

$$g(K_F, M) = \frac{K_F!}{M!(K_F - M)!} \tag{188}$$

representing the ways to distribute the M last fermions over K_F states, and its energy is

$$E(K_F, M) = \hbar \frac{K_F(K_F - 1)(2K_F - 1)}{6} + \hbar K_F M. \tag{189}$$

Clearly the vacua $(K_F, M = K_F)$ and $(K_F + 1, M = 0)$ are identical. Excitations over the Fermi sea come with energies in integer multiples of $\hbar$ and degeneracies according to the possible fermion arrangements.

For the bosonized system we choose polar phase space variables,

$$h_i = \frac{1}{2}(p_i^2 + x_i^2) \ , \ \theta_i = \arctan \frac{x_i}{p_i} \ , \ i = 1, 2 \tag{190}$$

in terms of which the single-particle Hamiltonian and Poisson structure is

$$\{\theta_i, h_j\}_{\mathrm{sp}} = \delta_{ij} \ , \ H_{\mathrm{sp}} = h_1 + h_2. \tag{191}$$

For the droplet description we can take $h_2 = R$ and $\theta_2 = \sigma$ which leaves $(h_1, \theta_1) \sim (x_1, p_1)$ as the residual phase space. The bosonic Hamiltonian is

$$H = \frac{1}{(2\pi\hbar)^2} \int d\sigma dh_1 d\theta_1 (\frac{1}{2}R^2 + h_1 R). \tag{192}$$

The ground state is a configuration with $R + h_1 = E_F$. The non-perturbative constraints $R > 0$, $h_1 > 0$ mean that the range of h_1 is $0 < h_1 < E_F$.

To obtain the matrix representation of R we define oscillator states $|a\rangle$, $a = 0, 1, 2, \ldots$ in the residual single-particle space (h_1, θ_1) satisfying $\hat{h}_1|a\rangle = \hbar(a+\frac{1}{2})|a\rangle$. The non-perturbative constraint for h_1 is implemented by restricting to the K_F-dimensional Hilbert space spanned by $a = 0, 1, \ldots, K_F$ with $E_F = \hbar K_F - 1$. In the matrix representation R^{ab} becomes a $U(K_F)$ current algebra. We also Fourier transform in σ as in (182) into discrete modes R_n^{ab}, $n = 0, \pm 1, \ldots$ (σ has a period 2π). The Hamiltonian (192) has no matrix ordering ambiguities (being quadratic in R and h_1) but it needs quantum ordering. Just as in the $1+1$-dimensional case, we normal order by pulling negative modes N to the left of positive ones. The result is

$$\frac{H}{\hbar} = \sum_{n>0} R_{-n}^{ab} R_n^{ba} + \frac{1}{2} R_0^{ab} R_0^{ba} + \left(a + \frac{1}{2}\right) R_0^{aa}. \tag{193}$$

To analyze the spectrum of (193) we perform the change of variables

$$\hat{R}_n^{ab} = R_{n-a+b}^{ab} + (a - K_F + 1)\delta^{ab}\delta_n. \tag{194}$$

The new fields $\hat{R}$ satisfy the same Kac-Moody algebra as R. The Hamiltonian (193) becomes

$$\frac{H}{\hbar} = \sum_{n>0} \hat{R}_{-n}^{ab}\hat{R}_n^{ba} + \frac{1}{2}\hat{R}_0^{ab}\hat{R}_0^{ba} + \left(K_F - \frac{1}{2}\right)\hat{R}_0^{aa} + \frac{K_F(K_F-1)(2K_F-1)}{6}. \tag{195}$$

The above is the standard quadratic form in $\hat{R}$ plus a constant and a term proportional to the $U(1)$ charge $\hat{R}_0^{aa} = N - K_F(K_F - 1)/2$.

The ground state consists of the vacuum multiplet $|K_F, M\rangle$, annihilated by all positive modes $\hat{R}_n$ and transforming in the M-fold fully antisymmetric irrep of $SU(K_F)$ ($0 \leq M \leq K_F - 1$), with degeneracy equal to the dimension of this representation $K_F!/M!(K_F - M)!$. The $U(1)$ charge of $\hat{R}$ is given by the number of boxes in the Young tableau of the irreps, so it is M. The fermion number is, then, $N = K_F(K_F - 1)/2 + M$. Overall, we have a full correspondence with the many-body fermion ground states found before; the state $M = K_F$ is absent, consistently with the fact that the corresponding many-body state is the state $M = 0$ for a shifted K_F.

The energy of the above states consists of a constant plus a dynamical contribution from the zero mode $\hat{R}_0$. The quadratic part contributes $\frac{1}{2}\hbar M$, while the linear part contributes $\hbar(K_F - \frac{1}{2})M$. Overall, the energy is $\hbar K_F(K_F - 1)(2K_F - 1)/6 + \hbar K_F M$, also in agreement with the many-body result.

Excited states are obtained by acting with creation operators $\hat{R}_{-n}$ on the vacuum. These will have integer quanta of energy. Due to the presence of zero-norm states, the corresponding Fock representation truncates in just the right way to reproduce the states of second-quantized fermions with an $SU(K_F)$ internal symmetry and fixed total fermion number. These particle-hole states are, again, into one-to-one correspondence with the excitation states of the many-body system, built as towers of one-dimensional excited Fermi seas over single-particle states $E_{m,n}$, one tower for each value of n, with the correct excitation energy. We have the non-perturbative constraint $0 \leq n < K_F$, as well as constraints related to the non-depletion of the Fermi sea for each value of n, just as in the one-dimensional case. The number of fermions for each tower can vary, the off-diagonal operators $\hat{R}_n^{ab}$ creating transitions between towers, with the total particle number fixed to N by the value of the $U(1)$ Casimir.

The above will suffice to give a flavor of the non-commutative bosonization method. There are clearly many issues that still remain open, not the least of which is the identification of a fermion creation operator in this framework. Putting the method to some good use would also be nice.

8. $\tau\acute{\alpha}$ $\pi\acute{\alpha}\nu\tau\alpha$ $\rho\varepsilon\tilde{\iota}$ ... (and it all keeps flowing...)

This was a lightning review of the more recent and current aspects of non-commutative fluids and their uses in many-body systems. There is a lot more to learn and do. If some of the readers are inspired and motivated into further study or research in this subject, then this narrative has served its purpose. We shall stop here.

References

[1] H.S. Snyder, *Quantized space-time*, Phys. Rev. **71** (1947), 38.

[2] T. Eguchi and H. Kawai, *Reduction of dynamical degrees of freedom in the large N gauge theory*, Phys. Rev. Lett. **48** (1982), 1063.

[3] A. Connes, *The action functional in non-commutative geometry*, Commun. Math. Phys. **117** (1988), 673; *Gravity coupled with matter and the foundation of non-commutative geometry*, Commun. Math. Phys. **182** (1996), 155, [arXiv:hep-th/9603053].

[4] A. Connes, M.R. Douglas and A.S. Schwarz, *Non-commutative geometry and matrix theory: Compactification on tori*, JHEP **9802** (1998), 003, [arXiv:hep-th/9711162].

[5] N. Seiberg and E. Witten, *String theory and non-commutative geometry*, JHEP **9909** (1999), 032, [arXiv:hep-th/9908142].

[6] J.A. Harvey, *Komaba lectures on non-commutative solitons and D-branes*, arXiv:hep-th/0102076.

[7] M.R. Douglas and N.A. Nekrasov, *Non-commutative field theory*, Rev. Mod. Phys. **73** (2001), 977, [arXiv:hep-th/0106048].

[8] R.J. Szabo, *Quantum field theory on non-commutative spaces*, Phys. Rept. **378** (2003), 207, [arXiv:hep-th/0109162].

[9] CK. Zachos, *Deformation quantization: Quantum mechanics lives and works in phase space*, Int. J. Mod. Phys. A **17** (2002), [arXiv:hep-th/0110114].

[10] H. Groenewold, Physica **12** (1946), 405; J. Moyal, Proc. Camb. Phil. Soc. **45** (1949), 99.

[11] J. Madore, S. Schraml, P. Schupp and J. Wess, *Gauge theory on non-commutative spaces*, Eur. Phys. J. C **16** (2000), 161, [arXiv:hep-th/0001203].

[12] C. Klimcik, *Gauge theories on the non-commutative sphere*, Commun. Math. Phys. **199** (1998), 257, [hep-th/9710153].

[13] U. Carow-Watamura and S. Watamura, *Differential calculus on fuzzy sphere and scalar field*, Int. J. Mod. Phys. A **13** (1998), 3235; [q-alg/9710034]; *Non-commutative geometry and gauge theory on fuzzy sphere*, Commun. Math. Phys. **212** (2000), 395, [hep-th/9801195].

[14] A.P. Polychronakos, *Flux tube solutions in non-commutative gauge theories*, Phys. Lett. B **495** (2000), 407, [arXiv:hep-th/0007043].

[15] D.J. Gross and N.A. Nekrasov, *Solitons in non-commutative gauge theory*, JHEP **0103** (2001), 044, [arXiv:hep-th/0010090].

[16] D.J. Gross, A. Hashimoto and N. Itzhaki, *Observables of non-commutative gauge theories*, Adv. Theor. Math. Phys. **4** (2000), 893, [arXiv:hep-th/0008075].

[17] A.P. Polychronakos, *Non-commutative Chern-Simons terms and the non-commutative vacuum*, JHEP **0011** (2000), 008, [arXiv:hep-th/0010264].

[18] S. Deser, R. Jackiw and S. Templeton, *Topologically massive gauge theories*, Annals Phys. **140** (1982), 372, [Erratum-ibid. **185**, 406.1988 APNYA, 281, 409 (1988 APNYA,281,409-449.2000)].

[19] O. Alvarez, *Topological Quantization And Cohomology*, Commun. Math. Phys. **100** (1985), 279.

[20] A.H. Chamseddine and J. Frohlich, *The Chern-Simons action in non-commutative geometry*, J. Math. Phys. **35** (1994), 5195, [arXiv:hep-th/9406013].

[21] T. Krajewski, *Gauge invariance of the Chern-Simons action in non-commutative geometry*, arXiv:math-ph/9810015.

[22] G.H. Chen and Y.S. Wu, *One-loop shift in non-commutative Chern-Simons coupling*, Nucl. Phys. B **593** (2001), 562, [arXiv:hep-th/0006114]; *On gauge invariance of non-commutative Chern-Simons theories*, Nucl. Phys. B **628** (2002), 473, [arXiv:hep-th/0111109].

[23] S. Mukhi and N.V. Suryanarayana, *Chern-Simons terms on non-commutative branes*, JHEP **0011** (2000), 006. [arXiv:hep-th/0009101].

[24] N.E. Grandi and G.A. Silva, *Chern-Simons action in non-commutative space*, Phys. Lett. B **507** (2001), 345, [arXiv:hep-th/0010113].

[25] A. Khare and M.B. Paranjape, *Solitons in $2 + 1$ dimensional non-commutative Maxwell Chern-Simons Higgs theories*, JHEP **0104** (2001), 002, [arXiv:hep-th/0102016].

[26] D. Bak, S.K. Kim, K.S. Soh and J.H. Yee, *Non-commutative Chern-Simons solitons*, Phys. Rev. D **64** (2001), 025018, [arXiv:hep-th/0102137].

[27] V.P. Nair and A.P. Polychronakos, *On level quantization for the non-commutative Chern-Simons theory*, Phys. Rev. Lett. **87** (2001), 030403, [arXiv:hep-th/0102181].

[28] D. Bak, K.M. Lee and J.H. Park, *Chern-Simons theories on non-commutative plane*, Phys. Rev. Lett. **87** (2001), 030402, [arXiv:hep-th/0102188].

[29] R. Jackiw, V.P. Nair, S.Y. Pi and A.P. Polychronakos, *Perfect fluid theory and its extensions*, J. Phys. A **37** (2004), R327, [arXiv:hep-ph/0407101].

[30] V. Arnold and B. Khesin, *Topological Methods in Hydrodynamics*, (Springer, Berlin, 1998).

[31] S. Bahcall and L. Susskind, *Fluid dynamics, Chern-Simons theory and the quantum Hall effect*, Int. J. Mod. Phys. B **5** (1991), 2735.

[32] D. Bak, K.M. Lee and J.H. Park, *Comments on non-commutative gauge theories*, Phys. Lett. B **501** (2001), 305, [arXiv:hep-th/0011244].

[33] L. Susskind, *The quantum Hall fluid and non-commutative Chern-Simons theory*, arXiv:hep-th/0101029.

[34] R. Jackiw, S.Y. Pi and A.P. Polychronakos, *Noncommuting gauge fields as a Lagrange fluid*, Annals Phys. **301** (2002), 157, [arXiv:hep-th/0206014].

[35] J. Goldstone, unpublished communication to R. Jackiw; M. Bordemann and J. Hoppe, *The Dynamics of relativistic membranes. 1. Reduction to two-dimensional*

fluid dynamics, Phys. Lett. B **317** (1993), 315, [arXiv:hep-th/9307036]; B. de Wit, J. Hoppe and H. Nicolai, *On the quantum mechanics of supermembranes*, Nucl. Phys. B **305** (1988), 545; R. Jackiw and A.P. Polychronakos, *Fluid dynamical profiles and constants of motion from d-branes*, Commun. Math. Phys. **207** (1999), 107, [arXiv:hep-th/9902024].

[36] H. Liu, **-Trek II: *n operations, open Wilson lines and the Seiberg-Witten map*, Nucl. Phys. B **614** (2001), 305, [arXiv:hep-th/0011125].

[37] Y. Okawa and H. Ooguri, *An exact solution to Seiberg-Witten equation of noncommutative gauge theory*, Phys. Rev. D **64** (2001), 046009, [arXiv:hep-th/0104036].

[38] A.P. Polychronakos, *Abelian Chern-Simons theories in $2 + 1$ dimensions*, Annals Phys. **203** (1990), 231.

[39] G.V. Dunne, R. Jackiw and C.A. Trugenberger, *Topological (Chern-Simons) Quantum Mechanics*, Phys. Rev. D **41** (1990), 661; G.V. Dunne and R. Jackiw, *'Peierls substitution' and Chern-Simons quantum mechanics*, Nucl. Phys. Proc. Suppl. **33C** (1993), 114, [arXiv:hep-th/9204057].

[40] R.B. Laughlin, 'The Quantum Hall Effect,' R.E. Prange and S.M. Girvin (Eds), p. 233.

[41] J.H. Brodie, L. Susskind and N. Toumbas, *How Bob Laughlin tamed the giant graviton from Taub-NUT space*, JHEP **0102** (2001), 003, [arXiv:hep-th/0010105]; I. Bena and A. Nudelman, *On the stability of the quantum Hall soliton*, JHEP **0012** (2000), 017, [arXiv:hep-th/0011155]; S. S. Gubser and M. Rangamani, *D-brane dynamics and the quantum Hall effect*, JHEP **0105** (2001), 041, [arXiv:hep-th/0012155] L. Cappiello, G. Cristofano, G. Maiella and V. Marotta, *Tunnelling effects in a brane system and quantum Hall physics*, Mod. Phys. Lett. A **17** (2002), 1281, [arXiv:hep-th/0101033].

[42] A.P. Polychronakos, *Quantum Hall states as matrix Chern-Simons theory*, JHEP **0104** (2001), 011, [arXiv:hep-th/0103013]; *Quantum Hall states on the cylinder as unitary matrix Chern-Simons theory*, JHEP **0106** (2001), 070, [arXiv:hep-th/0106011].

[43] F. Calogero, *Solution of the one-dimensional N body problems with quadratic and/or inversely quadratic pair potentials*, J. Math. Phys. **12** (1971), 419.

[44] B. Sutherland, *Exact Results For A Quantum Many Body Problem In One-Dimension*, Phys. Rev. A **4** (1971), 2019; *Exact Results For A Quantum Many Body Problem In One-Dimension. 2*, Phys. Rev. A **5** (1972), 1372.

[45] J. Moser, *Three integrable Hamiltonian systems connected with isospectral deformations*, Adv. Math. **16** (1975), 197.

[46] A.P. Polychronakos, *Nonrelativistic bosonization and fractional statistics*, Nucl. Phys. B **324** (1989), 597; *Exchange Operator Formalism For Integrable Systems of Particles*, Phys. Rev. Lett. **69** (1992), 703, [arXiv:hep-th/9202057]. *Exact anyonic states for a general quadratic Hamiltonian*, Phys. Lett. B **264** (1991), 362.

[47] L. Brink, T.H. Hansson, S. Konstein and M.A. Vasiliev, *The Calogero model: Anyonic representation, fermionic extension and supersymmetry*, Nucl. Phys. B **401** (1993), 591, [arXiv:hep-th/9302023].

[48] H. Azuma and S. Iso, *Explicit relation of quantum Hall effect and Calogero-Sutherland model*, Phys. Lett. B **331** (1994), 107, [arXiv:hep-th/9312001]; S. Iso and

S.J. Rey, *Collective Field Theory of the Fractional Quantum Hall Edge State and the Calogero-Sutherland Model*, Phys. Lett. B **352** (1995), 111, [arXiv:hep-th/9406192].

[49] S. Ouvry, *On the relation between the anyon and Calogero models*, arXiv:cond-mat/9907239.

[50] M.A. Olshanetsky and A.M. Perelomov, *Classical integrable finite dimensional systems related to Lie algebras*, Phys. Rept. **71** (1981), 313, and **94** (1983), 6.

[51] A.P. Polychronakos, *Integrable systems from gauged matrix models*, Phys. Lett. B **266** (1991), 29.

[52] For reviews of the Calogero model closest in spirit to the present discussion see A.P. Polychronakos, *Generalized statistics in one dimension*, published in 'Topological aspects of low-dimensional systems,' Les Houches Session LXIX (1998), Springer Ed. [hep-th/9902157]; *Physics and mathematics of Calogero particles*, J. Phys. A **39** (2006), 12793, [arXiv:hep-th/0607033].

[53] R. Gopakumar, S. Minwalla and A. Strominger, *non-commutative solitons*, JHEP **0005** (2000), 020, [arXiv:hep-th/0003160].

[54] D.P. Jatkar, G. Mandal and S.R. Wadia, *Nielsen-Olesen vortices in non-commutative Abelian Higgs model*, JHEP **0009** (2000), 018, [arXiv:hep-th/0007078].

[55] J.A. Harvey, P. Kraus and F. Larsen, *Exact non-commutative solitons*, JHEP **0012** (2000), 024, [arXiv:hep-th/0010060].

[56] X.G. Wen, *Gapless boundary excitations in the quantum Hall states and in the chiral spin states*, Phys. Rev. B **41** (1990), 12838.

[57] S. Iso, D. Karabali and B. Sakita, *Fermions in the lowest Landau level: Bosonization, W(infinity) algebra, droplets, chiral bosons*, Phys. Lett. B **296** (1992), 143, [arXiv:hep-th/9209003].

[58] A. Cappelli, C.A. Trugenberger and G.R. Zemba, *Infinite symmetry in the quantum Hall effect*, Nucl. Phys. B **396** (1993), 465, [arXiv:hep-th/9206027].

[59] A.P. Polychronakos, *Waves and solitons in the continuum limit of the Calogero-Sutherland model*, Phys. Rev. Lett. **74** (1995), 5153, [arXiv:hep-th/9411054].

[60] S. Hellerman and M. Van Raamsdonk, *Quantum Hall physics equals non-commutative field theory*, JHEP **0110** (2001), 039, [arXiv:hep-th/0103179].

[61] D. Karabali and B. Sakita, *Chern-Simons matrix model: Coherent states and relation to Laughlin wavefunctions*, Phys. Rev. B **64** (2001), 245316, [arXiv:hep-th/0106016]; *Orthogonal basis for the energy eigenfunctions of the Chern-Simons matrix model*, Phys. Rev. B **65** (2002), 075304, [arXiv:hep-th/0107168].

[62] B. Morariu and A.P. Polychronakos, *Finite non-commutative Chern-Simons with a Wilson line and the quantum Hall effect*, JHEP **0107** (2001), 006, [arXiv:hep-th/0106072]; *Fractional quantum Hall effect on the two-sphere: A matrix model proposal*, Phys. Rev. D **72** (2005), 125002, [arXiv:hep-th/0510034].

[63] C. Duval and P.A. Horvathy, *Exotic Galilean symmetry in the non-commutative plane, and the Hall effect*, J. Phys. A **34** (2001), 10097, [arXiv:hep-th/0106089]. P.A. Horvathy, *Non-commuting coordinates in vortex dynamics and in the Hall effect, related to 'exotic' Galilean symmetry*, arXiv:hep-th/0207075.

[64] T.H. Hansson, J. Kailasvuori and A. Karlhede, *Charge and current in the quantum Hall matrix model*, Phys. Rev. B **68** (2003), 035327.

[65] Y.X. Chen, M.D. Gould and Y.Z. Zhang, *Finite matrix model of quantum Hall fluids on S**2*, arXiv:hep-th/0308040.

[66] D. Berenstein, *A matrix model for a quantum Hall droplet with manifest particle-hole symmetry*, Phys. Rev. D **71** (2005), 085001, [arXiv:hep-th/0409115].

[67] A. Cappelli and M. Riccardi, *Matrix model description of Laughlin Hall states*, arXiv:hep-th/0410151.

[68] A. Ghodsi, A.E. Mosaffa, O. Saremi and M.M. Sheikh-Jabbari, *LLL vs. LLM: Half BPS sector of $N = 4$ SYM equals to quantum Hall system*, arXiv:hep-th/0505129.

[69] B. Sakita, *Collective variables of fermions and bosonization*, Phys. Lett. B **387** (1996), 118, [arXiv:hep-th/9607047].

[70] D. Karabali and V.P. Nair, *Quantum Hall effect in higher dimensions*, Nucl. Phys. B **641** (2002), 533, [arXiv:hep-th/0203264]; *The effective action for edge states in higher dimensional quantum Hall systems*, Nucl. Phys. B **679** (2004), 427, [arXiv:hep-th/0307281]; *Edge states for quantum Hall droplets in higher dimensions and a generalized WZW model*, Nucl. Phys. B **697** (2004), 513, [arXiv:hep-th/0403111].

[71] A.P. Polychronakos, *Chiral actions from phase space (quantum Hall) droplets*, Nucl. Phys. B **705** (2005), 457, [arXiv:hep-th/0408194]; *Kac-Moody theories for colored phase space (quantum Hall) droplets*, Nucl. Phys. B **711** (2005), 505, [arXiv:hep-th/0411065].

[72] F. Bloch, Z. Phys. **81** (1933), 363; S. Tomonaga, Prog. Theor. Phys. **5** (1950), 544. W. Thirring, Ann. Phys. (N.Y.) **3** (1958), 91; J.M. Luttinger, J. Math. Phys. 4 (1963), 1154; D. Mattis and E. Lieb, J. Math. Phys. **6** (1965), 304; S. R. Coleman, Phys. Rev. D **11** (1975), 2088; S. Mandelstam, Phys. Rept. **23** (1976), 307.

[73] E. Witten, *Nonabelian bosonization in two dimensions*, Commun. Math. Phys. **92** (1984), 455.

[74] A. Luther, *Bosonized Fermions In Three-Dimensions*, Phys. Rept. **49** (1979), 261; F.D.M. Haldane, *Luttinger's Theorem and Bosonization of the Fermi Surface*, Varenna 1992 Lectures [cond-mat/0505529]; D. Schmeltzer, *Bosonization in one and two dimensions*, Phys. Rev. B **47** (1993), 11980; A. Houghton and J.B. Marston, *Bosonization and fermion liquids in dimensions greater than one*, Phys. Rev. B **48** (1993), 7790; A.H. Castro Neto and E. Fradkin, *Bosonization of the Low Energy Excitations of Fermi Liquids*, Phys. Rev. Lett. **72** (1994), 1393; D.V. Khveshchenko, *Geometrical approach to bosonization of $D > 1$ dimensional (non) Fermi liquids*, Phys. Rev. B **52** (1995), 4833.

[75] A.P. Polychronakos, *Bosonization in higher dimensions via non-commutative field theory*, Phys. Rev. Lett. **96** (2006), 186401, [arXiv:hep-th/0502150]; A. Enciso and A.P. Polychronakos, *The fermion density operator in the droplet bosonization picture*, Nucl. Phys. B **751** (2006), 376, [arXiv:hep-th/0605040].

Alexios P. Polychronakos
Physics Department
City College of New York
160 Convent Avenue
New York, NY 10031, USA
e-mail: `alexios@sci.ccny.cuny.edu`

Quantum Spaces, 161–201
© 2007 Birkhäuser Verlag Basel

Heisenberg Spin Chains: From Quantum Groups to Neutron Scattering Experiments

Jean-Michel Maillet

Abstract. Heisenberg spin-1/2 chains are the archetype of quantum integrable one-dimensional models describing magnetic properties of a wide range of compounds (like the $KCuF_3$ crystal), which can be probed experimentally through neutron scattering experiments, while being at the same time at the root of the invention of Bethe ansatz and Yang-Baxter structures that led in turn to quantum groups discovery. The aim of this contribution is to describe these algebraic ingredients and to show how to obtain from them (using combined analytical and numerical analysis) dynamical correlation functions of integrable Heisenberg spin-1/2 chains, the Fourier transform of which, the so-called dynamical structure factors, being directly measured in inelastic neutron scattering experiments. Our method is based on the algebraic Bethe ansatz and the resolution of the quantum inverse scattering problem. It leads to recent progress in the computation of integrable Heisenberg spin-1/2 chains correlation functions that we review here.

1. Introduction

One of the main tasks of statistical mechanics is to understand macroscopic quantities such as specific heat, susceptibility, or transport properties for a fluid or a crystal in terms of microscopic elementary interactions between the constituents which are for example molecules, or ions. A fundamental theoretical quantity for this study is the so-called dynamical structure factor (the Fourier transform of the dynamical two-point correlation function). The importance of these functions originates from the following facts:

(i) They can be measured directly via scattering of neutrons or photons at the material to be studied [1–7], so that if we are able to compute these functions within a model given by some Hamiltonian describing microscopic interactions, we can compare this model with the reality.

(ii) From these quantities it is possible to compute other fundamental macroscopic quantities of statistical mechanics for systems in thermodynamical equilibrium and close to this equilibrium, like in particular transport coefficients (see, e.g., [7]).

Thus if one is interested in understanding, for example, magnetic properties of crystals one should find models describing the microscopic interactions between the spins of the constituent ions and develop methods to calculate within such models the dynamical spin-spin correlation functions. This is for a generically interacting quantum Hamiltonian a very involved problem, quite often out of reach of any treatment by perturbation theory. Hence, the strategy to attack this difficult task has been first to construct simple enough but representative models encoding the main features of magnetic properties of crystals. A serious but not senseless simplification in this process is the reduction of the dimensionality of the problem leading in particular to consider models defined in one dimension. Although drastic at first sight, this strategy proved to be quite successful. In fact there exists today an impressive list of magnetic materials where the interaction between the different constituents is mainly along one-dimensional chains whereas the energy exchange between the different chains is negligible [8]. Strong one-dimensional magnetic character is most usually produced by separating the chains carrying the dominant magnetic interaction by large non-magnetic complex ions, like in $CuCl_2 \cdot 2NC_5H_5$. Note however that in these systems, the three-dimensional character is usually recovered at sufficiently low temperature.

A very interesting example of such a compound is provided by the (rather exotic) $KCuF_3$ crystal which displays properties characteristic of one-dimensional antiferromagnets [8–12]. Although the $KCuF_3$ crystal is fully three-dimensional, its one-dimensional magnetic properties are attributed to the distortion of the octahedral environment of the Cu^{2+} ions due to the Jahn-Teller effect [13]. It leads to a spatial alignment of the 3d orbitals in Cu^{2+} resulting in a strong exchange interaction along one axis of the crystal (the chain axis) while in the perpendicular direction the exchange interaction is very small due to poor overlap of the corresponding orbitals. The ratio between the two interaction constants has been evaluated to be of the order 0.027 [9], making the magnetic behavior of the $KCuF_3$ compound effectively one-dimensional. Further, the Cu^{2+} ions provide [8] effective spin-1/2-dynamical variables in interaction which is well represented by the Heisenberg spin chain Hamiltonian [14]. The XXZ spin-$\frac{1}{2}$ Heisenberg chain in an external magnetic field h is a quantum interacting model defined on a one-dimensional lattice with Hamiltonian

$$H = H^{(0)} - hS_z, \tag{1}$$

$$H^{(0)} = \sum_{m=1}^{M} \left\{ \sigma_m^x \sigma_{m+1}^x + \sigma_m^y \sigma_{m+1}^y + \Delta(\sigma_m^z \sigma_{m+1}^z - 1) \right\}, \tag{2}$$

$$S_z = \frac{1}{2} \sum_{m=1}^{M} \sigma_m^z, \qquad [H^{(0)}, S_z] = 0. \tag{3}$$

Here Δ is the anisotropy parameter (essentially equal to 1 for KCuF$_3$), M is the number of sites of the chain (and here we assume for simplicity periodic boundary conditions), h denotes the uniform external magnetic field, and $\sigma_m^{x,y,z}$ are the local spin operators (here in the spin-$\frac{1}{2}$ representation) associated with each site m of the chain. The quantum space of states is $\mathcal{H} = \otimes_{m=1}^{M}\mathcal{H}_m$, where $\mathcal{H}_m \sim \mathbb{C}^2$ is called local quantum space, with $\dim \mathcal{H} = 2^M$. The operators $\sigma_m^{x,y,z}$ act as the corresponding Pauli matrices in the space $\mathcal{H}_m$ and as the identity operator elsewhere.

Following our discussion above, to be able to compare predictions of such a one-dimensional model to actual magnetic compounds such as KCuF$_3$, we need to compute various physical observable quantities such as the dynamical structure factors; they are the Fourier transform of the dynamical spin-spin correlation functions which at nonzero temperature T, lattice distance m and time difference t, are given as traces over the space of states,

$$S^{\alpha\beta}(m,t) = \frac{tr(\sigma_1^\alpha\, e^{iHt}\, \sigma_{m+1}^\beta\, e^{-iHt}\, e^{-\frac{H}{kT}})}{tr(e^{-\frac{H}{kT}})}. \tag{4}$$

At zero temperature, this expression reduces to an average value of the product of Heisenberg spin operators taken in the ground state $|\psi_g\rangle$, the normalized (non-degenerated in the disordered regime) state with lowest energy level of the Heisenberg chain,

$$S^{\alpha\beta}(m,t) = \langle\psi_g|\,\sigma_1^\alpha\, e^{iHt}\, \sigma_{m+1}^\beta\, e^{-iHt}\,|\psi_g\rangle. \tag{5}$$

The Fourier transform (in space and time) $S^{\alpha\beta}(q,\omega)$ of this dynamical correlation functions is related, at first order in the neutron-crystal interaction, to the differential magnetic cross sections for the inelastic scattering of unpolarized neutrons off a crystal (like KCuF$_3$), with energy transfer ω and momentum transfer q through the following formula [5]:

$$\frac{d\sigma}{d\Omega d\omega} \sim \left(\delta_{\alpha\beta} - \frac{q_\alpha q_\beta}{q^2}\right) S^{\alpha\beta}(q,\omega). \tag{6}$$

Hence, to compare the Heisenberg model to experimental measurements of the neutron scattering cross sections, we need to compute the dynamical spin-spin correlation functions (4) or (5).

This amounts first to determine the spectrum of the Heisenberg Hamiltonian. Further, we need to identify the action of the local spin operators in the corresponding eigenstate basis and obtain their matrix elements to be summed up to perform the trace, and the scalar products, necessary in the actual computation of (4) or (5).

The solution to these different steps turns out to be a fantastic challenge involving deep algebraic structures hidden in the original Bethe ansatz solution [15] of the Heisenberg Hamiltonian spectrum and unraveled along its extensions [16–20] in particular through the associated Yang-Baxter structures [21–27]; these led, in the search of an algebraic way to construct new integrable models [28–31], to the

164 J.-M. Maillet

discovery of quantum groups [32–35]; it was later realized that the underlying symmetry algebra of the Heisenberg model in the infinite lattice limit is the quantum affine algebra $U_q(\hat{sl}_2)$ [36, 37].

The aim of this lecture is to describe the methods used towards the solution of these successive steps. Some of them are already 75 years old and go back to H. Bethe [15], while others have been developed only in the last ten years. But before going into the historical developments and technical details about these tools, and as a motivation to eventually spend some time learning about them, we would like to give here one of the results that we obtained rather recently [38–40]: the graphical plot (as a function of q and ω) of the total dynamical structure factor at zero temperature $S(q, \omega)$ and its successful comparison to experimental neutron scattering measurements on the KCuF$_3$ crystal (the colors encode here the value of the function in the (q, ω) plane, from light gray corresponding to zero value to dark gray in the highest contributions), see Fig. 1.

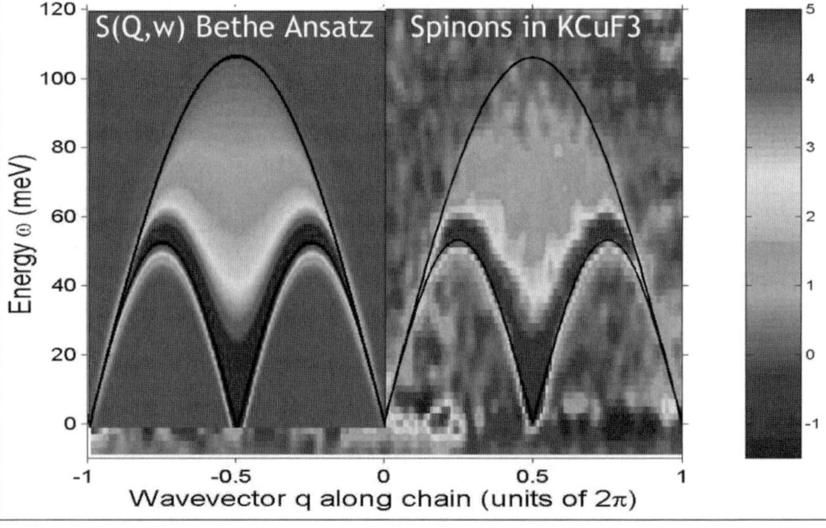

FIGURE 1. The dynamical structure factor $S(q, \omega)$, on the right computed using Bethe ansatz techniques and on the left from inelastic neutron scattering experiment on KCuF$_3$ [11] (experimental data and picture, courtesy A. Tennant)

This computation involves both analytical (exact) results about the spectrum of the Heisenberg Hamiltonian, the matrix elements of the local spin operators between eigenstates using Bethe ansatz techniques and numerical analysis used to perform the sums over these matrix elements to obtain the dynamical structure factor $S(q, \omega)$ (see Section 3).

What makes these results at all possible is the integrable nature of the Heisenberg Hamiltonian, namely in particular the possibility to determine its exact spectrum. This model, introduced by Heisenberg in 1928 [14], can in fact be considered as the archetype of a large class of integrable (called also exactly solvable) models in low dimensions in classical and quantum statistical mechanics and field theory. They already found many applications ranging from condensed matter physics (see, e.g., [26, 41–43]) to high energy physics (see, e.g., [44–46]).

The history of these integrable models of statistical physics started in fact a bit before the Heisenberg spin chain, with the proposal by Lenz and by Ising [47, 48]) of the Ising model to investigate ferromagnetic properties of solids. Ising first solved the one-dimensional case where there is no phase transition at any finite temperature to a ferromagnetic ordered state. It is rather unfortunate that Ising did not realize at that time that this failure was a peculiarity of the one-dimensional situation. However, this was taken by Heisenberg as a motivation to propose his own model in 1928 [14], based on a more sophisticated treatment of the interactions between the spins (using in particular their full quantum operator nature which was simplified drastically in the Ising case). In this way, the more complicated Heisenberg model was exploited (successfully) first, and only after theoretical physicists (and chemists!) did return to the somehow simpler Ising model [49–53].

The Bethe solution of the Heisenberg spin chain in 1931 gave the starting point for the development of the field of quantum integrable models in one-dimensional statistical mechanics, using his now famous Bethe ansatz and its further extensions [15, 17, 18, 20, 21].

The Ising model generated also a fantastic line of research, starting with the Onsager solution of the two-dimensional case in 1942 [51]. It had a major impact on the theory of critical phenomena and launched a series of studies of two-dimensional exactly solvable models of classical statistical mechanics (in fact related through their transfer matrices to the above quantum one-dimensional spin chains) culminating in the works of Baxter in the 1970s on the 6-vertex (related to the XXZ chain) and 8-vertex (related to the XYZ chain) models (see the book [26] and references therein).

This remarkable success (also with the works of Gaudin, Yang and many others; see [25–27] and references therein) led to apply these techniques to a quite interesting continuum model, the nonlinear Schrödinger model, which was also classically solvable through the inverse scattering methods using its Lax pair structure, see, e.g., [54, 55] and references therein. This led to the discovery of an algebraic version of the Bethe ansatz by Faddeev, Sklyanin and Taktadjan [23, 24]. The algebraic structure at work in this method has been coined since this pioneering work, the Yang-Baxter algebra. It is written as a quadratic algebra of quantum operators depending on a continuous parameter λ (the spectral parameter) and governed by an R-matrix which in the case of the Heisenberg XXZ chain is directly related to the Boltzman weights of the 6-vertex model. For that case, there are four operators, A, B, C, D that can be considered as forming the operator entries of a

2×2 matrix, the monodromy matrix,

$$T(\lambda) = \begin{pmatrix} A(\lambda) & B(\lambda) \\ C(\lambda) & D(\lambda) \end{pmatrix}. \tag{7}$$

This monodromy matrix is constructed from the R-matrix of the model as a specific ordered product all along the chain (see the next section). The quadratic commutation relations between the operators A, B, C, D can then be written in a compact way as

$$R_{12}(\lambda, \mu) \, T_1(\lambda) \, T_2(\mu) = T_2(\mu) \, T_1(\lambda) \, R_{12}(\lambda, \mu), \tag{8}$$

with the tensor notation $T_1(\lambda) = T(\lambda) \otimes \mathrm{Id}$ and $T_2(\mu) = \mathrm{Id} \otimes T(\mu)$. There the R-matrix appears as the structure constants of the Yang-Baxter algebra. It is a linear operator in the tensor product $V_1 \otimes V_2$, where each V_i is isomorphic to $\mathbf{C}^2$, and depends generically on two spectral parameters λ_1 and λ_2 associated to these two vector spaces. It is denoted by $R_{12}(\lambda_1, \lambda_2)$. Such an R-matrix satisfies the Yang-Baxter equation,

$$R_{12}(\lambda_1, \lambda_2) \, R_{13}(\lambda_1, \lambda_3) \, R_{23}(\lambda_2, \lambda_3) = R_{23}(\lambda_2, \lambda_3) \, R_{13}(\lambda_1, \lambda_3) \, R_{12}(\lambda_1, \lambda_2). \tag{9}$$

These commutation relations imply in particular that the transfer matrices, defined as

$$\mathcal{T}(\lambda) = \mathrm{tr}\, T(\lambda) = A(\lambda) + D(\lambda), \tag{10}$$

commute for different values of the spectral parameter $[\mathcal{T}(\lambda), \mathcal{T}(\mu)] = 0$ and also with S_z, $[\mathcal{T}(\lambda), S_z] = 0$. The Hamiltonian (2) at $h = 0$ is related to $\mathcal{T}(\lambda)$ by the 'trace identity'

$$H^{(0)} = 2 \sinh \eta \, \frac{d\mathcal{T}(\lambda)}{d\lambda} \mathcal{T}^{-1}(\lambda) \Big|_{\lambda = \frac{\eta}{2}} - 2M \cosh \eta. \tag{11}$$

Therefore, the spectrum of the Hamiltonian (1) is given by the common eigenvectors of the transfer matrices and of S_z. They can be constructed by the successive action of operator $B(\lambda_i)$ (or equivalently by the $C(\lambda_i)$) on a reference state provided the spectral parameters λ_i satisfy the original Bethe equations. The analysis of these equations leads to the determination of the Hamiltonian spectrum, and to the determination of the ground state, in particular in the limit of infinite chains.

It is interesting to mention that the above algebraic structures have nice classical limits that are related to Lie-Poisson structures (see [54, 55] and references therein). It enables us to construct the corresponding classical integrable models purely from the knowledge of a Lie algebra and its representations. The similar question for the quantum case was of great importance in constructing new quantum integrable models, not only on the lattice but also in the continuum [28–31]. In turn, the full solution of this problem led to the discovery of quantum groups [32–35].

After determining the spectrum, the next task is to consider the computation of correlation functions such as (5). There are two main routes to compute dynamical two-point correlation functions of this type, namely, depending on the lattice distance m and on the time variable t (we assume here translational invariance):

(i) Compute first the action of local operators on the ground state

$$\sigma_1^\alpha \, e^{iHt} \, \sigma_{m+1}^\beta \, e^{-iHt} \, |\,\psi_g\,\rangle = |\,\tilde{\psi}_g\,\rangle \tag{12}$$

and then calculate the resulting scalar product to get

$$S^{\alpha\beta}(m,t) \;=\; \langle\,\psi_g\,|\,\tilde{\psi}_g\,\rangle. \tag{13}$$

Note however that for dynamical correlation functions this amounts to evaluate the action of the exponential of the Hamiltonian operator not only on Hamiltonian eigenstates (which is easy) but also on general states resulting from the action of local operators on Hamiltonian eigenstates (which is rather complicated).

(ii) Insert the identity as a sum over a complete set of normalized states $|\,\psi_i\,\rangle$ (for instance, the basis constructed out of the eigenvectors of the Hamiltonian) between the local operators to obtain a representation for the correlation function as a sum over matrix elements of local operators,

$$S^{\alpha\beta}(m,t) \;=\; \sum_i \langle\,\psi_g\,|\,\sigma_1^\alpha\,|\,\psi_i\,\rangle\,\langle\,\psi_i\,|\,\sigma_1^\alpha\,|\,\psi_g\,\rangle\, e^{i(E_i-E_g)t}\, e^{im(P_i-P_g)}, \tag{14}$$

where, E_i, P_i and E_g, P_g are the energy and momentum eigenvalues of the states $|\,\psi_i\,\rangle$ and of the ground state $|\,\psi_g\,\rangle$ respectively. This amounts again to be able to act with local operators on eigenstates, to compute the resulting scalar products, and finally to perform the above sum containing in the XXZ spin-$\frac{1}{2}$ model case with M sites 2^M terms.

In both approaches, we need to obtain the action of local operators on Hamiltonian eigenstates in a compact and manageable form and then to evaluate the resulting scalar product. This problem turns out to be very involved due to the highly nonlocal nature of the Bethe eigenstates. Indeed, the creation operators of Bethe eigenstates (the operators $B(\lambda)$) are extremely nonlocal in terms of local spin operators σ_i^α. In fact (see next section) they are the sum of 2^M terms (M is the number of lattice sites in the chain), each term being some product of spin operators σ_i^α from the site one to the site M. As a consequence, A, B, C, D operators do not have a priori simple commutation relations with the local spin operators, which is the ingredient we would need to compute the action of the latter on Bethe eigenstates. It is a major problem that prevents for very long the computation of correlation functions. In fact, the first case to be understood was the free fermion point $\Delta = 0$ (a computation essentially equivalent to the one for the two-dimensional Ising model). In that case, thanks to a Jordan-Wigner transformation, it is possible to rewrite the Hamiltonian as a quadratic expression in the fermionic operators and hence to use them as creation operators for its eigenstates while the local spin operators have also a simple expression in terms of them. It is this property, namely the fact that all relevant quantities can be embedded inside the same Clifford algebra, that finally opened the possibility to compute the correlation functions in that case. Nevertheless tremendous work was necessary to achieve full answers [51–53, 56–60].

Going beyond the free Fermion case has been a major challenge for the last thirty years.

For integrable quantum spin chains [61–63] and lattice models [26], the first attempts to go beyond free Fermion models relied on the Bethe ansatz techniques [23,64] and were undertaken by A.G. Izergin and V.E. Korepin (see, e.g., [61] and references therein). Their approach yields formulae for the correlation functions [61, 65–67] written as vacuum expectation values of some determinants depending on so-called "dual fields" which were introduced to overcome the huge combinatorial sums arising in particular from the action of local operators on Bethe states. However these formulae are not completely explicit, since these "dual fields" cannot be eliminated from the final result.

In the last fifteen years, two main approaches to a more explicit computation of form factors and correlation functions have been developed, mainly for lattice models.

One of these approaches was initiated by M. Jimbo, T. Miwa and their collaborators [36, 37, 68, 69] and enables, using some (rather well-controlled) hypothesis, to compute form factors and correlation functions of quantum spin chains of infinite length (and in their massive regime) by expressing them in terms of traces of q-deformed vertex operators over an irreducible highest weight representation of the corresponding quantum affine algebra. This quantum affine algebra is conjectured to be the infinite-dimensional symmetry algebra of the Heisenberg infinite chain, and all relevant quantities can be embedded in this algebra, making the computation of correlation functions possible. The vertex operators traces turn out to satisfy an axiomatic system of equations called q-deformed Knizhnik-Zamolodchikov (q-KZ) equations, the solutions of which can be expressed in terms of multiple integral formulae. Using these equations similar formulae can be conjectured in the massless regime.

Recently, a more algebraic representation for the solution of these q-deformed Knizhnik-Zamolodchikov equations have been obtained for the XXX and XXZ (and conjectured for the XYZ) spin 1/2 chains; in these representations, all elementary blocks of the correlation functions can be expressed in terms of some transcendental functions [70–72]. A detailed review of the approach can be found in [62].

These results, their proofs, together with their extension to nonzero magnetic field have been obtained in 1999 [38, 73] using the algebraic Bethe ansatz framework [23–25] and the actual resolution of the so-called quantum inverse scattering problem [38, 74].

The main steps of this method are as follows. Let us first note that any n-point correlation function of the Heisenberg chain can be reconstructed as a sum of elementary building blocks defined in the following way:

$$F_m(\{\epsilon_j, \epsilon_j'\}) = \langle\, \psi_g \,|\, \prod_{j=1}^{m} E_j^{\epsilon_j', \epsilon_j} \,|\, \psi_g \,\rangle. \tag{15}$$

Here $|\psi_g\rangle$ is the normalized ground state of the chain and $E_j^{\epsilon'_j,\epsilon_j}$ denotes the elementary operator acting on the quantum space $\mathcal{H}_j$ at site j as the 2×2 matrix of elements $E_{lk}^{\epsilon',\epsilon}=\delta_{l,\epsilon'}\delta_{k,\epsilon}$.

A multiple integral representation for these building blocks was obtained for the first time in [68, 69]. We briefly recall how we derived them in the framework of algebraic Bethe Ansatz [38, 73] by solving the following successive problems:

(i) determination of the ground state $\langle\psi_g|$,
(ii) evaluation of action of the product of local operators on this ground state,
(iii) computation of the scalar product of the resulting state with $|\psi_g\rangle$,
(iv) thermodynamic limit.

The starting point of our method is to use in step (i) the description of the eigenstates obtained via algebraic Bethe ansatz [23, 61]. They are constructed in this framework in terms of generalized creation and annihilation operators which are themselves highly nonlocal. Acting with local operators on such states in step (ii) is therefore a-priori a nontrivial problem. One of the key-ingredients of our method, which enables us to compute this action explicitly, is the solution of the so-called quantum inverse scattering problem [38, 74]: local operators are reconstructed in terms of the generators of the so-called Yang-Baxter algebra, which contains in particular the creation/annihilation operators for the eigenstates. Hence, all computations can now be done in the Yang-Baxter algebra. In particular, the step (ii) is now completed using only the quadratic commutation relations satisfied by these generators [73]. The computation of the resulting scalar products in step (iii) may also present some technical difficulties. In the case of the XXZ Heisenberg chain, it has been solved using again the algebraic structure of the Yang-Baxter algebra [38, 83]. Finally, the step (iv) is obtained using the results of [19, 20].

Note that this procedure remains essentially the same in the case of the two-point correlation functions. The main difference is that, in step (ii), the reconstruction of the corresponding local operators from the solution of the inverse problem gives rise to a more complicated combination of the generators of the Yang-Baxter algebra, so that the use of their commutation relations to determine their action on the eigenstates involves more complicated combinatorics.

At zero magnetic field our method gives a complete proof of the multiple integral representations obtained in [37, 68, 69] both for massive and massless regimes. Hence, together with the works [68, 69], it also gives a proof that correlation functions of the XXZ (inhomogeneous) chain indeed satisfy (reduced) q-deformed Knizhnik-Zamolodchikov equations. Moreover, time or temperature dependent correlation functions can also be computed [63, 80, 81] using such techniques.

This method allows also for the computation of the matrix elements of the local spin operators and the above elementary blocks of the correlation functions for the finite chain. Hence, thermodynamic limit can be considered separately. In particular, using both analytical results from Bethe ansatz for these matrix elements of the spin operators [38, 73, 74, 84] and numerical methods to take the summation over intermediate states it has recently been possible to compute [39, 40] dynamical structure factors (i.e., Fourier transform of the dynamical spin-spin correlation functions) for finite XXZ Heisenberg spin chains in a magnetic field (with for example 500 or 1000 sites) and to compare successfully these theoretical results with actual neutron scattering experiments, for example on $KCuF_3$ as shown in Fig. 1.

This article is meant to be a rather brief review on the problem of correlation functions in quantum integrable models and more specifically in the XXZ Heisenberg model. More detailed account of the results sketched here together with their proofs can be found in the original articles [38, 63, 73–80, 84–86] and in [39, 40, 87]. This lecture is organized as follows. The space of states of the Heisenberg spin chain will be described in the next section. It includes a brief introduction to the algebraic Bethe ansatz and to various tools of importance in the computation of correlation functions, like in particular the solution of the quantum inverse scattering problem and the determinant representations of the scalar products of states. Section 3 is devoted to the correlation functions of the finite chain and the description of the method leading to Fig. 1. Correlation functions in the thermodynamic limit are studied in Section 4. In Section 5 we describe several exact and asymptotic results together with some open problems. Conclusions and some perspectives are given in the last section.

2. Heisenberg spin chain and algebraic Bethe ansatz

The space of states is of dimension 2^M as it follows from the definition of the Hamiltonian in (1). Apart from the completely ferromagnetic states with all spins up or down, the construction of the Hamiltonian eigenvectors is rather nontrivial. The purpose of this section is to briefly explain the basics of the knowledge of the space of states in the framework of the algebraic Bethe ansatz, leading in particular to the determination of the spectrum of (1).

2.1. Algebraic Bethe ansatz

The algebraic Bethe ansatz originated from the fusion of the original (coordinate) Bethe ansatz and of the inverse scattering method in its Hamiltonian formulation [23–25]. At the root of the algebraic Bethe ansatz method is the construction of the quantum monodromy matrix. In the case of the XXZ chain (1) the monodromy matrix is a 2×2 matrix,

$$T(\lambda) = \begin{pmatrix} A(\lambda) & B(\lambda) \\ C(\lambda) & D(\lambda) \end{pmatrix}, \tag{16}$$

with operator-valued entries A, B, C and D which depend on a complex parameter λ (spectral parameter) and act in the quantum space of states $\mathcal{H}$ of the chain. One of the main properties of these operators is that the trace of T, namely $A + D$, commutes with the Hamiltonian H, while operators B and C can be used as creation operators of respectively eigenvectors and dual eigenvectors of $A + D$

and hence of H itself. The monodromy matrix is defined as the following ordered product,

$$T(\lambda) = L_M(\lambda)\ldots L_2(\lambda)L_1(\lambda), \tag{17}$$

where $L_n(\lambda)$ denotes the quantum L-operator at the site n of the chain:

$$L_n(\lambda) = \begin{pmatrix} \sinh(\lambda + \frac{\eta}{2}\sigma_n^z) & \sinh\eta\,\sigma_n^- \\ \sinh\eta\,\sigma_n^+ & \sinh(\lambda - \frac{\eta}{2}\sigma_n^z) \end{pmatrix}. \tag{18}$$

The parameter η is related to the anisotropy parameter as $\Delta = \cosh\eta$. It follows from this definition that the monodromy matrix is a highly nonlocal operator in terms of the local spin operators $\sigma_n^{x,y,z}$. However, the commutation relations between the operators A, B, C, D can be computed in a simple way. They are given by the quantum R-matrix,

$$R(\lambda, \mu) = \begin{pmatrix} 1 & 0 & 0 & 0 \\ 0 & b(\lambda, \mu) & c(\lambda, \mu) & 0 \\ 0 & c(\lambda, \mu) & b(\lambda, \mu) & 0 \\ 0 & 0 & 0 & 1 \end{pmatrix} \tag{19}$$

where

$$b(\lambda, \mu) = \frac{\sinh(\lambda - \mu)}{\sinh(\lambda - \mu + \eta)}, \quad c(\lambda, \mu) = \frac{\sinh(\eta)}{\sinh(\lambda - \mu + \eta)}, \tag{20}$$

The R-matrix is a linear operator in the tensor product $V_1 \otimes V_2$, where each V_i is isomorphic to $\mathbf{C}^2$, and depends generically on two spectral parameters λ_1 and λ_2 associated to these two vector spaces. It is denoted by $R_{12}(\lambda_1, \lambda_2)$. Such an R-matrix satisfies the Yang-Baxter equation,

$$R_{12}(\lambda_1, \lambda_2)\, R_{13}(\lambda_1, \lambda_3)\, R_{23}(\lambda_2, \lambda_3) = R_{23}(\lambda_2, \lambda_3)\, R_{13}(\lambda_1, \lambda_3)\, R_{12}(\lambda_1, \lambda_2). \tag{21}$$

It gives the following commutation relations among the operators entries of the monodromy matrix,

$$R_{12}(\lambda, \mu)\, T_1(\lambda)\, T_2(\mu) = T_2(\mu)\, T_1(\lambda)\, R_{12}(\lambda, \mu), \tag{22}$$

with the tensor notations $T_1(\lambda) = T(\lambda) \otimes \mathrm{Id}$ and $T_2(\mu) = \mathrm{Id} \otimes T(\mu)$. These commutation relations imply in particular that the transfer matrices, defined as

$$\mathcal{T}(\lambda) = \mathrm{tr}\, T(\lambda) = A(\lambda) + D(\lambda), \tag{23}$$

commute for different values of the spectral parameter $[\mathcal{T}(\lambda), \mathcal{T}(\mu)] = 0$ and also with S_z, $[\mathcal{T}(\lambda), S_z] = 0$. The Hamiltonian (2) at $h = 0$ is related to $\mathcal{T}(\lambda)$ by the 'trace identity' (11).

Therefore, the spectrum of the Hamiltonian (1) is given by the common eigenvectors of the transfer matrices and of S_z.

For technical reasons, it is actually convenient to introduce a slightly more general object, the twisted transfer matrix

$$\mathcal{T}_\kappa(\lambda) = A(\lambda) + \kappa D(\lambda), \tag{24}$$

where κ is a complex parameter. The particular case of $\mathcal{T}_\kappa(\lambda)$ at $\kappa = 1$ corresponds to the usual (untwisted) transfer matrix $\mathcal{T}(\lambda)$. It will be also convenient to consider an inhomogeneous version of the XXZ chain, for which

$$T_{1\ldots M}(\lambda; \xi_1, \ldots, \xi_M) = L_M(\lambda - \xi_M + \eta/2) \ldots L_1(\lambda - \xi_1 + \eta/2). \tag{25}$$

Here, $\xi_1, \ldots, \xi_M$ are complex parameters (inhomogeneity parameters) attached to each site of the lattice. The homogeneous model (1) corresponds to the case where $\xi_j = \eta/2$ for $j = 1, \ldots, M$.

In the framework of algebraic Bethe ansatz, an arbitrary quantum state can be obtained from the vectors generated by multiple action of operators $B(\lambda)$ on the reference vector $|\,0\,\rangle$ with all spins up (respectively by multiple action of operators $C(\lambda)$ on the dual reference vector $\langle\,0\,|$),

$$|\,\psi\,\rangle = \prod_{j=1}^{N} B(\lambda_j)|\,0\,\rangle, \qquad \langle\,\psi\,| = \langle\,0\,| \prod_{j=1}^{N} C(\lambda_j), \qquad N = 0, 1, \ldots, M. \tag{26}$$

2.2. Description of the spectrum

Let us consider here the subspace $\mathcal{H}^{(M/2-N)}$ of the space of states $\mathcal{H}$ with a fixed number N of spins down. In this subspace, the eigenvectors $|\,\psi_\kappa(\{\lambda\})\,\rangle$ (respectively $\langle\,\psi_\kappa(\{\lambda\})\,|$) of the twisted transfer matrix $\mathcal{T}_\kappa(\mu)$ can be constructed in the form (26), where the parameters $\lambda_1, \ldots, \lambda_N$ satisfy the system of twisted Bethe equations

$$\mathcal{Y}_\kappa(\lambda_j|\{\lambda\}) = 0, \qquad j = 1, \ldots, N. \tag{27}$$

Here, the function $\mathcal{Y}_\kappa$ is defined as

$$\mathcal{Y}_\kappa(\mu|\{\lambda\}) = a(\mu) \prod_{k=1}^{N} \sinh(\lambda_k - \mu + \eta) + \kappa\, d(\mu) \prod_{k=1}^{N} \sinh(\lambda_k - \mu - \eta), \tag{28}$$

and $a(\lambda)$, $d(\lambda)$ are the eigenvalues of the operators $A(\lambda)$ and $D(\lambda)$ on the reference state $|\,0\,\rangle$. In the normalization (18) and for the inhomogeneous model (25), we have

$$a(\lambda) = \prod_{a=1}^{M} \sinh(\lambda - \xi_a + \eta), \qquad d(\lambda) = \prod_{a=1}^{M} \sinh(\lambda - \xi_a). \tag{29}$$

The corresponding eigenvalue of $\mathcal{T}_\kappa(\mu)$ on $|\,\psi_\kappa(\{\lambda\})\,\rangle$ (or on a dual eigenvector) is

$$\tau_\kappa(\mu|\{\lambda\}) = a(\mu) \prod_{k=1}^{N} \frac{\sinh(\lambda_k - \mu + \eta)}{\sinh(\lambda_k - \mu)} + \kappa\, d(\mu) \prod_{k=1}^{N} \frac{\sinh(\mu - \lambda_k + \eta)}{\sinh(\mu - \lambda_k)}. \tag{30}$$

The solutions of the system of twisted Bethe equations (27) have been analyzed in [88]. In general, not all of these solutions correspond to eigenvectors of $\mathcal{T}_\kappa(\mu)$.

Definition 2.1. *A solution $\{\lambda\}$ of the system* (27) *is called* admissible *if*

$$d(\lambda_j) \prod_{\substack{k=1 \\ k \neq j}}^{N} \sinh(\lambda_j - \lambda_k + \eta) \neq 0, \qquad j = 1, \ldots, N, \tag{31}$$

and un-admissible *otherwise. A solution is called* off-diagonal *if the corresponding parameters $\lambda_1, \ldots, \lambda_N$ are pairwise distinct, and* diagonal *otherwise.*

One of the main results of [88] is that, for generic parameters κ and $\{\xi\}$, the set of the eigenvectors corresponding to the admissible off-diagonal solutions of the system of twisted Bethe equations (27) forms a basis in the subspace $\mathcal{H}^{(M/2-N)}$. It has been proven in [80] that this result is still valid in the homogeneous case $\xi_j = \eta/2$, $j = 1, \ldots, N$, at least if κ is in a punctured vicinity of the origin (i.e., $0 < |\kappa| < \kappa_0$ for κ_0 small enough). Note however that, for specific values of κ and $\{\xi\}$, the basis of the eigenvectors in $\mathcal{H}^{(M/2-N)}$ may include some states corresponding to un-admissible solutions of (27) (in particular in the homogeneous limit at $\kappa = 1$).

At $\kappa = 1$, it follows from the trace identity (11) that the eigenvectors of the transfer matrix coincide, in the homogeneous limit, with the ones of the Hamiltonian (1). The corresponding eigenvalues in the case of zero magnetic field can be obtained from (11), (30):

$$H^{(0)} \, | \, \psi(\{\lambda\}) \rangle = \Big(\sum_{j=1}^{N} E(\lambda_j) \Big) \cdot | \, \psi(\{\lambda\}) \rangle, \tag{32}$$

where the (bare) one-particle energy $E(\lambda)$ is equal to

$$E(\lambda) = \frac{2 \sinh^2 \eta}{\sinh(\lambda + \frac{\eta}{2}) \sinh(\lambda - \frac{\eta}{2})}. \tag{33}$$

2.3. Drinfel'd twist and F-basis

As already noted, the operators A, B, C, D are highly nonlocal in terms of local spin operators. There exists however an interesting description of these operators by means of a change of basis of the space of states. In particular, this basis will provide a direct access to the scalar products of states. The root of this new basis is provided by the notion of Drinfel'd twist [35] associated to the R-matrix of the XXZ chain. It leads to the notion of factorizing F-matrices. To be essentially self-contained we briefly recall here their main properties and refer to [84] for more details and proofs.

Definition 2.2. *For inhomogeneity parameters ξ_j in generic positions and for any integer n one can associate to any element σ of the symmetric group S_n of order n a unique R-matrix $R^{\sigma}_{1 \ldots n}(\xi_1, \ldots, \xi_n)$, denoted for simplicity $R^{\sigma}_{1 \ldots n}$, constructed as an ordered product (depending on σ) of the elementary R-matrices $R_{ij}(\xi_i, \xi_j)$.*

We have the following property for arbitrary integer n:

Proposition 2.1.

$$R^{\sigma}_{1...n} \, T_{1...n}(\lambda; \xi_1, \ldots, \xi_n) = T_{\sigma(1)...\sigma(n)}(\lambda; \xi_{\sigma(1)}, \ldots, \xi_{\sigma(n)}) \, R^{\sigma}_{1...n}. \qquad (34)$$

We can now define the notion of factorizing F-matrix:

Definition 2.3. *A factorizing F-matrix associated to a given elementary R matrix is an invertible matrix $F_{1...n}(\xi_1, \ldots, \xi_n)$, defined for arbitrary integer n, satisfying the following relation for any element σ of S_n:*

$$F_{\sigma(1)...\sigma(n)}(\xi_{\sigma(1)}, \ldots, \xi_{\sigma(n)}) \, R^{\sigma}_{1...n}(\xi_1, \ldots, \xi_n) = F_{1...n}(\xi_1, \ldots, \xi_n). \qquad (35)$$

In other words, such an F-matrix factorizes the corresponding R-matrix for arbitrary integer n. Taking into account the fact that the parameters ξ_n are in one-to-one correspondence with the vector spaces $\mathcal{H}_n$, we can adopt simplified notations such that

$$F_{1...n}(\xi_1, \ldots, \xi_n) = F_{1...n},$$

$$F_{\sigma(1)...\sigma(n)}(\xi_{\sigma(1)}, \ldots, \xi_{\sigma(n)}) = F_{\sigma(1)...\sigma(n)}.$$

Theorem 2.1 ([84]). *For the XXZ model with inhomogeneity parameters ξ_n in generic positions, there exist a factorizing, triangular F-matrix. It is constructed explicitly from the R-matrix.*

It has two important properties:

Proposition 2.2 ([84]). *In the F-basis, the monodromy matrix $\widetilde{T}$*

$$\widetilde{T}_{1...M}(\lambda; \xi_1, \ldots, \xi_M) = F_{1...M} T_{1...M}(\lambda; \xi_1, \ldots, \xi_M) \, F^{-1}_{1...M} \qquad (36)$$

is totally symmetric under any simultaneous permutations of the lattice sites i and of the corresponding inhomogeneity parameters ξ_i.

The second property gives the explicit expressions of the monodromy matrix in the F-basis. For the XXZ-$\frac{1}{2}$ model, the quantum monodromy operator is a 2×2 matrix with entries A, B, C, D which are obtained as sums of 2^{M-1} operators which themselves are products of M local spin operators on the quantum chain. As an example, the B operator is given as

$$B_{1...M}(\lambda) = \sum_{i=1}^{N} \sigma_i^- \, \Omega_i + \sum_{i \neq j \neq k} \sigma_i^- \, (\sigma_j^- \, \sigma_k^+) \, \Omega_{ijk} + \text{higher terms}, \qquad (37)$$

where the matrices Ω_i, Ω_{ijk}, are diagonal operators acting respectively on all sites but i, on all sites but i, j, k, and the higher order terms involve more and more exchange spin terms like $\sigma_j^- \, \sigma_k^+$. It means that the B operator returns one spin somewhere on the chain, this operation being however dressed non-locally and with non-diagonal operators by multiple exchange terms of the type $\sigma_j^- \, \sigma_k^+$.

So, whereas these formulas in the original basis are quite involved, their expressions in the F-basis simplify drastically:

Proposition 2.3 ([84]). *The operators D, B and C in the F-basis are given by the formulas*

$$\widetilde{D}_{1...M}(\lambda;\xi_1,\ldots,\xi_M) = \overset{M}{\underset{i=1}{\otimes}} \begin{pmatrix} b(\lambda,\xi_i) & 0 \\ 0 & 1 \end{pmatrix}_{[i]}, \tag{38}$$

$$\widetilde{B}_{1...M}(\lambda) = \sum_{i=1}^{M} \sigma_i^- \, c(\lambda,\xi_i) \underset{j\neq i}{\otimes} \begin{pmatrix} b(\lambda,\xi_j) & 0 \\ 0 & b^{-1}(\xi_j,\xi_i) \end{pmatrix}_{[j]}, \tag{39}$$

$$\widetilde{C}_{1...M}(\lambda) = \sum_{i=1}^{M} \sigma_i^+ \, c(\lambda,\xi_i) \underset{j\neq i}{\otimes} \begin{pmatrix} b(\lambda,\xi_j)\, b^{-1}(\xi_i,\xi_j) & 0 \\ 0 & 1 \end{pmatrix}_{[j]}, \tag{40}$$

and the operator $\widetilde{A}$ can be obtained from quantum determinant relations.

We wish first to stress that while the operators $\widetilde{A}$, $\widetilde{B}$, $\widetilde{C}$, $\widetilde{D}$ satisfy the same quadratic commutation relations as A, B, C, D, they are completely symmetric under simultaneous exchange of the inhomogeneity parameters of the spaces $\mathcal{H}_n$. It really means that the factorizing F-matrices we have constructed solve the combinatorial problem induced by the nontrivial action of the permutation group S_M given by the R-matrix. In the F-basis the action of the permutation group on the operators $\widetilde{A}$, $\widetilde{B}$, $\widetilde{C}$, $\widetilde{D}$ is trivial.

Further, it can be shown that the pseudo-vacuum vector is left invariant, namely, it is an eigenvector of the total F-matrix with eigenvalue 1; in particular, the algebraic Bethe ansatz can be carried out also in the F-basis. Hence, a direct computation of Bethe eigenvectors and of their scalar products in this F-basis is made possible, while it was a priori very involved in the original basis. There, only commutation relations between the operators A, B, C, D can be used, leading (see [61]) to very intricate sums over partitions.

2.4. Solution of the quantum inverse problem

The very simple expressions of the monodromy matrix operators entries D, B, C in the F-basis suggests that any local operator $E_j^{\epsilon'_j,\epsilon_j}$, acting in a local quantum space $\mathcal{H}_j$ at site j, can be expressed in terms of the entries of the monodromy matrix. This is the so-called quantum inverse scattering problem. The solution to this problem was found in [38, 74]:

Theorem 2.2.

$$E_j^{\epsilon'_j,\epsilon_j} = \prod_{\alpha=1}^{j-1} \mathcal{T}(\xi_\alpha) \cdot T_{\epsilon_j,\epsilon'_j}(\xi_j) \cdot \prod_{\alpha=1}^{j} \mathcal{T}^{-1}(\xi_\alpha). \tag{41}$$

The proof of this theorem is elementary (see [38, 74]) and hence it can be obtained for a large class of lattice integrable models. It relies essentially on the property that the R-matrix $R(\lambda,\mu)$ reduces to the permutation operator for $\lambda = \mu$.

An immediate consequence of this theorem is that the operators A, B, C, and D generate the space of all operators acting in $\mathcal{H}$.

2.5. Scalar products

We give here the expressions for the scalar product of an eigenvector of the twisted transfer matrix with any arbitrary state of the form (26). These scalar products can be expressed as determinant of rather simple matrices. The root of all these determinants is in fact the determinant representation for the partition function of the 6-vertex model with domain wall boundary conditions [89]. Let us first define, for arbitrary positive integers n, n' ($n \le n'$) and arbitrary sets of variables $\lambda_1, \ldots, \lambda_n$, $\mu_1, \ldots, \mu_n$ and $\nu_1, \ldots, \nu_{n'}$ such that $\{\lambda\} \subset \{\nu\}$, the $n \times n$ matrix $\Omega_\kappa(\{\lambda\}, \{\mu\}|\{\nu\})$ as

$$(\Omega_\kappa)_{jk}(\{\lambda\}, \{\mu\}|\{\nu\}) = a(\mu_k)\, t(\lambda_j, \mu_k) \prod_{a=1}^{n'} \sinh(\nu_a - \mu_k + \eta)$$

$$- \kappa\, d(\mu_k)\, t(\mu_k, \lambda_j) \prod_{a=1}^{n'} \sinh(\nu_a - \mu_k - \eta), \quad (42)$$

with

$$t(\lambda, \mu) = \frac{\sinh \eta}{\sinh(\lambda - \mu)\sinh(\lambda - \mu + \eta)}. \quad (43)$$

Proposition 2.4. [38,63,83] *Let* $\{\lambda_1, \ldots, \lambda_N\}$ *be a solution of the system of twisted Bethe equations (27), and* $\mu_1, \ldots, \mu_N$ *be generic complex numbers. Then,*

$$\langle 0| \prod_{j=1}^{N} C(\mu_j)\, |\psi_\kappa(\{\lambda\})\rangle = \langle \psi_\kappa(\{\lambda\})|\prod_{j=1}^{N} B(\mu_j)|0\rangle$$

$$= \frac{\prod\limits_{a=1}^{N} d(\lambda_a) \prod\limits_{a,b=1}^{N} \sinh(\mu_b - \lambda_a)}{\prod\limits_{a>b}^{N} \sinh(\lambda_a - \lambda_b)\sinh(\mu_b - \mu_a)} \cdot \det_N\left(\frac{\partial}{\partial \lambda_j}\tau_\kappa(\mu_k|\{\lambda\})\right) \quad (44)$$

$$= \frac{\prod\limits_{a=1}^{N} d(\lambda_a)}{\prod\limits_{a>b}^{N} \sinh(\lambda_a - \lambda_b)\sinh(\mu_b - \mu_a)} \cdot \det_N \Omega_\kappa(\{\lambda\}, \{\mu\}|\{\lambda\}). \quad (45)$$

These equations are valid for any arbitrary complex parameter κ, in particular at $\kappa = 1$. In this case we may omit the subscript κ and denote $(\psi, \tau, \mathcal{Y}, \Omega) = (\psi_\kappa, \tau_\kappa, \mathcal{Y}_\kappa, \Omega_\kappa)|_{\kappa=1}$.

If the sets $\{\lambda\}$ and $\{\mu\}$ are different, the eigenvector $|\psi_\kappa(\{\lambda\})\rangle$ is orthogonal to the dual eigenvector $\langle \psi_\kappa(\{\mu\})|$. Otherwise we obtain a formula for the norm of the corresponding vector [38, 82, 90],

$$\langle \psi_\kappa(\{\lambda\}) \,|\, \psi_\kappa(\{\lambda\}) \rangle = \frac{\prod\limits_{a=1}^{N} d(\lambda_a)}{\prod\limits_{\substack{a,b=1 \\ a\neq b}}^{N} \sinh(\lambda_a - \lambda_b)} \cdot \det_N \Omega_\kappa(\{\lambda\}, \{\lambda\}|\{\lambda\})$$

$$= (-1)^N \frac{\prod\limits_{a=1}^{N} d(\lambda_a)}{\prod\limits_{\substack{a,b=1 \\ a\neq b}}^{N} \sinh(\lambda_a - \lambda_b)} \cdot \det_N \left(\frac{\partial}{\partial \lambda_k} \mathcal{Y}_\kappa(\lambda_j|\{\lambda\}) \right).$$

2.6. Action of operators A, B, C, D on a general state

An important step of the computation of correlation function is to express the action of any product of local operators on any Bethe eigenvector. From the solution of the quantum inverse scattering problem, this is given by the successive action of A, B, C, D operators on a vector constructed by action of C operators on the reference vector. Action of A, B, C, D on such a vector are well known (see for example [61]). They can be written in the following form:

$$\langle 0 | \prod_{k=1}^{N} C(\lambda_k)\, A(\lambda_{N+1}) = \sum_{a'=1}^{N+1} a(\lambda_{a'}) \frac{\prod\limits_{k=1}^{N} \sinh(\lambda_k - \lambda_{a'} + \eta)}{\prod\limits_{\substack{k=1 \\ k\neq a'}}^{N+1} \sinh(\lambda_k - \lambda_{a'})} \langle 0 | \prod_{\substack{k=1 \\ k\neq a'}}^{N+1} C(\lambda_k); \quad (46)$$

$$\langle 0 | \prod_{k=1}^{N} C(\lambda_k)\, D(\lambda_{N+1}) = \sum_{a=1}^{N+1} d(\lambda_a) \frac{\prod\limits_{k=1}^{N} \sinh(\lambda_a - \lambda_k + \eta)}{\prod\limits_{\substack{k=1 \\ k\neq a}}^{N+1} \sinh(\lambda_a - \lambda_k)} \langle 0 | \prod_{\substack{k=1 \\ k\neq a}}^{N+1} C(\lambda_k). \quad (47)$$

The action of the operator $B(\lambda)$ can be obtained similarly,

$$\langle 0 | \prod_{k=1}^{N} C(\lambda_k)\, B(\lambda_{N+1}) = \sum_{a=1}^{N+1} d(\lambda_a) \frac{\prod\limits_{k=1}^{N} \sinh(\lambda_a - \lambda_k + \eta)}{\prod\limits_{\substack{k=1 \\ k\neq a}}^{N+1} \sinh(\lambda_a - \lambda_k)}$$

$$\times \sum_{\substack{a'=1 \\ a'\neq a}}^{N+1} \frac{a(\lambda_{a'})}{\sinh(\lambda_{N+1} - \lambda_{a'} + \eta)} \frac{\prod\limits_{\substack{j=1 \\ j\neq a}}^{N+1} \sinh(\lambda_j - \lambda_{a'} + \eta)}{\prod\limits_{\substack{j=1 \\ j\neq a,a'}}^{N+1} \sinh(\lambda_j - \lambda_{a'})} \langle 0 | \prod_{\substack{k=1 \\ k\neq a,a'}}^{N+1} C(\lambda_k), \quad (48)$$

and the action of C is obvious.

3. Correlation functions: Finite chain

To compute correlation functions of some product of local operators, the following successive problems have to be addressed:

(i) determination of the ground state $\langle\,\psi_g\,|$,
(ii) evaluation of the action of the product of the local operators on it, and
(iii) computation of the scalar product of the resulting state with $|\,\psi_g\,\rangle$.

Using the solution of the quantum inverse scattering problem together with the explicit determinant formulas for the scalar products and the norm of the Bethe state, one sees that matrix elements of local spin operators and correlation functions can be expressed as (multiple) sums of determinants [73]. It should be stressed that this result is purely algebraic and is valid for finite chains of arbitrary length M.

3.1. Matrix elements of local operators

We begin with the calculation of the one-point functions. These results follow directly from the solution of the quantum inverse scattering problem, the above action of operators A, B, C and D, and the determinant representation of the scalar products. We consider

$$F_N^-(m, \{\mu_j\}, \{\lambda_k\}) = \langle\,0\,|\ \prod_{j=1}^{N+1} C(\mu_j)\ \sigma_m^-\ \prod_{k=1}^{N} B(\lambda_k)\,|\,0\,\rangle \tag{49}$$

and

$$F_N^+(m, \{\lambda_k\}, \{\mu_j\}) = \langle\,0\,|\ \prod_{k=1}^{N} C(\lambda_k)\ \sigma_m^+\ \prod_{j=1}^{N+1} B(\mu_j)\,|\,0\,\rangle, \tag{50}$$

where $\{\lambda_k\}_n$ and $\{\mu_j\}_{n+1}$ are solutions of Bethe equations.

Proposition 3.1. *For two Bethe states with spectral parameters $\{\lambda_k\}_N$ and $\{\mu_j\}_{N+1}$, the matrix element of the operator σ_m^- can be represented as a determinant,*

$$F_N^-(m, \{\mu_j\}, \{\lambda_k\}) = \frac{\phi_{m-1}(\{\mu_j\})}{\phi_{m-1}(\{\lambda_k\})} \frac{\displaystyle\prod_{j=1}^{N+1} \sinh(\mu_j - \xi_m + \eta)}{\displaystyle\prod_{k=1}^{N} \sinh(\lambda_k - \xi_m + \eta)}$$

$$\cdot \frac{\det_{N+1} H^-(m, \{\mu_j\}, \{\lambda_k\})}{\displaystyle\prod_{N+1\geq k>j\geq 1} \sinh(\mu_k - \mu_j) \prod_{1\leq\beta<\alpha\leq N} \sinh(\lambda_\beta - \lambda_\alpha)}, \tag{51}$$

$$\phi_m(\{\lambda_k\}) = \prod_{k=1}^{N} \prod_{j=1}^{m} b^{-1}(\lambda_k, \xi_j), \tag{52}$$

and the $(N+1) \times (N+1)$ matrix H^- is defined as

$$H^-_{ab}(m) = \frac{\varphi(\eta)}{\varphi(\mu_a - \lambda_b)} \left(a(\lambda_b) \prod_{\substack{j=1 \\ j \neq a}}^{N+1} \varphi(\mu_j - \lambda_b + \eta) - d(\lambda_b) \prod_{\substack{j=1 \\ j \neq a}}^{N+1} \varphi(\mu_j - \lambda_b - \eta) \right)$$
for $b < N+1,$ $\qquad\qquad\qquad\qquad\qquad\qquad\qquad\qquad\qquad\qquad$ (53)

$$H^-_{aN+1}(m) = \frac{\varphi(\eta)}{\varphi(\mu_a - \xi_m + \eta)\varphi(\mu_a - \xi_m)}. \qquad\qquad (54)$$

For the matrix element $F_N^+(m, \{\lambda_k\}, \{\mu_j\})$ we get

$$F_N^+(m, \{\lambda_k\}, \{\mu_j\}) = \frac{\phi_m(\lambda_k)\,\phi_{m-1}(\lambda_k)}{\phi_{m-1}(\mu_j)\,\phi_m(\mu_j)} F_N^-(m, \{\mu_j\}, \{\lambda_k\}). \qquad (55)$$

The matrix elements of the operator σ_m^z between two Bethe states have been obtained similarly [38].

3.2. Elementary blocks of correlation functions

In this section we consider a more general case of correlation functions: the ground state mean value of any product of the local elementary 2×2 matrices $E_{lk}^{\epsilon',\epsilon} = \delta_{l,\epsilon'}\delta_{k,\epsilon}$:

$$F_m(\{\epsilon_j, \epsilon_j'\}) = \frac{\langle \psi_g | \prod_{j=1}^{m} E_j^{\epsilon_j',\epsilon_j} | \psi_g \rangle}{\langle \psi_g | \psi_g \rangle}. \qquad (56)$$

An arbitrary n-point correlation function can be obtained as a sum of such mean values. Using the solution of the quantum inverse scattering problem, we reduce this problem to the computation of the ground state mean value of an arbitrary ordered product of monodromy matrix elements,

$$F_m(\{\epsilon_j, \epsilon_j'\}) = \phi_m^{-1}(\{\lambda\}) \frac{\langle \psi_g | T_{\epsilon_1,\epsilon_1'}(\xi_1) \dots T_{\epsilon_m,\epsilon_m'}(\xi_m) | \psi_g \rangle}{\langle \psi_g | \psi_g \rangle}. \qquad (57)$$

To calculate these mean values we first describe generically the product of the monodromy matrix elements. For that purpose, one should consider the two following sets of indices, $\alpha^+ = \{j : 1 \leq j \leq m,\ \epsilon_j = 1\}$, $\mathrm{card}(\alpha^+) = s'$, $\max_{j \in \alpha^+}(j) \equiv j'_{\max}$, $\min_{j \in \alpha^+}(j) \equiv j'_{\min}$, and similarly $\alpha^- = \{j : 1 \leq j \leq m,\ \epsilon_j' = 2\}$, $\mathrm{card}(\alpha^-) = s$, $\max_{j \in \alpha^-}(j) \equiv j_{\max}$, $\min_{j \in \alpha^-}(j) \equiv j_{\min}$. The intersection of these two sets is not empty and corresponds to the operators $B(\xi_j)$. Consider now the action

$$\langle 0 | \prod_{k=1}^{N} C(\lambda_k) T_{\epsilon_1,\epsilon_1'}(\lambda_{N+1}) \dots T_{\epsilon_m,\epsilon_m'}(\lambda_{N+m}), \qquad (58)$$

applying one by one the formulae (46)–(48). For all the indices j from the sets α^+ and α^- one obtains a summation on the corresponding indices a_j' (for $j \in \alpha^+$, corresponding to the action of the operators $A(\lambda)$ or $B(\lambda)$) or a_j (for $j \in \alpha^-$, corresponding to the action of the operators $D(\lambda)$ or $B(\lambda)$). As the product of the monodromy matrix elements is ordered these summations are also ordered and the corresponding indices should be taken from the following sets, $\mathbf{A}_j = \{b : 1 \le b \le N + m, \ b \ne a_k, a'_k, \ k < j\}$ and $\mathbf{A}'_j = \{b : 1 \le b \le N + m, \ b \ne a'_k, \ k < j, \ b \ne a_k, \ k \le j\}$. Thus,

$$\langle 0 | \prod_{k=1}^{N} C(\lambda_k) \, T_{\epsilon_1, \epsilon'_1}(\lambda_{N+1}) \dots T_{\epsilon_m, \epsilon'_m}(\lambda_{N+m})$$

$$= \sum_{\{a_j, a'_j\}} G_{\{a_j, a'_j\}}(\lambda_1, \dots, \lambda_{N+m}) \langle 0 | \prod_{b \in \mathbf{A}_{m+1}} C(\lambda_b). \quad (59)$$

The summation is taken over the indices a_j for $j \in \alpha^-$ and a'_j for $j \in \alpha^+$ such that $1 \le a_j \le N + j$, $a_j \in \mathbf{A}_j$, $1 \le a'_j \le N + j$, $a'_j \in \mathbf{A}'_j$. The functions $G_{\{a_j, a'_j\}}(\lambda_1, \dots, \lambda_{N+m})$ can then be easily obtained from the formulae (46)–(48) taking into account that $\lambda_a = \xi_{a-N}$ for $a > N$:

$$G_{\{a_j, a'_j\}}(\lambda_1, \dots, \lambda_{N+m}) = \prod_{j \in \alpha^-} d(\lambda_{a_j}) \frac{\displaystyle\prod_{\substack{b=1 \\ b \in \mathbf{A}_j}}^{N+j-1} \sinh(\lambda_{a_j} - \lambda_b + \eta)}{\displaystyle\prod_{\substack{b=1 \\ b \in \mathbf{A}'_j}}^{N+j} \sinh(\lambda_{a_j} - \lambda_b)}$$

$$\times \prod_{j \in \alpha^+} a(\lambda_{a'_j}) \frac{\displaystyle\prod_{\substack{b=1 \\ b \in \mathbf{A}'_j}}^{N+j-1} \sinh(\lambda_b - \lambda_{a'_j} + \eta)}{\displaystyle\prod_{\substack{b=1 \\ b \in \mathbf{A}_{j+1}}}^{N+j} \sinh(\lambda_b - \lambda_{a'_j})}. \quad (60)$$

Now to calculate the normalized mean value (57) we apply the determinant representation for the scalar product. It should be mentioned that the number of operators $C(\lambda)$ has to be equal to the number of the operators $B(\lambda)$, as otherwise the mean value is zero, and hence the total number of elements in the sets α^+ and α^- is $s + s' = m$. Taking into account that in (57), for $b > N$, $\lambda_b = \xi_{b-N}$ one has to consider the following scalar products,

$$\frac{\langle 0 | \displaystyle\prod_{b \in \mathbf{A}_{m+1}} C(\lambda_b) \prod_{k=1}^{N} B(\lambda_k) | 0 \rangle}{\langle 0 | \displaystyle\prod_{k=1}^{N} C(\lambda_k) \prod_{k=1}^{N} B(\lambda_k) | 0 \rangle},$$

for all the permitted values of a_j, a'_j. Finally we obtain:

$$F_m(\{\epsilon_j, \epsilon'_j\}) = \frac{1}{\displaystyle\prod_{k<l} \sinh(\xi_k - \xi_l)} \sum_{\{a_j, a'_j\}} H_{\{a_j, a'_j\}}(\lambda_1, \dots, \lambda_{N+m}), \quad (61)$$

the sum being taken on the same set of indices a_j, a_j' as in (59). The functions $H_{\{a_j, a_j'\}}(\{\lambda\})$ can be obtained using (60) and the determinant representations for the scalar products.

3.3. Two-point functions

The method presented in the last section is quite straightforward and gives formally the possibility to compute any correlation function. However, it has been developed for the computation of the average values of monomials in the monodromy matrix operators entries, leading to the elementary building blocks, whereas the study of the two-point functions involves big sums of such blocks. Indeed, let us consider for example the correlation function $\langle \sigma_1^z \sigma_{m+1}^z \rangle$. Then, according to the solution of the inverse scattering problem (41), we need to calculate the expectation value

$$\langle \psi(\{\lambda\}) | (A - D)(\xi_1) \cdot \prod_{a=2}^{m} \mathcal{T}(\xi_a) \cdot (A - D)(\xi_{m+1}) \cdot \prod_{b=1}^{m+1} \mathcal{T}^{-1}(\xi_b) | \psi(\{\lambda\}) \rangle. \quad (62)$$

Since $| \psi(\{\lambda\}) \rangle$ is an eigenvector, the action of $\prod_{b=1}^{m+1} \mathcal{T}^{-1}(\xi_b)$ on this state merely produces a numerical factor. However, it is much more complicated to evaluate the action of $\prod_{a=2}^{m} \mathcal{T}(\xi_a)$. Indeed, we have to act first with $(A - D)(\xi_1)$ on $\langle \psi(\{\lambda\}) |$ (or with $(A - D)(\xi_{m+1})$ on $| \psi(\{\lambda\}) \rangle$), which gives a sum of states which are no longer eigenvectors of the transfer matrix, and on which the multiple action of $\mathcal{T}$ is not simple. In fact, the product $\prod_{a=2}^{m}(A + D)(\xi_a)$ would lead to a sum of 2^{m-1} elementary blocks. This is not very convenient, in particular at large distance m. Therefore, to obtain manageable expressions for such correlation functions, it is of great importance to develop an alternative and compact way to express the multiple action of the transfer matrix on arbitrary states or, in other words, to make an effective re-summation of the corresponding sum of the 2^{m-1} terms. This can be achieved in the following way:

Proposition 3.2. *Let κ, $x_1, \ldots, x_m$ and $\mu_1, \ldots, \mu_N$ be generic parameters. Then the action of $\prod_{a=1}^{m} \mathcal{T}_\kappa(x_a)$ on a state of the form $\langle 0 | \prod_{j=1}^{N} C(\mu_j)$ can be formally written as*

$$\langle 0 | \prod_{j=1}^{N} C(\mu_j) \prod_{a=1}^{m} \mathcal{T}_\kappa(x_a) = \frac{1}{N!} \oint_{\Gamma\{x\} \cup \Gamma\{\mu\}} \prod_{j=1}^{N} \frac{dz_j}{2\pi i} \cdot \prod_{a=1}^{m} \tau_\kappa(x_a | \{z\}) \cdot \prod_{a=1}^{N} \frac{1}{\mathcal{Y}_\kappa(z_a | \{z\})}$$

$$\times \prod_{\substack{j,k=1 \\ j<k}}^{N} \frac{\sinh(z_j - z_k)}{\sinh(\mu_j - \mu_k)} \cdot \det_N \Omega_\kappa(\{z\}, \{\mu\} | \{z\}) \cdot \langle 0 | \prod_{j=1}^{N} C(z_j), \quad (63)$$

where the integration contour $\Gamma\{x\} \cup \Gamma\{\mu\}$ surrounds the points[1] $x_1, \ldots, x_m$ and $\mu_1, \ldots, \mu_N$ and does not contain any other pole of the integrand.

---

[1] More precisely, for a set of complex variables $\{\nu_1, \ldots, \nu_l\}$, the notation $\Gamma\{\nu\}$ should be understood in the following way: $\Gamma\{\nu\}$ is the boundary of a set of poly-disks $\mathcal{D}_a(r)$ in $\mathbb{C}^N$, i.e., $\Gamma\{\nu\} = \cup_{a=1}^{l} \bar{\mathcal{D}}_a(r)$ with $\bar{\mathcal{D}}_a(r) = \{z \in \mathbb{C}^N : |z_k - \nu_a| = r, \quad k = 1, \ldots, N\}$.

One of the simplest applications concerns the generating function of the two-point correlation function of the third components of spin, which is defined as the normalized expectation value $\langle Q_{l,m}^{\kappa} \rangle$ of the operator

$$Q_{l,m}^{\kappa} = \prod_{n=l}^{m} \left(\frac{1+\kappa}{2} + \frac{1-\kappa}{2} \cdot \sigma_n^z \right) = \prod_{j=1}^{l-1} \mathcal{T}(\xi_j) \cdot \prod_{j=l}^{m} \mathcal{T}_\kappa(\xi_j) \cdot \prod_{j=1}^{m} \mathcal{T}^{-1}(\xi_j), \quad (64)$$

where $|\psi(\{\lambda\})\rangle$ is an eigenvector of $\mathcal{T}(\mu)$ in the subspace $\mathcal{H}^{(M/2-N)}$. The two-point correlation function of the third components of local spins in the eigenvector $|\psi(\{\lambda\})\rangle$ can be obtained in terms of the second 'lattice derivative' and the second derivative with respect to κ of the generating function $\langle Q_{l,m}^{\kappa} \rangle$ at $\kappa = 1$:

$$\langle \sigma_l^z \, \sigma_{l+m}^z \rangle = \langle \sigma_l^z \rangle + \langle \sigma_{l+m}^z \rangle - 1$$

$$+ 2\frac{\partial^2}{\partial\kappa^2} \langle Q_{l,l+m}^{\kappa} - Q_{l,l+m-1}^{\kappa} - Q_{l+1,l+m}^{\kappa} + Q_{l+1,l+m-1}^{\kappa} \rangle\Big|_{\kappa=1}. \quad (65)$$

Due to the translational invariance of the correlation functions in the homogeneous model, we will simply consider the expectation value $\langle Q_{1,m}^{\kappa} \rangle$. For any given eigenvector, we obtain the following result:

Theorem 3.1. *Let $\{\lambda\}$ be an admissible off-diagonal solution of the system of untwisted Bethe equations, and let us consider the corresponding expectation value $\langle Q_{1,m}^{\kappa} \rangle$ in the inhomogeneous finite XXZ chain. Then there exists $\kappa_0 > 0$ such that, for $|\kappa| < \kappa_0$, the following representations hold:*

$$\langle Q_{1,m}^{\kappa} \rangle = \frac{1}{N!} \oint_{\Gamma\{\xi\}\cup\Gamma\{\lambda\}} \prod_{j=1}^{N} \frac{dz_j}{2\pi i} \cdot \prod_{a=1}^{m} \frac{\mathcal{T}_\kappa(\xi_a|\{z\})}{\mathcal{T}(\xi_a|\{\lambda\})} \cdot \prod_{a=1}^{N} \frac{1}{\mathcal{Y}_\kappa(z_a|\{z\})}$$

$$\times \det_N \Omega_\kappa(\{z\}, \{\lambda\}|\{z\}) \cdot \frac{\det_N \Omega(\{\lambda\}, \{z\}|\{\lambda\})}{\det_N \Omega(\{\lambda\}, \{\lambda\}|\{\lambda\})}. \quad (66)$$

The integration contours are such that the only singularities of the integrand which contribute to the integral are the points $\xi_1, \ldots, \xi_m$ and $\lambda_1 \ldots, \lambda_N$.

From this result, we can extract a compact representation for the two-point function of σ^z [79]. Similar expressions exist for other correlation functions of the spin operators, and in particular for the time dependent case [63, 79]. Moreover, this multiple contour integral representation permits to relate two very different ways to compute two-point correlation functions of the type, $g_{12} = \langle \omega | \theta_1 \theta_2 | \omega \rangle$, namely,

(i) to compute the action of local operators on the ground state $\theta_1 \theta_2 | \omega \rangle = | \tilde{\omega} \rangle$ and then to calculate the resulting scalar product $g_{12} = \langle \omega | \tilde{\omega} \rangle$ as was explained in the previous sections.

(ii) to insert a sum over a complete set of states $| \omega_i \rangle$ (for instance, a complete set of eigenvectors of the Hamiltonian) between the local operators θ_1 and θ_2

and to obtain the representation for the correlation function as a sum over matrix elements of local operators,

$$g_{12} = \sum_i \langle \omega | \theta_1 | \omega_i \rangle \cdot \langle \omega_i | \theta_2 | \omega \rangle. \tag{67}$$

In fact the above representation as multiple contour integrals contains both expansions. Indeed there are two ways to evaluate the corresponding integrals: either to compute the residues in the poles inside Γ, or to compute the residues in the poles within strips of the width $i\pi$ outside Γ.

The first way leads to a representation of the correlation function $\langle \sigma_1^z \sigma_{m+1}^z \rangle$ in terms of the previously obtained [75] m-multiple sums. Evaluation of the above contour integral in terms of the poles outside the contour Γ gives us the expansion (ii) of the correlation function (i.e., an expansion in terms of matrix elements of σ^z between the ground state and all excited states). This relation holds also for the time dependent case [63, 79].

3.4. Towards the comparison with neutron scattering experiments

In this section, we first briefly review all elements necessary for the computation of the dynamical spin-spin correlation functions of the anisotropic Heisenberg model, following [39,40] and leading in particular to the successful comparison with neutron scattering experiments, see Fig. 1. We start by giving our notation and discussing the eigenstates in some details. The reference state is taken to be the state with all spins up, $|0\rangle = \otimes_{i=1}^{M} | \uparrow \rangle_i$. Since the total magnetization commutes with the Hamiltonian, the Hilbert space separates into subspaces of fixed magnetization, determined from the number of reversed spins N. We take the number of sites M to be even, and $2N \leq M$, the other sector being accessible through a change in the reference state.

Eigenstates in each subspace are completely characterized for $2N \leq M$ by a set of rapidities $\{\lambda_j\}$, $j = 1, \ldots, N$, solution to the Bethe equations

$$\left[\frac{\sinh(\lambda_j + i\zeta/2)}{\sinh(\lambda_j - i\zeta/2)} \right]^M = \prod_{k \neq j}^{N} \frac{\sinh(\lambda_j - \lambda_k + i\zeta)}{\sinh(\lambda_j - \lambda_k - i\zeta)}, \qquad j = 1, \ldots, N, \tag{68}$$

where $\Delta = \cos \zeta$. In view of the periodicity of the sinh function in the complex plane, we can restrict the possible values that the rapidities can take to the strip $-\pi/2 < \mathrm{Im}\lambda \leq \pi/2$, or alternately define an extended zone scheme in which λ and $\lambda + i\pi\mathbb{Z}$ are identified.

A more practical version of the Bethe equations is obtained by writing them in logarithmic form,

$$\mathrm{atan}\left[\frac{\tanh(\lambda_j)}{\tan(\zeta/2)} \right] - \frac{1}{M} \sum_{k=1}^{N} \mathrm{atan}\left[\frac{\tanh(\lambda_j - \lambda_k)}{\tan \zeta} \right] = \pi \frac{I_j}{M}. \tag{69}$$

Here, I_j are distinct half-integers which can be viewed as quantum numbers: each choice of a set $\{I_j\}$, $j = 1, \ldots, N$ (with I_j defined $\mathrm{mod}(M)$) uniquely specifies a set of rapidities, and therefore an eigenstate. The energy of a state is given as a function of the rapidities by

$$E = J \sum_{j=1}^{N} \frac{-\sin^2 \zeta}{\cosh 2\lambda_j - \cos \zeta} - h\left(\frac{M}{2} - N\right), \tag{70}$$

whereas the momentum has a simple representation in terms of the quantum numbers,

$$q = \sum_{j=1}^{N} i \ln\left[\frac{\sinh(\lambda_j + i\zeta/2)}{\sinh(\lambda_j - i\zeta/2)}\right] = \pi N + \frac{2\pi}{M} \sum_{j=1}^{N} I_j \quad \mod 2\pi. \tag{71}$$

The ground state is given by $I_j^0 = -\frac{N+1}{2} + j$, $j = 1, \ldots, N$, and all excited states are in principle obtained from the different choices of sets $\{I_j\}$.

To study dynamics, some ingredients have to be added to the Bethe Ansatz: the matrix elements of spin operators between eigenstates (form factors). In terms of form factors for the Fourier-transformed spin operators $S_q^a = \frac{1}{\sqrt{M}} \sum_{j=1}^{M} e^{iqj} S_j^a$, the structure factor can be written as a sum

$$S^{a\bar{a}}(q, \omega) = 2\pi \sum_{\alpha \neq GS} |\langle GS | S_q^a | \alpha \rangle|^2 \delta(\omega - \omega_\alpha) \tag{72}$$

over the whole set of intermediate eigenstates $|\alpha\rangle$ (distinct from the ground state $|GS\rangle$) in a fixed magnetization subspace. Each term in (72) can be obtained [38] as a product of determinants of specific matrices, which are fully determined for given *bra* and *ket* eigenstates by a knowledge of the corresponding sets of rapidities. The analytical summation of this series remains for the moment out of reach, but numerically, for chains of length a few hundred sites, quite feasible. Moreover, we know that the correlation functions of the finite chain approach their thermodynamic limit with errors of order $\frac{1}{M}$, hence if $M = 200$ for example the error is usually quite acceptable to make comparison with experiments.

The strategy to follow is now clear. We compute the S^{zz} and S^{-+} structure factors by directly summing the terms on the right-hand side of equation (72) over a judiciously chosen subset of eigenstates. The momentum delta functions are broadened to width $\epsilon \sim 1/M$ using $\delta_\epsilon(x) = \frac{1}{\sqrt{\pi}\epsilon} e^{-x^2/\epsilon^2}$ in order to obtain smooth curves. We scan through the eigenstates in the following order. First, we observe that the form factors of the spin operators between the ground state and an eigenstate $\{\lambda\}$ are extremely rapidly decreasing functions of the number of holes that need to be inserted in the configuration of the lowest-energy state (in the same base) in order to obtain the configuration $\{I\}$ corresponding to $\{\lambda\}$. We therefore scan through all bases and configurations for increasing number of holes, starting from one-hole states for S^{zz}, and zero-hole states for S^{-+}. Although the number of possible configurations for fixed base and number of holes is a rapidly increasing function of the number of holes, we find that the total contributions for fixed bases also rapidly decrease for increasing hole numbers. We therefore limit ourselves to states with up to three holes, corresponding to up to six-particle excitations. We can quantify the quality of the present computational method by evaluating the summation rules for the longitudinal and transverse form factors. Namely, by integrating over momentum and frequency, we should saturate the values

$$\int_{-\infty}^{\infty} \frac{d\omega}{2\pi} \frac{1}{M} \sum_q S^{zz}(q,\omega) = \frac{1}{4} - \langle S^z \rangle^2 = \frac{1}{4}\left[1 - (1 - \frac{2N}{M})^2\right], \qquad (73)$$

$$\int_{-\infty}^{\infty} \frac{d\omega}{2\pi} \frac{1}{M} \sum_q S^{-+}(q,\omega) = \frac{1}{2} - \langle S^z \rangle = \frac{N}{M}. \qquad (74)$$

In Fig. 2, we plot the longitudinal structure factor as a function of momentum and frequency for anisotropy $\Delta = 0.75$, for four values of the magnetization. Fig. 3 contains the transverse structure factor for the same anisotropy and magnetizations.

For all intermediate states involving strings, we explicitly check that the deviations from the string hypothesis are small. We find in general that states involving strings of length higher than two are admissible solutions to the Bethe equations for high enough magnetizations. At zero field, only two-string states have exponentially small deviations δ, and all higher-string states must be discarded.

The relative contributions to the structure factors from different bases is very much dependent on the system size, the anisotropy, and the magnetization. In general, we find that two- and four-particle contributions are sufficient to saturate well over 90% of the summation rules in all cases, for system sizes up to $M = 200$. Interestingly, however, we find that string states also contribute noticeably in many cases. For example, in Fig. 4, we plot the zero-field transverse structure factor contributions coming from intermediate states with one string of length two and up to three holes. Around six or seven percent of the weight is accounted for by these states, and similar or somewhat lower figures are found in other cases. Strings of length higher than two do not contribute significantly. For example, we find only around $5.7 \cdot 10^{-8}\%$ of the summation rule from states with one string of length three, for the longitudinal structure factor for $\Delta = 0.25$ at $M = N/4$ with $N = 128$. For $\Delta = 0.75$, we find $6.3 \cdot 10^{-7}\%$. For the transverse correlators, the figures are $2.3 \cdot 10^{-12}\%$ and $3.1 \cdot 10^{-12}\%$. Even though these numbers would increase if we could go to larger system sizes, we do not expect them to ever become numerically significant.

The imperfect saturation of the summation rules that we obtain in general can be ascribed either to higher states in the hierarchy which are not included in our partial summations, or states that are in principle included, but which are rejected in view of their deviations from the string hypothesis. As the proportion of excluded string states can be rather large (ranging anywhere from zero to fifty percent), we believe the latter explanation to be the correct one. In any case, these results are precise enough to be compared successfully to different data from neutron scattering experiments for several magnetic compounds. From our results covering the whole Brillouin zone and frequency space, it is straightforward to

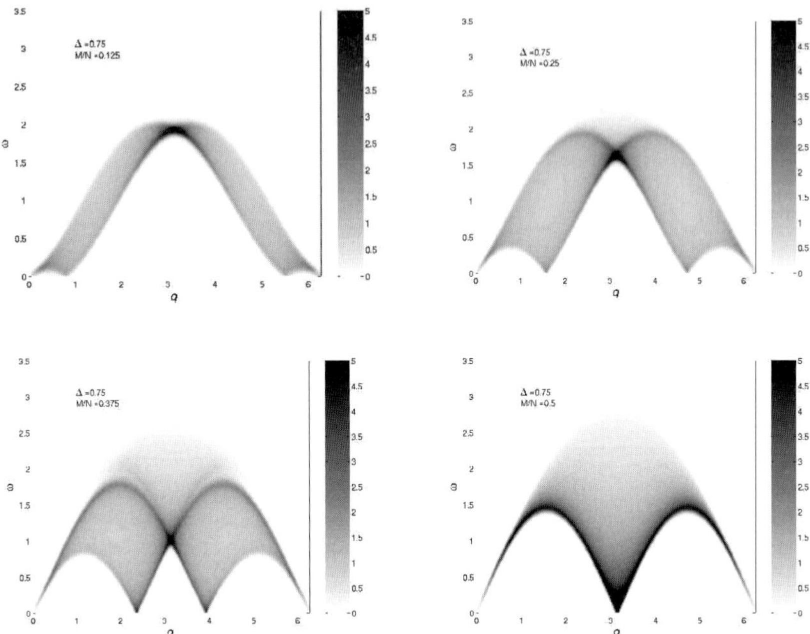

FIGURE 2. Longitudinal structure factor as a function of momentum q and frequency ω, for $\Delta = 0.75$, and $N = M/8, M/4, 3M/8$, and $M/2$. Here, $M = 200$ and all contributions up to two holes are taken into account. The summation rule is thereby saturated to 98.6%, 97.0%, 95.4% and 97.8%.

obtain space-time dependent correlation functions by inverse Fourier transform:

$$\langle S^a_{j+1}(t) S^{\bar{a}}_1(0)\rangle_c = \frac{1}{M} \sum_{\alpha \neq GS} |\langle GS| S^a_{q_\alpha} |\alpha\rangle|^2 e^{-iq_\alpha j - i\omega_\alpha t}. \tag{75}$$

It is possible to compare these results to known exact results for equal-time correlators at short distance, and to the large-distance asymptotic form obtained from conformal field theory. This comparison can only be made at zero field, where both sets of results are known exactly. The comparison turns out to be extremely good, as can be expected from the high saturation of the summation rules [40].

4. Correlation functions: Infinite chain

In the thermodynamic limit, $M \to \infty$ and at zero magnetic field, the model exhibits three different regimes depending on the value of Δ [26]. For $\Delta < -1$, the

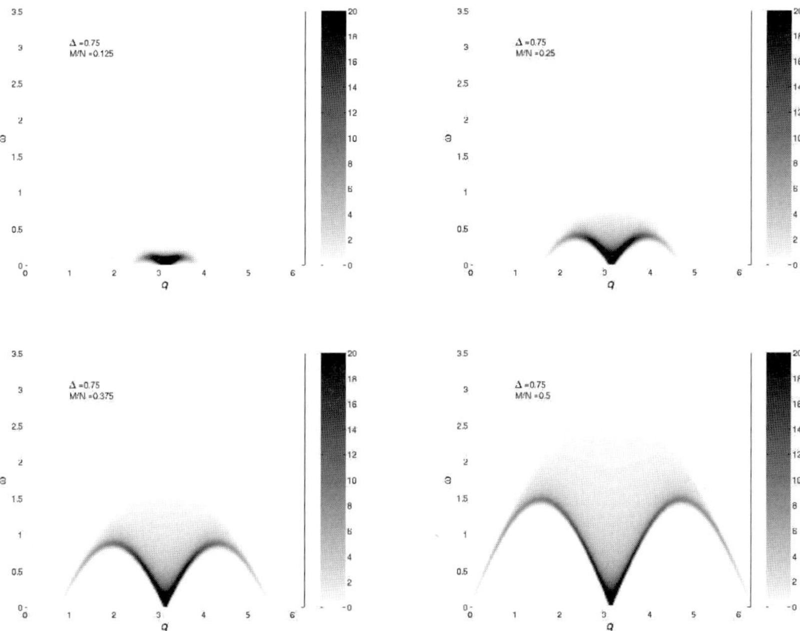

FIGURE 3. Transverse structure factor as a function of momentum q and frequency ω, for $\Delta = 0.75$, and $N = M/8, M/4, 3M/8$, and $M/2$. Here, $M = 200$ and all contributions up to two holes are taken into account. The summation rule is thereby saturated to 99.3%, 97.8%, 96.5% and 98.8%.

model is ferromagnetic, for $-1 < \Delta < 1$, the model has a non-degenerated antiferromagnetic ground state, and no gap in the spectrum (massless regime), while for $\Delta > 1$, the ground state is twice degenerated with a gap in the spectrum (massive regime). In both cases, the ground state has spin zero. Hence the number of parameters λ in the ground state vectors is equal to half the size M of the chain. For $M \to \infty$, these parameters will be distributed in some continuous interval according to a density function ρ.

4.1. The thermodynamic limit

In this limit, the Bethe equations for the ground state, written in their logarithmic form, become a linear integral equation for the density distribution of these λ's,

$$\rho_{\text{tot}}(\alpha) + \int_{-\Lambda}^{\Lambda} K(\alpha - \beta)\rho_{\text{tot}}(\beta)\, d\beta = \frac{p'_{0_{\text{tot}}}(\alpha)}{2\pi}, \tag{76}$$

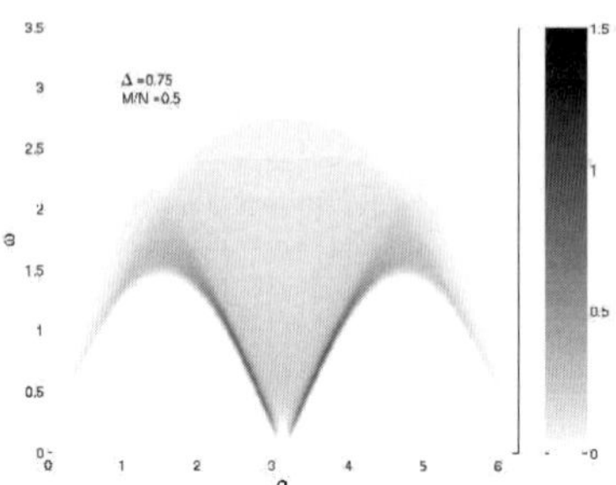

FIGURE 4. The two-string contributions to the transverse structure factor at zero magnetic field, as a function of momentum q and frequency ω, and for anisotropy 0.75. The density scale has been enhanced as compared to that used in the previous figures. Here, $M = 200$ and contributions up to three holes are taken into account. The summation rule contributions from these states is 6.3%.

where the new real variables α are defined in terms of general spectral parameters λ differently in the two domains. From now on, we only describe the massless regime (see [73] for the other case) $-1 < \Delta < 1$ where $\alpha = \lambda$. The density ρ is defined as the limit of the quantity $\frac{1}{M(\lambda_{j+1}-\lambda_j)}$, and the functions $K(\lambda)$ and $p'_{0_{\text{tot}}}(\lambda)$ are the derivatives with respect to λ of the functions $-\frac{\theta(\lambda)}{2\pi}$ and $p_{0_{\text{tot}}}(\lambda)$:

$$K(\alpha) = \frac{\sin 2\zeta}{2\pi \sinh(\alpha + i\zeta)\sinh(\alpha - i\zeta)}$$
$$p'_0(\alpha) = \frac{\sin \zeta}{\sinh(\alpha + i\frac{\zeta}{2})\sinh(\alpha - i\frac{\zeta}{2})} \qquad \text{for } -1 < \Delta < 1, \text{ with } \zeta = i\eta, \quad (77)$$

$$\text{with} \quad p'_{0_{\text{tot}}}(\alpha) = \frac{1}{M}\sum_{i=1}^{M} p'_0(\alpha - \beta_k - i\frac{\zeta}{2}), \tag{78}$$

where $\beta_k = \xi_k$. The integration limit Λ is equal to $+\infty$ for $-1 < \Delta < 1$. The solution for the equation (76) in the homogeneous model where all parameters ξ_k are equal to $\eta/2$, that is the density for the ground state of the Hamiltonian in the thermodynamic limit, is given by the following function [19]:

$$\rho(\alpha) = \frac{1}{2\zeta \cosh(\frac{\pi\alpha}{\zeta})}.$$

For technical convenience, we will also use the function

$$\rho_{\text{tot}}(\alpha) = \frac{1}{M}\sum_{i=1}^{M} \rho(\alpha - \beta_k - i\frac{\zeta}{2}).$$

It will be also convenient to consider, without any loss of generality, that the inhomogeneity parameters are contained in the region $-\zeta < \mathrm{Im}\beta_j < 0$. Using these results, for any $\mathcal{C}^\infty$ function f (π-periodic in the domain $\Delta > 1$), sums over all the values of f at the point α_j, $1 \leq j \leq N$, parameterizing the ground state, can be replaced in the thermodynamic limit by an integral:

$$\frac{1}{M}\sum_{j=1}^{N} f(\alpha_j) = \int_{-\Lambda}^{\Lambda} f(\alpha)\rho_{\mathrm{tot}}(\alpha)\,d\alpha + O(M^{-1}).$$

Thus, multiple sums obtained in correlation functions will become multiple integrals. Similarly, it is possible to evaluate the behavior of the determinant formulas for the scalar products and the norm of Bethe vectors (and in particular their ratios) in the limit $M \to \infty$.

4.2. Elementary blocks

From the representations as multiple sums of these elementary blocks in the finite chain we can obtain their multiple integral representations in the thermodynamic limit. Let us now consider separately the two regimes of the XXZ model. In the massless regime $\eta = -i\zeta$ is imaginary, the ground state parameters λ are real and the limit of integration is infinity $\Lambda = \infty$. In this case we consider the inhomogeneity parameters ξ_j such that $0 > \mathrm{Im}(\xi_j) > -\zeta$. For the correlation functions in the thermodynamic limit one obtains the following result in this regime:

Proposition 4.1.

$$F_m(\{\epsilon_j, \epsilon'_j\}) = \prod_{k<l} \frac{\sinh\frac{\pi}{\zeta}(\xi_k - \xi_l)}{\sinh(\xi_k - \xi_l)} \prod_{j=1}^{s'} \int_{-\infty-i\zeta}^{\infty-i\zeta} \frac{d\lambda_j}{2i\zeta} \prod_{j=s'+1}^{m} \int_{-\infty}^{\infty} i\frac{d\lambda_j}{2\zeta}$$

$$\times \prod_{a=1}^{m}\prod_{k=1}^{m} \frac{1}{\sinh\frac{\pi}{\zeta}(\lambda_a - \xi_k)} \prod_{j\in\alpha^-} \left(\prod_{k=1}^{j-1} \sinh(\mu_j - \xi_k - i\zeta) \prod_{k=j+1}^{m} \sinh(\mu_j - \xi_k) \right)$$

$$\times \prod_{j\in\alpha^+} \left(\prod_{k=1}^{j-1} \sinh(\mu'_j - \xi_k + i\zeta) \prod_{k=j+1}^{m} \sinh(\mu'_j - \xi_k) \right) \prod_{a>b} \frac{\sinh\frac{\pi}{\zeta}(\lambda_a - \lambda_b)}{\sinh(\lambda_a - \lambda_b - i\zeta)},$$

where the parameters of integration are ordered in the following way: $\{\lambda_1, \ldots \lambda_m\} = \{\mu'_{j'_{\max}}, \ldots, \mu'_{j'_{\min}}, \mu_{j_{\min}}, \ldots, \mu_{j_{\max}}\}$.

The homogeneous limit ($\xi_j = -i\zeta/2, \forall j$) of the correlation function $F_m(\{\epsilon_j, \epsilon'_j\})$ can then be taken in an obvious way. We have obtained similar representations for the massive regime, and also in the presence of a nonzero magnetic field [73]. For zero magnetic field, these results agree exactly with the ones obtained by Jimbo and Miwa in [69], using spin particular q-KZ equations. It means that for zero magnetic field, the elementary blocks of correlation functions indeed satisfy q-KZ equations. Recently, more algebraic representations of solutions of the q-KZ equations have been obtained that correspond to the above correlation functions [70, 71]. From the finite chain representation for the two-point function, it is also possible to obtain multiple integral representations for that case as well, in particular for their generating function [75, 76]. They correspond to different huge re-summations and symmetrization of the corresponding elementary blocks, as in the finite chain situation [75]. Moreover, the case of time dependent correlation functions has also been obtained [63, 79]. Finally, let us note that at the free fermion point, all the results presented here lead, in a very elementary way, to already known results [63, 76, 80].

5. Exact and asymptotic results

5.1. Exact results at $\Delta = 1/2$

Up to now, two exact results have been obtained for the case of anisotropy $\Delta = 1/2$: the exact value of the emptiness formation probability for arbitrary distance m [77] and the two-point function of the third component of spin [85]. These two results follow from the above multiple integral representations for which, due to the determinant structure of the integrand, the corresponding multiple integrals can be separated and hence explicitly computed for this special value of the anisotropy.

5.1.1. The emptiness formation probability. This correlation function $\tau(m)$ (the probability to find in the ground state a ferromagnetic string of length m) is defined as the expectation value

$$\tau(m) = \langle \psi_g | \prod_{k=1}^{m} \frac{1 - \sigma_k^z}{2} | \psi_g \rangle \tag{79}$$

where $|\psi_g\rangle$ denotes the normalized ground state. In the thermodynamic limit $(M \to \infty)$, this quantity can be expressed as a multiple integral with m integrations [37, 38, 68, 69, 73].

Proposition 5.1. *For* $\Delta = \cos\zeta$, $0 < \zeta < \pi$, $\tau(m) = \lim_{\xi_1, \dots \xi_m \to -\frac{i\zeta}{2}} \tau(m, \{\xi_j\})$, *where*

$$\tau(m, \{\xi_j\}) = \frac{1}{m!} \int_{-\infty}^{\infty} \frac{Z_m(\{\lambda\}, \{\xi\})}{\prod\limits_{a<b}^{m} \sinh(\xi_a - \xi_b)} \det_m \left(\frac{i}{2\zeta \sinh\frac{\pi}{\zeta}(\lambda_j - \xi_k)} \right) d^m\lambda, \tag{80}$$

$$Z_m(\{\lambda\}, \{\xi\}) = \prod_{a=1}^{m} \prod_{b=1}^{m} \frac{\sinh(\lambda_a - \xi_b)\sinh(\lambda_a - \xi_b - i\zeta)}{\sinh(\lambda_a - \lambda_b - i\zeta)}$$

$$\times \frac{\det_m \left(\frac{-i \sin\zeta}{\sinh(\lambda_j - \xi_k)\sinh(\lambda_j - \xi_k - i\zeta)} \right)}{\prod\limits_{a>b}^{m} \sinh(\xi_a - \xi_b)}. \tag{81}$$

The proof is given in [75]. Due to the determinant structure of the integrand, the integrals can be separated and computed for the special case $\Delta = \frac{1}{2}$ ($\zeta = \pi/3$):

Proposition 5.2. *Let $\xi_k = \varepsilon_k - i\pi/6$ and $\varepsilon_{ab} = \varepsilon_a - \varepsilon_b$. Then*

$$\tau(m, \{\varepsilon_j\}) = \frac{(-1)^{\frac{m^2-m}{2}}}{2^{m^2}} \prod_{a>b}^{m} \frac{\sinh 3\varepsilon_{ba}}{\sinh \varepsilon_{ba}} \prod_{\substack{a,b=1 \\ a \neq b}}^{m} \frac{1}{\sinh \varepsilon_{ab}} \cdot \det{}_m \left(\frac{3 \sinh \frac{\varepsilon_{jk}}{2}}{\sinh \frac{3\varepsilon_{jk}}{2}} \right), \quad (82)$$

$$\tau(m) = \left(\frac{1}{2} \right)^{m^2} \prod_{k=0}^{m-1} \frac{(3k+1)!}{(m+k)!}. \quad (83)$$

Observe that the quantity $A_m = \prod_{k=0}^{m-1}(3k+1)!/(m+k)!$ is the number of alternating sign matrices of size m. This result was conjectured in [91].

5.1.2. The two-point function of σ^z. The two-point functions can be obtained, as in the finite chain situation, from a generating function $\langle Q_\kappa(m) \rangle$; in the thermo-dynamic limit, we use the following multiple integral representation [79]:

$$\langle Q_\kappa(m) \rangle = \sum_{n=0}^{m} \frac{\kappa^{m-n}}{n!(m-n)!} \oint_{\Gamma\{-i\zeta/2\}} \frac{d^m z}{(2\pi i)^m} \int_{\mathbb{R}-i\zeta} d^n\lambda \int_{\mathbb{R}} d^{m-n}\lambda \cdot \prod_{j=1}^{m} \frac{\varphi^m(z_j)}{\varphi^m(\lambda_j)}$$

$$\times \prod_{j=1}^{n} \left\{ t(z_j, \lambda_j) \prod_{k=1}^{m} \frac{\sinh(z_j - \lambda_k - i\zeta)}{\sinh(z_j - z_k - i\zeta)} \right\} \prod_{j=n+1}^{m} \left\{ t(\lambda_j, z_j) \prod_{k=1}^{m} \frac{\sinh(\lambda_k - z_j - i\zeta)}{\sinh(z_k - z_j - i\zeta)} \right\}$$

$$\times \prod_{j=1}^{m} \prod_{k=1}^{m} \frac{\sinh(\lambda_k - z_j - i\zeta)}{\sinh(\lambda_k - \lambda_j - i\zeta)} \cdot \det{}_m \left(\frac{i}{2\zeta \sinh \frac{\pi}{\zeta}(\lambda - z)} \right). \quad (84)$$

Here,

$$\Delta = \cos\zeta, \quad t(z, \lambda) = \frac{-i\sin\zeta}{\sinh(z - \lambda)\sinh(z - \lambda - i\zeta)}, \quad \varphi(z) = \frac{\sinh(z - i\frac{\zeta}{2})}{\sinh(z + i\frac{\zeta}{2})}, \quad (85)$$

and the integrals over the variables z_j are taken with respect to a closed contour Γ which surrounds the point $-i\zeta/2$ and does not contain any other singularities of the integrand. The equation (84) is valid for the homogeneous XXZ chain with arbitrary $-1 < \Delta < 1$. If we consider the inhomogeneous XXZ model with inhomogeneities $\xi_1, \ldots, \xi_m$, then one should replace in the representation (84) the function φ^m in the following way:

$$\varphi^m(z) \to \prod_{b=1}^{m} \frac{\sinh(z - \xi_b - i\zeta)}{\sinh(z - \xi_b)}, \qquad \varphi^{-m}(\lambda) \to \prod_{b=1}^{m} \frac{\sinh(\lambda - \xi_b)}{\sinh(\lambda - \xi_b - i\zeta)}. \quad (86)$$

In order to come back to the homogeneous case, one should set $\xi_k = -i\zeta/2$, $k = 1, \ldots, m$ in (86). In the inhomogeneous model, the integration contour Γ surrounds the points $\xi_1, \ldots, \xi_m$, and the integrals over z_j are therefore equal to the sum of the residues of the integrand in these simple poles. It turns out that again for the special case $\Delta = \frac{1}{2}$ integrals can be separated and computed to give [85]:

192   J.-M. Maillet

Proposition 5.3.

$$\langle Q_\kappa(m)\rangle = \frac{3^m}{2^{m^2}} \prod_{a>b}^{m} \frac{\sinh 3(\xi_a - \xi_b)}{\sinh^3(\xi_a - \xi_b)} \sum_{n=0}^{m} \kappa^{m-n} \sum_{\substack{\{\xi\}=\{\xi_{\gamma_+}\}\cup\{\xi_{\gamma_-}\} \\ |\gamma_+|=n}} \det_m \hat{\Phi}^{(n)}$$

$$\times \prod_{a\in\gamma_+} \prod_{b\in\gamma_-} \frac{\sinh(\xi_b - \xi_a - \frac{i\pi}{3})\sinh(\xi_a - \xi_b)}{\sinh^2(\xi_b - \xi_a + \frac{i\pi}{3})},$$

$$\hat{\Phi}^{(n)}(\{\xi_{\gamma_+}\}, \{\xi_{\gamma_-}\}) = \left(\begin{array}{c|c} \Phi(\xi_j - \xi_k) & \Phi(\xi_j - \xi_k - \frac{i\pi}{3}) \\ \hline \Phi(\xi_j - \xi_k + \frac{i\pi}{3}) & \Phi(\xi_j - \xi_k) \end{array} \right), \quad \Phi(x) = \frac{\sinh \frac{x}{2}}{\sinh \frac{3x}{2}}.$$

Here the sum is taken with respect to all partitions of the set $\{\xi\}$ into two disjoint subsets $\{\xi_{\gamma_+}\} \cup \{\xi_{\gamma_-}\}$ of cardinality n and $m - n$ respectively. The first n lines and columns of the matrix $\hat{\Phi}^{(n)}$ are associated with the parameters $\xi \in \{\xi_{\gamma_+}\}$. The remaining lines and columns are associated with $\xi \in \{\xi_{\gamma_-}\}$.

Thus, we have obtained an explicit answer for the generating function $\langle Q_\kappa(m)\rangle$ of the inhomogeneous XXZ model. It is also possible to check that the above sum over partitions remains indeed finite in the homogeneous limit $\xi_k \to 0$. Finally, for small distances it is possible to compute the above expressions explicitly as polynomial functions of the variable κ of degree m. Interestingly, it turns out that all coefficients are integer numbers divided by 2^{m^2} [85], meaning a possible combinatorial interpretation of these numbers as for the emptiness formation probability computed in the previous section.

5.2. Asymptotic results

An important issue is the analysis of the multiple integral representations of correlation functions for large distances. There it means analyzing asymptotic behavior of m-fold integrals for m large. An interesting example to study in this respect is provided by the emptiness formation probability. This correlation function reduces to a single elementary block. Moreover, we already described its exact value for an anisotropy $\Delta = \frac{1}{2}$ in the previous section. In fact, it is possible to obtain the asymptotic behavior of $\tau(m)$ using the saddle-point method for arbitrary values of the anisotropy $\Delta > -1$. This was performed for the first time in [76] in the case of free fermions ($\Delta = 0$), but it can be applied to the general case as well. We present here the results in the massless and massive regimes [63, 78].

To apply the saddle-point method to the emptiness formation probability, it is convenient to express its integral representation in the following form:

$$\tau(m) = \int_{\mathcal{D}} d^m\lambda \, G_m(\{\lambda\}) \, e^{m^2 S_m(\{\lambda\})}, \tag{87}$$

with

$$S_m(\{\lambda\}) = - \frac{1}{m^2} \sum_{a>b}^{m} \log[\sinh(\lambda_a - \lambda_b + \eta) \sinh(\lambda_a - \lambda_b - \eta)]$$

$$+ \frac{1}{m} \sum_{a=1}^{m} \log[\sinh(\lambda_a + \eta/2) \sinh(\lambda_a - \eta/2)]$$

$$+ \frac{1}{m^2} \lim_{\xi_1 \ldots \xi_m \to \eta/2} \log \left[\left(\frac{-2i\pi}{\sinh \eta} \right)^m \frac{\left(\det \rho(\lambda_j, \xi_k) \right)^2}{\prod_{a \neq b} \sinh(\xi_a - \xi_b)} \right] \tag{88}$$

and

$$G_m(\{\lambda\}) = \lim_{\xi_1 \ldots \xi_m \to \eta/2} \frac{\det_m \left[\frac{i}{2\pi} t(\lambda_j, \xi_k) \right]}{\det_m \rho(\lambda_j, \xi_k)}. \tag{89}$$

In (87), the integration domain $\mathcal{D}$ is such that the variable of integration $\lambda_1, \ldots, \lambda_m$ are ordered in the interval $\mathcal{C} = [-\Lambda_h, \Lambda_h]$ (i.e., $-\Lambda_h < \lambda_1 < \cdots < \lambda_m < \Lambda_h$ in the massless case, and $-i\Lambda_h < i\lambda_1 < \cdots < i\lambda_m < i\Lambda_h$ in the massive case).

The main problem in the saddle point analysis is that, a priori, we do not know any asymptotic equivalent of the quantity $G_m(\lambda)$ when $m \to \infty$. Nevertheless, in the case of zero magnetic field, it is still possible to compute the asymptotic behavior of (87) in the leading order, provided we make the following hypothesis: we assume that the integrand of (87) admits a maximum for a certain value $\lambda'_1, \ldots, \lambda'_m$ of the integration variables $\lambda_1, \ldots, \lambda_m$, that, for large m, the distribution of these parameters $\lambda'_1, \ldots, \lambda'_m$ can be described by a density function $\rho_s(\lambda')$ of the form

$$\rho_s(\lambda'_j) = \lim_{m \to \infty} \frac{1}{m(\lambda'_{j+1} - \lambda'_j)}, \tag{90}$$

on the symmetric interval $[-\Lambda, \Lambda]$ and that, at the leading order in m, we can replace the sums over the set of parameters $\{\lambda'\}$ by integrals weighted with the density $\rho_s(\lambda')$.

First, it is easy to determine the maximum of the function $S_m(\{\lambda\})$. Indeed, let $\{\tilde{\lambda}\}$ be solution of the system

$$\partial_{\lambda_j} S_m(\{\tilde{\lambda}\}) = 0, \qquad 1 \leq j \leq m. \tag{91}$$

In the limit $m \to \infty$, if we suppose again that the parameters $\tilde{\lambda}_1, \ldots, \tilde{\lambda}_m$ become distributed according to a certain density $\tilde{\rho}_s(\lambda)$ and that sums over the $\tilde{\lambda}_j$ become integrals over this density, the system (91) turns again into a single integral equation for $\tilde{\rho}_s$, that can be solved explicitly in the case of zero magnetic field. It gives the maximum of $S_m(\{\lambda\})$ when $m \to \infty$.[2]

The second step is to show that the factor $G_m(\{\lambda\})$ gives always a negligible contribution compared to $S_m(\{\tilde{\lambda}\})$ at this order in m, at least for any distribution

---

[2] At this main order in m, there exists a unique solution of the integral equation for $\tilde{\rho}_s$, and we know it corresponds to a maximum because $S_m(\{\lambda\}) \to -\infty$ on the boundary of $\mathcal{D}$.

of the variables λ_j satisfying the previous hypothesis of regularity. We obtain

$$\lim_{m \to \infty} \frac{1}{m^2} \log G_m(\{\lambda\}) = 0 \tag{92}$$

for any distribution of $\{\lambda\}$ with good properties of regularity, in particular for the saddle point. This means that, at the main order in m, the factor $G_m(\{\lambda\})$ does not contribute to the value of the maximum of the integrand.

Finally we obtain the following result concerning the asymptotic behaviour of $\tau(m)$ for $m \to \infty$ (see [63, 78]):

$$S^{(0)}(\Delta) = \lim_{m \to \infty} \frac{\log \tau(m)}{m^2}, \tag{93}$$

$$= -\frac{\zeta}{2} - \sum_{n=1}^{\infty} \frac{e^{-n\zeta}}{n} \frac{\sinh(n\zeta)}{\cosh(2n\zeta)}, \qquad (\Delta = \cosh \zeta > 1), \tag{94}$$

$$= \log \frac{\pi}{\zeta} + \frac{1}{2} \int_{\mathbb{R}-i0} \frac{d\omega}{\omega} \frac{\sinh \frac{\omega}{2}(\pi - \zeta) \cosh^2 \frac{\omega\zeta}{2}}{\sinh \frac{\pi\omega}{2} \sinh \frac{\omega\zeta}{2} \cosh \omega\zeta}, \qquad (-1 < \Delta = \cos \zeta < 1). \tag{95}$$

It coincides with the exact known results obtained in [76, 92] at the free fermion point and in [77, 91] at $\Delta = 1/2$, and is in agreement with the expected (infinite) value in the Ising limit. Similar techniques can be applied to the two-point function. However, the result that has been extracted so far is only the absence of the gaussian term. Unfortunately, we do not know up to now how to extract the expected power law corrections to the gaussian behavior from this saddle point analysis. More powerful methods will certainly be needed to go further.

5.3. Asymptotic behavior of the two-point functions

The long-distance asymptotic behavior of physical correlation functions, such as the two-point functions, have attracted long-standing interest. In the case of the XXZ model, some predictions were made already a long time ago. These predictions are confirmed by the numerical summation over the exact form factors that we performed for the XXZ model in the disordered regime [40].

In the massive regime ($\Delta > 1$), spin-spin correlation functions are expected to decay exponentially with the distance and the exact value of the correlation length was proposed in [94]. For the XXZ chain in the massless regime ($-1 < \Delta \leq 1$), zero temperature is a critical point and the correlation length becomes infinite in units of the lattice spacing. The leading long-distance effects can be predicted by conformal field theory and the correlation functions are expected to decay as a power of the distance. In particular, one expects that, at the leading order,

$$\langle \sigma_j^x \, \sigma_{j+n}^x \rangle = (-1)^n \frac{A}{n^{\pi-\zeta}} + \cdots, \tag{96}$$

$$\langle \sigma_j^z \, \sigma_{j+n}^z \rangle = -\frac{1}{\pi(\pi - \zeta)} \frac{1}{n^2} + (-1)^n \frac{A_z}{n^{\frac{\pi}{\pi-\zeta}}} + \cdots. \tag{97}$$

A conjecture for the non-universal correlation amplitudes A and A_z can be found in [95–97]. The exact value of the critical exponents in (96)–(97) was proposed for the first time in [98].

However, at the moment there does not exist any direct derivation of these predictions from the exact expressions of the correlation functions on the lattice. In the last subsection we have shown how to determine, at least in the main order, the asymptotic behavior of the emptiness formation probability using the saddle-point method. We could expect to be able to apply the same technique to the new multiple integral representation of the two-point function.

In particular, one can notice immediately that each term of the representation of the generating functional $\langle Q_{1,m}^{\kappa} \rangle$ has a structure very similar to the one for the emptiness formation probability. Indeed, it is possible to apply to the whole sum a slight modification of the saddle-point technique presented here. It shows that, as it should be, there is no contribution of order $\exp(\alpha m^2)$ when $m \to \infty$.

However, to obtain the precise asymptotic behavior of the two-point function, one should be able to analyze sub-leading corrections to this saddle-point method, which is technically quite difficult. It is not obvious in particular from these expressions that, in the massless regime, the leading asymptotic behavior of the two-point function is only of power-law order.

It is also quite interesting (and relevant experimentally) to consider other lattice models such as spin chains with magnetic or non-magnetic impurities [99–101] or models with electrons (carrying both spin and charge) like the Hubbard model and to compute in particular their transport properties.

6. Conclusion and perspectives

In this article, we have reviewed recent results concerning the computation of correlation functions in the XXZ chain by the methods of the inverse scattering problem and the algebraic Bethe ansatz. In conclusion, we would like to discuss some perspectives and problems to be solved.

One of the most interesting open problems is to prove the conformal field theory predictions [93, 98] concerning the asymptotic behavior of the correlation functions. This is certainly a very important issue not only for physical applications but also from a theoretical viewpoint. Moreover, it also would open the route towards the generalization of the methods presented here to quantum integrable models of field theory. We have seen that in particular cases, the multiple integral representations enable for a preliminary asymptotic analysis. Nevertheless, this problem remains one of the main challenges in the topics that have been described in this article.

A possible way to solve this problem would be to find the thermodynamic limit of the master equations (like the one obtained for the two-point correlation functions). It is natural to expect that, in this limit, one should obtain a representation for these correlation functions in terms of a functional integral, which could eventually be estimated for large time and distance.

Note that the master equation shows a direct analytic relation between the multiple integral representations and the form factor expansions for the correlation functions. It seems likely that similar representations exist for other models solvable by algebraic Bethe ansatz. It would be in particular very interesting to obtain an analogue of this master equation in the case of the field theory models, which could provide an analytic link between short distance and long distance expansions of their correlation functions. Other models of interest include models with magnetic [99] or non-magnetic impurities, meaning different integrable boundary conditions [100,101], and also the Hubbard model the transport properties of which have high experimental interest, see, e.g., [102] and references therein.

Acknowledgments

It is a great pleasure to thank N. Kitanine, N.A. Slavnov and V. Terras for their longstanding collaboration on the difficult but exiting topics presented here, and J.-S. Caux for his energy and skills that make our collaboration successful in comparing Bethe ansatz results to actual neutron scattering experiments.

References

[1] F. Bloch, *On the magnetic scattering of neutrons*, Phys. Rev. **50** (1936), 259–260.

[2] J.S. Schwinger, *On the magnetic scattering of neutrons*, Phys. Rev. **51** (1937), 544–552.

[3] O. Halpern and M.H. Johnson, *On the magnetic scattering of neutrons*, Phys. Rev. **55** (1938), 898–923.

[4] L. Van Hove, *Correlations in space and time and Born approximation scattering in systems of interacting particles*, Phys. Rev. **95** (1954), 249–262.

[5] L. Van Hove, *Time dependent correlations between spins and neutron scattering in ferromagnetic crystals*, Phys. Rev. **95** (1954), 1374–1384.

[6] W. Marshall and S.W. Lovesey, *Theory of thermal neutron scattering*, Academic Press, Oxford, 1971.

[7] R. Balescu, *Equilibrium and nonequilibrium statistical mechanics*, J. Wiley, New York, 1975.

[8] M. Steiner, J. Villain, and C.G. Windsor, *Theoretical and experimental studies on one-dimensional magnetic systems*, Adv. Phys. **25** (1976), 87–209.

[9] S.K. Satija, J.D. Axe, G. Shirane, H. Yoshizawa, and K. Hirakawa, *Neutron scattering study of spin waves in one-dimensional antiferromagnet* $KCuF_3$, Phys. Rev. **B21** (1980), 2001–2007.

[10] S.E. Nagler, D.A. Tennant, R.A. Cowley, T.G. Perring, and S.K. Satija, *Spin dynamics in the quantum antiferromagnetic chain compound* $KCuF_3$, Phys. Rev. **B44** (1991), 12361–12368.

[11] D.A. Tennant, R.A. Cowley, S.E. Nagler, and A.M. Tsvelik, *Measurement of the spin-excitation continuum in one-dimensional* $KCuF_3$ *using neutron scattering*, Phys. Rev. **B52** (1995), 13368–13380.

[12] D.A. Tennant, S.E. Nagler D. Welz, G. Shirane, and K. Yamada, *Effects of coupling between chains on the magnetic excitation spectrum of* $KCuF_3$, Phys. Rev. **B52** (1995), 13381–13389.

[13] H.A. Jahn and E. Teller, *Stability of polyatomic molecules in degenerate electronic states. I. Orbital degeneracy*, Proc. Roy. Soc. London Ser. **A161** (1937), 220–235.

[14] W. Heisenberg, *Zur Theorie der Ferromagnetismus*, Z. Phys. **49** (1928), 619.

[15] H. Bethe, *Zur Theorie der Metalle I. Eigenwerte und Eigenfunktionen Atomkete*, Z. Phys. **71** (1931), 205.

[16] L. Hulthen, *Über das Austauschproblem eines Kristalls*, Arkiv. Mat. Astron. Fys. **26** A(11) (1938), 1–106.

[17] R. Orbach, *Linear antiferromagnetic chain with anisotropic coupling*, Phys. Rev. **112** (1958), 309.

[18] L.R. Walker, *Antiferromagnetic linear chain*, Phys. Rev. **116** (1959), 1089.

[19] C.N. Yang and C.P. Yang, *One-dimensional chain of anisotropic spin-spin interactions. I. Proof of Bethe's hypothesis for ground state in a finite system*, Phys. Rev. **150** (1966), 321.

[20] C.N. Yang and C.P. Yang, *One-dimensional chain of anisotropic spin-spin interactions. II. Properties of the ground state energy per lattice site for an infinite system*, Phys. Rev. **150** (1966), 327.

[21] C.-N. Yang, *Some exact results for the many body problems in one dimension with repulsive delta function interaction*, Phys. Rev. Lett. **19** (1967), 1312–1314.

[22] E.H. Lieb and D.C. Mattis, *Mathematical Physics in One Dimension* (Academic Press, New-York, 1966).

[23] L.D. Faddeev, E.K. Sklyanin, and L.A. Takhtajan, *Quantum inverse problem method I*, Theor. Math. Phys. **40** (1980), 688.

[24] L.A. Takhtajan and L.D. Faddeev, *The Quantum method of the inverse problem and the Heisenberg XYZ model*, Russ. Math. Surveys **34** (1979), 11.

[25] H.B. Thacker, *Exact integrability in quantum field theory and statistical systems*, Rev. Mod. Phys. **53** (1981), 253.

[26] R. Baxter, *Exactly solved models in statistical mechanics*, Academic Press, London (1982).

[27] M. Gaudin, *La fonction d'onde de Bethe*, Masson, 1983.

[28] P.P. Kulish, N.Yu. Reshetikhin, and E.K. Sklyanin, *Yang-Baxter equation and representation theory I*, Lett. Math. Phys. **5** (1981), 393–403.

[29] P.P. Kulish and N.Yu. Reshetikhin, *Quantum linear problem for the sine-Gordon equation and higher representations*, Zap. Nauch. Sem. LOMI **101** (1981), 101–110, Translation in J. Sov. Math. **23** (1983), 2435–41.

[30] P.P. Kulish and N.Yu. Reshetikhin, Gl_3-*invariant solutions of the Yang-Baxter equation and associated quantum systems*, J. Sov. Math. **34** (1986), 1948–1971, translated from Zap. Nauch. Sem. LOMI **120** (1982), 92–121.

[31] P.P. Kulish and E.K. Sklyanin, *Quantum spectral transform method. Recent developments*, Lectures Notes in Physics **151** (1982), 61–119.

[32] M. Jimbo, *A q-difference analogue of $U(g)$ and the Yang-Baxter equation*, Lett. Math. Phys. **10** (1985), 63–69.

[33] M. Jimbo, *A q-analogue of $U(gl(N+1))$, Hecke algebra, and the Yang-Baxter equation*, Lett. Math. Phys. **11** (1986), 247–252.

[34] V.G. Drinfeld, *Hopf algebras and the quantum Yang-Baxter equation*, Soviet Math. Dokl. **32** (1985) (1), 254–258.

[35] V.G. Drinfel'd, *Quantum groups*, In the proceedings of the ICM 1986, Berkeley, USA, 1986, AMS, (1987).

[36] O. Foda, M. Jimbo, T. Miwa, K. Miki, and A. Nakayashiki. *Vertex operators in solvable lattice models*, J. Math. Phys. **35** (1994), 1346.

[37] M. Jimbo and T. Miwa, *Algebraic analysis of solvable lattice models*, AMS, Providence, RI (1995).

[38] N. Kitanine, J.M. Maillet, and V. Terras, *Form Factors of the XXZ Heisenberg spin-1/2 finite chain*, Nucl. Phys. **B554** (1999), 647–678.

[39] J.-S. Caux and J.M. Maillet, *Computation of dynamical correlation functions of Heisenberg chains in a field*, Phys. Rev. Lett. **95** (2005), 077201.

[40] J.-S. Caux, R. Hagemans, and J.M. Maillet, *Computation of dynamical correlation functions of Heisenberg chains: the gapless anysotropic regime*, J. Stat. Mech., P09003 (2005).

[41] N. Andrei, K. Furuya, and J. Lowenstein, *Solution of the Kondo problem*, Rev. Mod. Phys. **55** (1983), 331–402.

[42] A.M. Tvselik and P.B. Wiegmann, *Exact results in the theory of magnetic alloys*, Adv. Phys. **32** (1983), 453–713.

[43] I. Affleck, in "Fields, strings and critical phenomena", Les Houches Lectures 1988, North-Holland, (1990).

[44] L.D. Faddeev and G.P. Korchemsky, *High energy QCD as a completely integrable model*, Phys. Lett. **B342** (1995), 311–322.

[45] J.A. Minahan and K. Zarembo, *The Bethe-ansatz for $N = 4$ super Yang-Mills*, JHEP, 013 (2003).

[46] G. Arutyunov, S. Frolov, and M. Staudacher, *Bethe ansatz for quantum strings*, JHEP, 016 (2004).

[47] W. Lenz, *Beitrag zum Verständnis der magnetischen Erscheinungen in festen Körpern*, Zeit. Phys. **21** (1920), 613–615.

[48] E. Ising, *Beitrag zur Theorie des Ferromagnetismus*, Zeit. Phys. **31** (1925), 253–258.

[49] H.A. Kramers and G.H. Wannier, *Statistics of the Two-Dimensional Ferromagnet Part I*, Phys. Rev. **60** (1941), 252–262.

[50] H.A. Kramers and G.H. Wannier, *Statistics of the Two-Dimensional Ferromagnet Part II*, Phys. Rev. **60** (1941), 263–277.

[51] L. Onsager, *Crystal statistics I: A two-dimensional model with an order-disorder transition*, Phys. Rev. **65** (1944), 117–149.

[52] B. Kaufman, *Crystal Statistics. II. Partition Function Evaluated by Spinor Analysis*, Phys. Rev. **76** (1949), 1232–1243.

[53] C.N. Yang, *The Spontaneous Magnetization of a Two-Dimensional Ising Model*, Phys. Rev. **85** (1952), 808–816.

[54] L.D. Faddeev and L. Takhtajan, *Hamiltonian methods in the theory of solitons*, Springer Verlag (1986).

[55] O. Babelon, D. Bernard and M. Talon *Introduction to classical integrable systems*, Cambridge Univ. Press, (2003).

[56] E. Lieb, T. Schultz, and D. Mattis, *Two soluble models of an antiferromagnetic chain*, Ann. Phys. **16** (1961), 407–466.

[57] B.M. McCoy, *Spin correlation functions of the XY model*, Phys. Rev. **173** (1968), 531–541.

[58] T.T. Wu, B.M. McCoy, C.A. Tracy, and E. Barouch, *Spin-spin correlation functions of the two-dimensional Ising model: exact theory in the scaling region*, Phys. Rev. **B13** (1976), 316–374.

[59] B.M. McCoy, C.A. Tracy, and T.T. Wu, *Two-Dimensional Ising Model as an Exactly Solvable Relativistic Quantum Field Theory: Explicit Formulas for n-Point Functions*, Phys. Rev. Lett. **38** (1977), 793–796.

[60] M. Sato, T. Miwa, and M. Jimbo, *Holonomic quantum fields*, IV Publ. Res. Inst. Math. Sci. IV Publ. Res. Inst. Math. Sci. **16** (1980), 531–584, **15** (1979), 871–972, **15** (1979), 577–629, **15** (1979), 201–278, **14** (1978), 223–267.

[61] V.E. Korepin, N. Bogoliugov, and A.G. Izergin, *Quantum inverse scattering method and correlation functions*, Cambridge University Press, Cambridge (1993).

[62] H. Boos, M. Jimbo, T. Miwa, F. Smirnov, and Y. Takeyama, *Algebraic representation of correlation functions in integrable spin chains*, hep-th/0601132.

[63] N. Kitanine, J.M. Maillet, N.A. Slavnov, and V. Terras, *On the algebraic Bethe ansatz approach to the correlation functions of the XXZ spin-1/2 Heisenberg chain*, in "Solvable Lattice Models 2004 – Recent Progress on Solvable Lattice Models", RIMS, Kyoto, Kokyuroku, No.1480 (2006) 14, hep-th/0505006.

[64] L.D. Faddeev, *How Algebraic Bethe Ansatz works for integrable model*, Les Houches School of Physics: Relativistic gravitation and gravitational radiation (1995), 149–219.

[65] A.G. Izergin and V.E. Korepin, *The quantum inverse scattering method approach to correlation functions*, Commun. Math. Phys. **94** (1984), 67–92.

[66] A.G. Izergin and V.E. Korepin, *Correlation functions for the Heisenberg XXZ-antiferromagnet*, Commun. Math. Phys. **99** (1985), 271–302.

[67] F.H.L. Eßler, H. Frahm, A.G. Izergin, and V.E. Korepin, *Determinant representation for correlation functions of spin 1/2 XXX and XXZ Heisenberg magnets*, Commun. Math. Phys. **174** (1995), 191–214.

[68] M. Jimbo, K. Miki, T. Miwa, and A. Nakayashiki, *Correlation functions of the XXZ model for $\Delta < -1$*, Phys. Lett. **A168** (1992), 256–263.

[69] M. Jimbo and T. Miwa, *Quantum KZ equation with $|q| = 1$ and correlation functions of the XXZ model in the gapless regime*, J. Phys. **A29** (1996), 2923–2958.

[70] H. Boos, M. Jimbo, T. Miwa, F. Smirnov, and Y. Takeyama, *A recursion formula for the correlation functions of an inhomogeneous XXX model*, Algebra and analysis **17** (2005), 115–159.

[71] H. Boos, M. Jimbo, T. Miwa, F. Smirnov, and Y. Takeyama, *Reduced qKZ equation and correlation functions of the XXZ model*, Commun. Math. Phys. **261** (2006), 245–276.

[72] H. Boos, M. Jimbo, T. Miwa, F. Smirnov, and Y. Takeyama, *Traces on the Sklyanin algebra and correlation functions of the eight-vertex model*, J. Phys. **A38** (2005), 7629–7660.

[73] N. Kitanine, J.M. Maillet, and V. Terras, *Correlation functions of the XXZ Heisemberg spin-1/2 chain in a magnetic field*, Nucl. Phys. **B567** (2000), 554–582.

[74] J.M. Maillet and V. Terras, *On the quantum inverse scattering problem*, Nucl. Phys. **B575** (2000), 627–644.

[75] N. Kitanine, J.M. Maillet, N. Slavnov and V. Terras, *Spin-spin correlation functions of the XXZ-1/2 Heisenberg chain in a magnetic field*, Nucl. Phys. **B641** (2002), 487–518.

[76] N. Kitanine, J.M. Maillet, N.A. Slavnov, and V. Terras, *Correlation functions of the XXZ spin-1/2 Heisenberg chain at the free Fermion point from their multiple integral representations*, Nucl. Phys. **B642** (2002), 433–455.

[77] N. Kitanine, J.M. Maillet, N.A. Slavnov, and V. Terras, *Emptiness formation probability of the XXZ spin-1/2 Heisenberg chain at Delta = 1/2*, J. Phys. **A35** (2002), L385.

[78] N. Kitanine, J.M. Maillet, N.A. Slavnov, and V. Terras, *Large distance asymptotic behaviour of the emptiness formation probability of the XXZ spin-1/2 Heisenberg chain*, J.Phys. **A35** (2002), L753.

[79] N. Kitanine, J.M. Maillet, N.A. Slavnov, and V. Terras, *Master equation for spin-spin correlation functions of the XXZ chain*, Nucl. Phys. **B712** [FS] (2005), 600.

[80] N. Kitanine, J.M. Maillet, N.A. Slavnov, and V. Terras, *Dynamical correlation functions of the XXZ spin-1/2 chain*, Nucl. Phys. **B729** [FS] (2005) , 558.

[81] F. Göhmann, A. Klümper, and A. Seel, *Integral representation of the density matrix of the XXZ chain at finite temperatures*, J. Phys. **A38** (2005), 1833–1842.

[82] V.E. Korepin, *Calculation of norms of Bethe wave functions*, Commun. Math. Phys. **86** (1982), 391–418.

[83] N.A. Slavnov, *Calculation of scalar products of wave functions and form factors in the framework of the algebraic Bethe Ansatz*, Theor. Math. Phys. **79** (1989), 502.

[84] J.M. Maillet and J. Sánchez de Santos, *Drinfel'd twists and algebraic Bethe ansatz*, Amer. Math. Soc. Transl. **201** (2000), 137–178.

[85] N. Kitanine, J.M. Maillet, N.A. Slavnov, and V. Terras, *Exact results for the sigma-z two-point function of the XXZ chain at Delta = 1/2*, J. Stat. Mech. (2005), L09002.

[86] N. Kitanine, J.M. Maillet, N.A. Slavnov, and V. Terras, *On the spin-spin correlation functions of the XXZ spin-1/2 infinite chain*, J. Phys. A: Math. Gen. **38** (2005) , 74417460.

[87] R.G. Pereira, J. Sirker, J.-S. Caux, R. Hagemans, J.M. Maillet, S.R. White, and I. Affleck, *Dynamical Spin Structure Factor for the Anisotropic Spin-1/2 Heisenberg Chain*, Phys. Rev. Lett. **96** (2006), 257202.

[88] V. Tarasov and A. Varchenko, *Completeness of Bethe vectors and difference equations with regular singular points*, Int. Math. Res. Notices **13** (1996), 637.

[89] A.G. Izergin, *Partition function of the six-vertex model in a finite volume*, Sov. Phys. Dokl. **32** (1987), 878.

[90] M. Gaudin, B.M. Mc Coy and T.T. Wu, *Normalization sum for the Bethe's hypothesis wave functions of the Heisenberg-Ising model*, Phys. Rev. **D23** (1981), 417.

[91] A.V. Razumov and Y.G. Stroganov, *Spin chains and combinatorics*, J. Phys. **A34** (2001), 3185.

[92] M. Shiroishi, M. Takahashi, and Y. Nishiyama, *Emptiness Formation Probability for the One-Dimensional Isotropic XY Model*, J. Phys. Soc. Jap. **70** (2001), 3535.

[93] A.A. Belavin, A.M. Polyakov, and A.B. Zamolodchikov, *Infinite conformal symmetry in two-dimensional quantum field theory*, Nucl. Phys. **B241** (1984), 333.

[94] J.D. Johnson, S. Krinsky, and B.M. McCoy, *Vertical-Arrow Correlation Length in the Eight-Vertex Model and the Low-Lying Excitations of the XYZ Hamiltonian*, Phys. Rev. **A8** (1973), 2526.

[95] S. Lukyanov and A. Zamolodchikov, *Exact expectation values of local fields in the quantum sine-Gordon model*, Nucl. Phys. **B493** (1997), 571.

[96] S. Lukyanov, *Correlation amplitude for the XXZ spin chain in the disordered regime*, Phys. Rev. **B59** (1999), 11163.

[97] S. Lukyanov and V. Terras, *Long-distance asymptotics of spin-spin correlation functions for the XXZ spin chain*, Nucl. Phys. **B654** (2003), 323.

[98] A. Luther and I. Peschel, *Calculation of critical exponents in two dimensions from quantum field theory in one dimension*, Phys. Rev. **B12** (1975), 3908.

[99] O. Castro Alvaredo and J.M. Maillet *Form factors of integrable Heisenberg (higher) spin chains*, arXiv:hep-th/0702186.

[100] N. Kitanine, K. Kozlowski, J.M. Maillet, G. Niccoli, N.A. Slavnov, and V. Terras, *Correlation functions of open Heisenberg chain I*, to appear.

[101] N. Kitanine, K. Kozlowski, J.M. Maillet, G. Niccoli, N.A. Slavnov, and V. Terras, *Correlation functions of open Heisenberg chain II*, to appear.

[102] F.H.L. Essler, H. Frahm, F. Goehmann, A. Klumper, and V.E. Korepin, *The One-Dimensional Hubbard Model*, Cambridge Univ. Press (2005).

Jean-Michel Maillet
Laboratoire de Physique
ENS Lyon et CNRS
46, allée d'Italie
69364 Lyon, France
e-mail: `maillet@ens-lyon.fr`

Quantum Spaces, 203–227
© 2007 Birkhäuser Verlag Basel

Non-commutative Geometry and the Spectral Model of Space-time

Alain Connes

Abstract. This is a report on our joint work with A. Chamseddine and M. Marcolli. This essay gives a short introduction to a potential application in physics of a new type of geometry based on spectral considerations which is convenient when dealing with non-commutative spaces, *i.e.*, spaces in which the simplifying rule of commutativity is no longer applied to the coordinates. Starting from the phenomenological Lagrangian of gravity coupled with matter one infers, using the spectral action principle, that space-time admits a fine structure which is a subtle mixture of the usual 4-dimensional continuum with a finite discrete structure F. Under the (unrealistic) hypothesis that this structure remains valid (*i.e.*, one does not have any "hyperfine" modification) until the unification scale, one obtains a number of predictions whose approximate validity is a basic test of the approach.

1. Background

Our knowledge of space-time can be summarized by the transition from the flat Minkowski metric

$$ds^2 = - dt^2 + dx^2 + dy^2 + dz^2 \tag{1}$$

to the Lorentzian metric

$$ds^2 = g_{\mu\nu} dx^\mu \, dx^\nu \tag{2}$$

of curved space-time with gravitational potential $g_{\mu\nu}$. The basic principle is the Einstein-Hilbert action principle

$$S_E[g_{\mu\nu}] = \frac{1}{G} \int_M r \sqrt{g} \, d^4x \tag{3}$$

where r is the scalar curvature of the space-time manifold M. This action principle only accounts for the gravitational forces and a full account of the forces observed so far requires the addition of new fields, and of corresponding new terms S_{SM} in the action, which constitute the Standard Model so that the total action is of the form

$$S = S_E + S_{SM}. \tag{4}$$

Passing from classical to quantum physics is achieved by the recipe of Dirac and Feynman so that the probability amplitude of a classical configuration A is

$$e^{i\,\frac{S(A)}{\hbar}}. \tag{5}$$

When combined with perturbative renormalization this recipe agrees remarkably well with experiment, but meets (at least) two basic problems:

- One cannot maintain both unitarity and renormalizability at arbitrary scales for the gravitational potential $g_{\mu\nu}$.
- The action S_{SM} is complicated beyond reason and thus only appears as "phenomenological".

To appreciate the second statement we give the explicit form of $S_{SM} = \int_M \mathcal{L}_{SM} \sqrt{g}\, d^4x$ below (*cf.* [26]):

$$
\begin{aligned}
\mathcal{L}_{SM} =\ & -\tfrac{1}{2}\partial_\nu g_\mu^a \partial_\nu g_\mu^a - g_s f^{abc}\partial_\mu g_\nu^a g_\mu^b g_\nu^c - \tfrac{1}{4}g_s^2 f^{abc} f^{ade} g_\mu^b g_\nu^c g_\mu^d g_\nu^e - \partial_\nu W_\mu^+ \partial_\nu W_\mu^- - \\
& M^2 W_\mu^+ W_\mu^- - \tfrac{1}{2}\partial_\nu Z_\mu^0 \partial_\nu Z_\mu^0 - \tfrac{1}{2c_w^2}M^2 Z_\mu^0 Z_\mu^0 - \tfrac{1}{2}\partial_\mu A_\nu \partial_\mu A_\nu - igc_w(\partial_\nu Z_\mu^0(W_\mu^+ W_\nu^- - \\
& W_\nu^+ W_\mu^-) - Z_\nu^0(W_\mu^+ \partial_\nu W_\mu^- - W_\mu^- \partial_\nu W_\mu^+) + Z_\mu^0(W_\nu^+ \partial_\nu W_\mu^- - W_\nu^- \partial_\nu W_\mu^+)) - \\
& igs_w(\partial_\nu A_\mu(W_\mu^+ W_\nu^- - W_\nu^+ W_\mu^-) - A_\nu(W_\mu^+ \partial_\nu W_\mu^- - W_\mu^- \partial_\nu W_\mu^+) + A_\mu(W_\nu^+ \partial_\nu W_\mu^- - \\
& W_\nu^- \partial_\nu W_\mu^+)) - \tfrac{1}{2}g^2 W_\mu^+ W_\mu^- W_\nu^+ W_\nu^- + \tfrac{1}{2}g^2 W_\mu^+ W_\nu^- W_\mu^+ W_\nu^- + g^2 c_w^2(Z_\mu^0 W_\mu^+ Z_\nu^0 W_\nu^- - \\
& Z_\mu^0 Z_\mu^0 W_\nu^+ W_\nu^-) + g^2 s_w^2(A_\mu W_\mu^+ A_\nu W_\nu^- - A_\mu A_\mu W_\nu^+ W_\nu^-) + g^2 s_w c_w(A_\mu Z_\nu^0(W_\mu^+ W_\nu^- - \\
& W_\nu^+ W_\mu^-) - 2A_\mu Z_\mu^0 W_\nu^+ W_\nu^-) - \tfrac{1}{2}\partial_\mu H \partial_\mu H - \tfrac{1}{2}m_h^2 H^2 - \partial_\mu \phi^+ \partial_\mu \phi^- - M^2 \phi^+ \phi^- - \\
& \tfrac{1}{2}\partial_\mu \phi^0 \partial_\mu \phi^0 - \tfrac{1}{2c_w^2}M^2 \phi^0 \phi^0 - \beta_h\left(\frac{2M^2}{g^2} + \frac{2M}{g}H + \tfrac{1}{2}(H^2 + \phi^0\phi^0 + 2\phi^+\phi^-)\right) + \frac{2M^4}{g^2}\alpha_h - \\
& g\alpha_h M\left(H^3 + H\phi^0\phi^0 + 2H\phi^+\phi^-\right) - \\
& \tfrac{1}{8}g^2\alpha_h\left(H^4 + (\phi^0)^4 + 4(\phi^+\phi^-)^2 + 4(\phi^0)^2\phi^+\phi^- + 4H^2\phi^+\phi^- + 2(\phi^0)^2 H^2\right) - \\
& gMW_\mu^+ W_\mu^- H - \tfrac{1}{2}g\frac{M}{c_w^2}Z_\mu^0 Z_\mu^0 H - \\
& \tfrac{1}{2}ig\left(W_\mu^+(\phi^0\partial_\mu\phi^- - \phi^-\partial_\mu\phi^0) - W_\mu^-(\phi^0\partial_\mu\phi^+ - \phi^+\partial_\mu\phi^0)\right) + \\
& \tfrac{1}{2}g\left(W_\mu^+(H\partial_\mu\phi^- - \phi^-\partial_\mu H) + W_\mu^-(H\partial_\mu\phi^+ - \phi^+\partial_\mu H)\right) + \tfrac{1}{2}g\frac{1}{c_w}Z_\mu^0(H\partial_\mu\phi^0 - \phi^0\partial_\mu H) - \\
& ig\frac{s_w^2}{c_w}MZ_\mu^0(W_\mu^+\phi^- - W_\mu^-\phi^+) + igs_w MA_\mu(W_\mu^+\phi^- - W_\mu^-\phi^+) - ig\frac{1-2c_w^2}{2c_w}Z_\mu^0(\phi^+\partial_\mu\phi^- - \\
& \phi^-\partial_\mu\phi^+) + igs_w A_\mu(\phi^+\partial_\mu\phi^- - \phi^-\partial_\mu\phi^+) - \tfrac{1}{4}g^2 W_\mu^+ W_\mu^-\left(H^2 + (\phi^0)^2 + 2\phi^+\phi^-\right) - \\
& \tfrac{1}{8}g^2\frac{1}{c_w^2}Z_\mu^0 Z_\mu^0\left(H^2 + (\phi^0)^2 + 2(2s_w^2 - 1)^2\phi^+\phi^-\right) - \tfrac{1}{2}g^2\frac{s_w^2}{c_w}Z_\mu^0\phi^0(W_\mu^+\phi^- + W_\mu^-\phi^+) - \\
& \tfrac{1}{2}ig^2\frac{s_w^2}{c_w}Z_\mu^0 H(W_\mu^+\phi^- - W_\mu^-\phi^+) + \tfrac{1}{2}g^2 s_w A_\mu\phi^0(W_\mu^+\phi^- + W_\mu^-\phi^+) + \tfrac{1}{2}ig^2 s_w A_\mu H(W_\mu^+\phi^- - \\
& W_\mu^-\phi^+) - g^2\frac{s_w}{c_w}(2c_w^2 - 1)Z_\mu^0 A_\mu\phi^+\phi^- - g^2 s_w^2 A_\mu A_\mu\phi^+\phi^- + \tfrac{1}{2}igs\lambda_{ij}^a(\bar{q}_i^\sigma\gamma^\mu q_j^\sigma)g_\mu^a - \bar{e}^\lambda(\gamma\partial + \\
& m_e^\lambda)e^\lambda - \bar{\nu}^\lambda\gamma\partial\nu^\lambda - \bar{u}_j^\lambda(\gamma\partial + m_u^\lambda)u_j^\lambda - \bar{d}_j^\lambda(\gamma\partial + m_d^\lambda)d_j^\lambda + \\
& igs_w A_\mu\left(-(\bar{e}^\lambda\gamma^\mu e^\lambda) + \tfrac{2}{3}(\bar{u}_j^\lambda\gamma^\mu u_j^\lambda) - \tfrac{1}{3}(\bar{d}_j^\lambda\gamma^\mu d_j^\lambda)\right) + \frac{ig}{4c_w}Z_\mu^0\{(\bar{\nu}^\lambda\gamma^\mu(1+\gamma^5)\nu^\lambda) + \\
& (\bar{e}^\lambda\gamma^\mu(4s_w^2 - 1 - \gamma^5)e^\lambda) + (\bar{d}_j^\lambda\gamma^\mu(\tfrac{4}{3}s_w^2 - 1 - \gamma^5)d_j^\lambda) + (\bar{u}_j^\lambda\gamma^\mu(1 - \tfrac{8}{3}s_w^2 + \gamma^5)u_j^\lambda)\} + \\
& \frac{ig}{2\sqrt{2}}W_\mu^+\left((\bar{\nu}^\lambda\gamma^\mu(1+\gamma^5)e^\lambda) + (\bar{u}_j^\lambda\gamma^\mu(1+\gamma^5)C_{\lambda\kappa}d_j^\kappa)\right) + \\
& \frac{ig}{2\sqrt{2}}W_\mu^-\left((\bar{e}^\lambda\gamma^\mu(1+\gamma^5)\nu^\lambda) + (\bar{d}_j^\kappa C_{\kappa\lambda}^\dagger\gamma^\mu(1+\gamma^5)u_j^\lambda)\right) + \\
& \frac{ig}{2\sqrt{2}}\frac{m_e^\lambda}{M}\left(-\phi^+(\bar{\nu}^\lambda(1-\gamma^5)e^\lambda) + \phi^-(\bar{e}^\lambda(1+\gamma^5)\nu^\lambda)\right) - \frac{g}{2}\frac{m_e^\lambda}{M}\left(H(\bar{e}^\lambda e^\lambda) + i\phi^0(\bar{e}^\lambda\gamma^5 e^\lambda)\right) + \\
& \frac{ig}{2M\sqrt{2}}\phi^+\left(-m_d^\kappa(\bar{u}_j^\lambda C_{\lambda\kappa}(1-\gamma^5)d_j^\kappa) + m_u^\lambda(\bar{u}_j^\lambda C_{\lambda\kappa}(1+\gamma^5)d_j^\kappa)\right) +
\end{aligned}
$$

$$\frac{ig}{2M\sqrt{2}}\phi^-\left(m_d^\lambda(\bar{d}_j^\lambda C_{\lambda\kappa}^\dagger(1+\gamma^5)u_j^\kappa)-m_u^\kappa(\bar{d}_j^\lambda C_{\lambda\kappa}^\dagger(1-\gamma^5)u_j^\kappa)\right)-\frac{g}{2}\frac{m_u^\lambda}{M}H(\bar{u}_j^\lambda u_j^\lambda)-$$

$$\frac{g}{2}\frac{m_d^\lambda}{M}H(\bar{d}_j^\lambda d_j^\lambda)+\frac{ig}{2}\frac{m_u^\lambda}{M}\phi^0(\bar{u}_j^\lambda\gamma^5 u_j^\lambda)-\frac{ig}{2}\frac{m_d^\lambda}{M}\phi^0(\bar{d}_j^\lambda\gamma^5 d_j^\lambda)+\bar{G}^a\partial^2 G^a+g_s f^{abc}\partial_\mu\bar{G}^a G^b g_\mu^c+$$

$$\bar{X}^+(\partial^2-M^2)X^+ + \bar{X}^-(\partial^2-M^2)X^- + \bar{X}^0(\partial^2-\tfrac{M^2}{c_w^2})X^0 + \bar{Y}\partial^2 Y+$$

$$igc_w W_\mu^+(\partial_\mu\bar{X}^0 X^- - \partial_\mu\bar{X}^+ X^0)+igs_w W_\mu^+(\partial_\mu\bar{Y}X^- - \partial_\mu\bar{X}^+ Y)+igc_w W_\mu^-(\partial_\mu\bar{X}^- X^0-$$

$$\partial_\mu\bar{X}^0 X^+)+igs_w W_\mu^-(\partial_\mu\bar{X}^- Y-\partial_\mu\bar{Y}X^+)+igc_w Z_\mu^0(\partial_\mu\bar{X}^+ X^+-\partial_\mu\bar{X}^- X^-)+$$

$$igs_w A_\mu(\partial_\mu\bar{X}^+ X^+-\partial_\mu\bar{X}^- X^-)-\tfrac{1}{2}gM\left(\bar{X}^+ X^+ H+\bar{X}^- X^- H+\tfrac{1}{c_w^2}\bar{X}^0 X^0 H\right)+$$

$$\frac{1-2c_w^2}{2c_w}igM\left(\bar{X}^+ X^0\phi^+-\bar{X}^- X^0\phi^-\right)+\frac{1}{2c_w}igM\left(\bar{X}^0 X^-\phi^+-\bar{X}^0 X^+\phi^-\right)+$$

$$igMs_w\left(\bar{X}^0 X^-\phi^+-\bar{X}^0 X^+\phi^-\right)+\tfrac{1}{2}igM\left(\bar{X}^+ X^+\phi^0-\bar{X}^- X^-\phi^0\right).$$

This action functional was expressed in flat space-time and needs of course to be minimally coupled with gravity. One also needs to take into account the experimental discovery of neutrino oscillations and add the corresponding new terms.

2. Why non-commutative spaces

The natural group of symmetries of the total action (4) is the semi-direct product

$$\mathcal{G}=\mathrm{Map}(M,G)\rtimes\mathrm{Diff}(M)\tag{6}$$

of the group $\mathrm{Map}(M,G)$ of gauge transformations of second kind by the group $\mathrm{Diff}(M)$ of diffeomorphisms. Here G is the gauge group, inferred from experiment

$$G=U(1)\times SU(2)\times SU(3)\,.\tag{7}$$

Since the symmetry group of the Einstein-Hilbert action of pure gravity is simply $\mathrm{Diff}(M)$ it is natural to ask wether there is a space X whose group of diffeomorphisms is directly of the form (6). The answer is:

No: for ordinary spaces.

Yes: for non-commutative spaces.

A "non-commutative space" is one in which the usual coordinates x^μ no longer satisfy the simplifying commutative rule saying that the order of the terms is irrelevant in a product. They are familiar to physicists since Heisenberg's discovery of the nontrivial commutation rules for the natural coordinates in the phase space of a microscopic mechanical system. In first approximation the group of diffeomorphisms of such a space is the group of automorphisms $\mathrm{Aut}(\mathcal{A})$ of the algebra $\mathcal{A}$ of coordinates. The new feature that arises in the non-commutative case is that there are "easy" automorphisms, namely those of the form

$$f\in\mathcal{A}\mapsto ufu^{-1}$$

where $u\in\mathcal{A}$ is an invertible element. Such automorphisms are called "inner" or "internal" and form a normal subgroup $\mathrm{Inn}(\mathcal{A})$ of the group $\mathrm{Aut}(\mathcal{A})$ so that one has the general exact sequence

$$1\to\mathrm{Inn}(\mathcal{A})\to\mathrm{Aut}(\mathcal{A})\to\mathrm{Out}(\mathcal{A})\to 1\,.\tag{8}$$

This exact sequence remains valid when taking into account the compatibility with the adjoint $f \mapsto f^*$ (one restricts to $\star$-automorphisms while $u \in \mathcal{A}$ is now a unitary element $uu^* = u^*u = 1$).

For an ordinary manifold X results from topology ($cf.$ [22]) preclude the existence of a space whose group of diffeomorphisms is the group $\mathcal{G}$ of (6). To understand how passing to non-commutative spaces adds the missing part $\mathrm{Map}(M, G)$, let us consider the simplest example where the algebra

$$\mathcal{A} = C^\infty(M, M_n(\mathbb{C})) = C^\infty(M) \otimes M_n(\mathbb{C})$$

consists of smooth maps from a manifold M to the algebra $M_n(\mathbb{C})$ of $n \times n$ matrices. One then shows that the group $\mathrm{Inn}(\mathcal{A})$ in that case is locally isomorphic to the group $\mathrm{Map}(M, G)$ of smooth maps from M to the small gauge group $G = PSU(n)$ (quotient of $SU(n)$ by its center) and that the general exact sequence (8) becomes identical to the exact sequence governing the structure of the group $\mathcal{G}$, namely

$$1 \to \mathrm{Map}(M, G) \to \mathcal{G} \to \mathrm{Diff}(M) \to 1. \tag{9}$$

Moreover the physics terminology of "internal symmetries" matches the mathematical one perfectly. We refer to Proposition 3.4 of [7] for the more involved case of the group (6).

3. What is a non-commutative geometry?

A refined notion of geometry (suitable in particular to deal with spaces whose coordinates do not commute) is obtained by focussing not on the traditional $g_{\mu\nu}$ but on the Dirac operator D. In extracting the square root of the Laplacian using a spin structure the Dirac operator enables us to talk about the line element $ds = D^{-1}$ instead of its square (2). The new paradigm for a geometric space is of *spectral* nature. A spectral geometry $(\mathcal{A}, \mathcal{H}, D)$ is given by an involutive unital algebra $\mathcal{A}$ represented as operators in a Hilbert space $\mathcal{H}$ and a self-adjoint operator D with compact resolvent such that all commutators $[D, a]$ are bounded for $a \in \mathcal{A}$. A spectral geometry is *even* if the Hilbert space $\mathcal{H}$ is endowed with a $\mathbb{Z}/2$- grading γ which commutes with any $a \in \mathcal{A}$ and anticommutes with D.

This notion extends the Riemannian paradigm as follows. A spin Riemannian manifold M gives rise in a canonical manner to a spectral geometry. The Hilbert space $\mathcal{H}$ is the Hilbert space $L^2(M, S)$ of square integrable spinors on M and the algebra $\mathcal{A} = C^\infty(M)$ of smooth functions on M acts in $\mathcal{H}$ by multiplication operators:

$$(f\,\xi)(x) = f(x)\,\xi(x), \quad \forall\, x \in M. \tag{10}$$

The operator D is the Dirac operator,

$$\not{\partial}_M = \sqrt{-1}\,\gamma^\mu\,\nabla_\mu. \tag{11}$$

The grading γ is given by the chirality operator which we denote by γ_5 in the four-dimensional case.

As it turns out this way of defining a geometry by specifying the Dirac operator is meaningful both in mathematical terms (where the Dirac operator specifies the fundamental class in KO-homology) and in physics terms (where, modulo a chiral gauge transformation, the Dirac operator is the inverse of the Euclidean propagator of fermions). From both sides (KO-homology and physics) a further "decoration" is needed in the form of a real structure. A real structure of KO-dimension $n \in \mathbb{Z}/8$ on a spectral geometry $(\mathcal{A}, \mathcal{H}, D)$ is an antilinear isometry $J : \mathcal{H} \to \mathcal{H}$, with the property that

$$J^2 = \varepsilon, \qquad JD = \varepsilon'DJ, \quad \text{and} \quad J\gamma = \varepsilon''\gamma J. \tag{12}$$

The numbers $\varepsilon, \varepsilon', \varepsilon'' \in \{-1, 1\}$ are a function of $n \mod 8$ given by

n	0	1	2	3	4	5	6	7
ε	1	1	−1	−1	−1	−1	1	1
ε'	1	−1	1	1	1	−1	1	1
ε''	1		−1		1		−1	

From the mathematical side the role of J is twofold, it embodies the crucial nuance between complex K-homology and "real" KO-homology which plays a key role in the conceptual understanding of homotopy types of manifolds. It also embodies the discovery by Tomita of the general structure of representations of non-commutative algebras. This corresponds to the commutation relation

$$[a, b^0] = 0 \quad \forall\, a, b \in \mathcal{A}, \tag{13}$$

where

$$b^0 = Jb^*J^{-1} \qquad \forall b \in \mathcal{A}. \tag{14}$$

From the physics side the operator J corresponds to the charge conjugation operator. The change from the Riemannian paradigm to the spectral one already occurred in geodesy. The notion of geometry is intimately tied up with the measurement of length and it was never completely obvious how to reach some agreement on a physical unit of length which would unify the numerous existing choices. Since the French revolution the concrete "mètre-étalon" (realized in the form of a platinum bar which is approximately 10^{-7} times the quarter of the meridian of the earth) was taken as unit of length in the metric system. Already in 1927, at the seventh conference on the metric system, in order to take into account the inevitable natural variations of the concrete "mètre-étalon", the idea emerged to compare it with a reference wave length (the red line of Cadmium). Around 1960 the reference to the "mètre-étalon" was finally abandoned and a new definition of the "mètre" was adopted as $1650763, 73$ times the wave length of the radiation corresponding to the transition between the levels 2p10 and 5d5 of the Krypton 86Kr. In 1967 the second was defined as the duration of 9192631770 periods of the radiation corresponding to the transition between the two hyperfine levels of Caesium-133. Finally in 1983 the "mètre" was defined as the distance travelled by light in $1/299792458$ second. In fact the speed of light is just a conversion factor and to define the "mètre" one gives it the specific value of

$$c = 299792458 \, m/s.$$

In other words the "mètre" is defined as a certain fraction $\frac{9192631770}{299792458} \sim 30.6633\ldots$ of the wave length of the radiation coming from the transition between the above hyperfine levels of the Caesium atom. The advantages of the new standard of length are many. By not being tied up with any specific location it is in fact available anywhere where Caesium is (the choice of Caesium as opposed to Helium or Hydrogen which are much more common in the universe is of course still debatable [2]).

In non-commutative geometry the Riemannian formula for the geodesic distance

$$d(x, y) = \inf \int_\gamma \sqrt{g_{\mu\nu} dx^\mu \, dx^\nu} \tag{15}$$

where the infimum is taken over all paths from x to y, is replaced by

$$d(x, y) = \sup\{|f(x) - f(y)| : f \in \mathcal{A}, \ \|[D, f]\| \le 1\}, \tag{16}$$

which gives the same answer in the Riemannian case but continues to make sense for spectral geometries where the algebra $\mathcal{A}$ is no longer commutative (x and y are then states on $\mathcal{A}$).

The traditional notions of geometry all have natural analogues in the spectral framework. We refer to [9] for more details. The dimension of a non-commutative geometry is not a number but a spectrum, the *dimension spectrum* (*cf.* [14]) which is the subset Π of the complex plane $\mathbb{C}$ at which the spectral functions have singularities. Under the hypothesis that the dimension spectrum is *simple*, *i.e.*, that the spectral functions have at most simple poles, the residue at the pole defines a far reaching extension (*cf.* [14]) of the fundamental integral in non-commutative geometry given by the Dixmier trace (*cf.* [9]). This extends the Wodzicki residue from pseudodifferential operators on a manifold to the general framework of spectral triples, and gives meaning to $\fint T$ in that context. It is simply given by

$$\fint T = \operatorname{Res}_{s=0} \operatorname{Tr}\left(T \, |D|^{-s}\right). \tag{17}$$

4. Inner fluctuations of a spectral geometry

The non-commutative world is rich in phenomena which have no commutative counterpart. We already saw above the role of inner automorphisms (as internal symmetries) which decompose the full automorphism group into equivalence classes modulo inner. In a similar manner the non-commutative metrics admit a natural foliation, the metrics on the same leaf are obtained as inner fluctuations. The corresponding transformation on the operator D is simply the addition $D \mapsto D_A = D + A + \varepsilon' \, J A J^{-1}$ where $A = A^*$ is an arbitrary selfadjoint element of Ω_D^1 with

$$\Omega_D^1 = \{\sum_j a_j\,[D, b_j] \mid a_j,\, b_j \in \mathcal{A}\}\,, \tag{18}$$

which is by construction a bimodule over $\mathcal{A}$.

The inner fluctuations in non-commutative geometry are generated by the existence of Morita equivalences (*cf.* [24]). Given an algebra $\mathcal{A}$, a Morita equivalent algebra $\mathcal{B}$ is the algebra of endomorphisms of a finite projective (right) module $\mathcal{E}$ over $\mathcal{A}$,

$$\mathcal{B} = \mathrm{End}_{\mathcal{A}}(\mathcal{E}). \tag{19}$$

Transferring the metric from $\mathcal{A}$ to $\mathcal{B}$ requires the choice of a *hermitian connection* ∇ on $\mathcal{E}$. A connection is a linear map $\nabla : \mathcal{E} \to \mathcal{E} \otimes_{\mathcal{A}} \Omega_D^1$ satisfying the Leibniz rule

$$\nabla(\xi a) = (\nabla \xi)a + \xi \otimes da\,, \quad \forall\, \xi \in \mathcal{E}\,,\ a \in \mathcal{A},$$

with $da = [D, a]$. Taking the obvious Morita equivalence between $\mathcal{A}$ and itself generates the inner fluctuations $D \mapsto D + A + \varepsilon'\, J\,A\,J^{-1}$.

By (14) one gets a right $\mathcal{A}$-module structure on $\mathcal{H}$,

$$\xi b = b^0 \xi\,, \quad \forall\, \xi \in \mathcal{H}\,, \quad b \in \mathcal{A}. \tag{20}$$

The unitary group of the algebra $\mathcal{A}$ then acts by the "adjoint representation" in $\mathcal{H}$ in the form

$$\xi \in \mathcal{H} \to \mathrm{Ad}(u)\,\xi = u\,\xi\,u^*\,, \quad \forall\, \xi \in \mathcal{H}\,, \quad u \in \mathcal{A}\,, \quad u\,u^* = u^*\,u = 1\,. \tag{21}$$

The order one condition

$$[[D, a], b^0] = 0 \qquad \forall\, a, b \in \mathcal{A} \tag{22}$$

ensures that for any $A \in \Omega_D^1$ with $A = A^*$ and any unitary $u \in \mathcal{A}$, one has

$$\mathrm{Ad}(u)(D + A + \varepsilon'\, J\,A\,J^{-1})\mathrm{Ad}(u^*) = D + \gamma_u(A) + \varepsilon'\, J\,\gamma_u(A)\,J^{-1},$$

where $\gamma_u(A) = u\,[D, u^*] + u\,A\,u^*$.

The above parallel between inner automorphisms and internal symmetries extends to a parallel between the inner fluctuations and the gauge potentials.

5. The spectral action principle

We shall recall in this section our joint work with Ali Chamseddine on the spectral action principle [3–6]. The starting point is the discussion of observables in gravity. By the principle of gauge invariance the only quantities which have a chance to be observable in gravity are those which are invariant under the gauge group, *i.e.*, the group of diffeomorphisms of the space-time M. Assuming first that we deal with a classical manifold (and Wick rotate to Euclidean signature for simplicity), one can form a number of such invariants (under suitable convergence conditions) as the integrals of the form

$$\int_M F(K)\,\sqrt{g}\,d^4x \tag{23}$$

where $F(K)$ is a scalar invariant function (the scalar curvature is one example of such a function but there are many others) of the Riemann curvature K. We refer to [16] for other more complicated examples of such invariants, where those of the form (23) appear as the *single integral* observables, *i.e.*, those which add up when evaluated on the direct sum of geometric spaces. Now while in theory a quantity like (23) is observable it is almost impossible to evaluate since it involves the knowledge of the entire space-time and is in that way highly non localized. On the other hand, spectral data are available in localized form anywhere, and are (asymptotically) of the form (23) when they are of the additive form

$$\text{Trace}\,(f(D/\Lambda)), \tag{24}$$

where D is the Dirac operator and f is a positive even function of the real variable while the parameter Λ fixes the mass scale. The spectral action principle asserts that the fundamental action functional S that allows to compare different geometric spaces at the classical level and is used in the functional integration to go to the quantum level, is itself of the form (24). The detailed form of the function f is largely irrelevant since the spectral action (24) can be expanded in decreasing powers of the scale Λ in the form

$$\text{Trace}\,(f(D/\Lambda)) \sim \sum_{k \in \Pi^+} f_k\, \Lambda^k \fint |D|^{-k} + f(0)\, \zeta_D(0) + o(1), \tag{25}$$

where Π^+ is the positive part of the dimension spectrum, the integral $\fint$ is defined in (17), and the function f only appears through the scalars

$$f_k = \int_0^\infty f(v)\, v^{k-1}\, dv. \tag{26}$$

The term independent of the parameter Λ is the value at $s = 0$ (regularity at $s = 0$ is assumed) of the zeta function,

$$\zeta_D(s) = \text{Tr}\,(|D|^{-s}). \tag{27}$$

The main result of our joint work with A. Chamseddine [3], [4] is that, when applied to the inner fluctuations of the product geometry $M \times F$ the spectral action gives the standard model coupled with gravity. Here M is a Riemannian compact spin 4-manifold, the standard model coupled with gravity is in the Euclidean form, and the geometry of the finite space F is encoded (as in the general framework of NCG) by a spectral geometry $(\mathcal{A}_F, \mathcal{H}_F, D_F)$.

For M the spectral geometry is given by (10), (11). For the non-commutative geometry F used in [4] to obtain the standard model coupled to gravity, all the ingredients are *finite-dimensional*. The algebra $\mathcal{A}_F = \mathbb{C} \oplus \mathbb{H} \oplus M_3(\mathbb{C})$ (*i.e.*, the direct sum of the algebras $\mathbb{C}$ of complex numbers, $\mathbb{H}$ of quaternions, and $M_3(\mathbb{C})$ of 3×3 matrices) encodes the gauge group. The Hilbert space $\mathcal{H}_F$ encodes the elementary quarks and leptons. The operator D_F encodes those free parameters of the standard model related to the Yukawa couplings.

The above work [4] had several shortcomings:

1. The finite geometry F is put in "by hand" with no conceptual understanding of the representation of $\mathcal{A}_F$ in $\mathcal{H}_F$.
2. There is a fermion doubling problem (*cf.* [21]) in the Fermionic part of the action.
3. It does not incorporate the neutrino mixing and see-saw mechanism for neutrino masses.

We showed in [12] and [7] how to solve these three problems (the first only partly since the number of generations is put by hand) simply by keeping the distinction between the following two notions of dimension of a non-commutative space,

- the metric dimension,
- the KO-dimension.

The metric dimension manifests itself by the growth of the spectrum of the Dirac operator and gives an upper bound to the dimension spectrum. In a (compact) space of dimension k the line element $ds = D^{-1}$ is an infinitesimal of order $1/k$ which means that the nth characteristic value of ds is of the order of $n^{-1/k}$ (in the non-compact case one replaces ds by $a\,ds$ for $a \in \mathcal{A}$). As far as space-time goes it appears that the situation of interest will be the four-dimensional one. In particular the metric dimension of the finite geometry F will be zero.

The KO-dimension is only well defined modulo 8 and it takes into account both the $\mathbb{Z}/2$-grading γ of $\mathcal{H}$ as well as the real structure J according to (12). The real surprise is that in order for things to work the only needed change (besides the easy addition of a right-handed neutrino) is to change the $\mathbb{Z}/2$ grading of the finite geometry F to its opposite in the "antiparticle" sector. It is only thanks to this that the Fermion doubling problem pointed out in [21] can be successfully handled. Moreover it will automatically generate the *full standard model, i.e.,* the model with neutrino mixing and the see-saw mechanism as follows from the full classification of Dirac operators: Theorem 6.7.

When one looks at the above table giving the KO-dimension of the finite space F one then finds that its KO-dimension is now equal to 6 modulo 8 (!). As a result we see that the KO-dimension of the product space $M \times F$ is in fact equal to $10 \sim 2$ modulo 8. Of course the above 10 is very reminiscent of string theory, in which the finite space F might be a good candidate for an "effective" compactification at least for low energies[1]. But 10 is also 2 modulo 8 which might be related to the observations of [20] about gravity.

It is also remarkable that the non-commutative spheres arising from quantum groups, such as the Podleś spheres already exhibit the situation where the metric dimension (0 in that case) is distinct from the KO-dimension (2 in that case) as pointed out in the work of L. Dąbrowski and A. Sitarz on Podleś quantum spheres [15].

---

[1]Note however that we are dealing with the standard model, not its supersymmetrized version.

6. The finite non-commutative geometry F

In this section we shall first describe in a conceptual manner the representation of $\mathcal{A}_F$ in $\mathcal{H}_F$ and classify the Dirac operators D_F. The only small nuance with [11] is that we incorporate a right-handed neutrino ν_R and change the $\mathbb{Z}/2$ grading in the antiparticle sector to its opposite. This, innocent as it looks, allows for a better conceptual understanding of the representation of $\mathcal{A}_F$ in $\mathcal{H}_F$ and also will completely alter the classification of Dirac operators (Theorem 6.7).

6.1. The representation of $\mathcal{A}_F$ in $\mathcal{H}_F$

We start from the involutive algebra (with $\mathbb{H}$ the quaternions with involution $q \to \bar{q}$)

$$\mathcal{A}_{LR} = \mathbb{C} \oplus \mathbb{H}_L \oplus \mathbb{H}_R \oplus M_3(\mathbb{C}). \tag{28}$$

We construct a natural representation $(\mathcal{A}_{LR}, \mathcal{H}_F, J_F, \gamma_F)$ fulfilling (12) and (13) in dimension 6 modulo 8. The commutation relation (13) shows that there is an underlying structure of $\mathcal{A}_{LR}$-bimodule on $\mathcal{H}_F$ and we shall use that structure as a guide. One uses the bimodule structure to define $\mathrm{Ad}(u)$ by (21).

Definition 6.1. *Let $\mathcal{M}$ be an $\mathcal{A}_{LR}$-bimodule. Then $\mathcal{M}$ is* odd *iff the adjoint action* (21) *of $s = (1, -1, -1, 1)$ fulfills* $\mathrm{Ad}(s) = -1$.

Such a bimodule is a representation of the reduction of $\mathcal{A}_{LR} \otimes_{\mathbb{R}} \mathcal{A}_{LR}^0$ by the projection $\frac{1}{2}(1 - s \otimes s^0)$. This subalgebra is an algebra over $\mathbb{C}$ and we restrict to complex representations. One defines the contragredient bimodule of a bimodule $\mathcal{M}$ as the complex conjugate space

$$\mathcal{M}^0 = \{\bar{\xi} \,;\, \xi \in \mathcal{M}\}, \quad a\,\bar{\xi}\,b = \overline{b^*\xi\,a^*}, \quad \forall\, a, b \in \mathcal{A}_{LR}. \tag{29}$$

We can now give the following characterization of the $\mathcal{A}_{LR}$-bimodule $\mathcal{M}_F$ and the real structure J_F for one generation.

Proposition 6.2.
- *The $\mathcal{A}_{LR}$-bimodule $\mathcal{M}_F$ is the direct sum of all inequivalent irreducible odd $\mathcal{A}_{LR}$-bimodules.*
- *The dimension of $\mathcal{M}_F$ is* 32.
- *The real structure J_F is given by the isomorphism with the contragredient bimodule.*

We define the $\mathbb{Z}/2$-grading γ_F by

$$\gamma_F = c - J_F\,c\,J_F\,, \quad c = (0, -1, 1, 0) \in \mathcal{A}_{LR}. \tag{30}$$

One then checks that the following holds:

$$J_F^2 = 1\,, \quad J_F\,\gamma_F = -\,\gamma_F\,J_F, \tag{31}$$

which together with the commutation of J_F with the Dirac operators, is characteristic of KO-dimension equal to 6 modulo 8.

The equality $\iota(\lambda, q, m) = (\lambda, q, \lambda, m)$ defines a homomorphism ι of involutive algebras from $\mathcal{A}_F$ to $\mathcal{A}_{LR}$ so that we view $\mathcal{A}_F$ as a subalgebra of $\mathcal{A}_{LR}$.

Definition 6.3. *The real representation* $(\mathcal{A}_F, \mathcal{H}_F, J_F, \gamma_F)$ *is the restriction to* $\mathcal{A}_F \subset \mathcal{A}_{LR}$ *of the direct sum* $\mathcal{M}_F \otimes \mathbb{C}^3$ *of three copies of* $\mathcal{M}_F$.

It has dimension $32 \times 3 = 96$, needless to say this 3 is the number of generations and it is put in by hand here. A conceptual explanation for the restriction to $\mathcal{A}_F \subset \mathcal{A}_{LR}$ is given in [7].

6.2. The unimodular unitary group $\mathrm{SU}(\mathcal{A}_F)$

Using the action of $\mathcal{A}_F$ in $\mathcal{H}_F$ one defines the unimodular subgroup $\mathrm{SU}(\mathcal{A}_F)$ of the unitary group $\mathrm{U}(\mathcal{A}_F) = \{u \in \mathcal{A}_F,\ uu^* = u^*u = 1\}$ as follows.

Definition 6.4. *We let* $\mathrm{SU}(\mathcal{A}_F)$ *be the subgroup of* $\mathrm{U}(\mathcal{A}_F)$ *defined by*

$$\mathrm{SU}(\mathcal{A}_F) = \{u \in \mathrm{U}(\mathcal{A}_F)\ :\ \mathrm{Det}(u) = 1\}$$

where $\mathrm{Det}(u)$ *is the determinant of the action of* u *in* $\mathcal{H}_F$.

One obtains both the standard model gauge group and its action on fermions from the adjoint action of $\mathrm{SU}(\mathcal{A}_F)$ in the following way:

Proposition 6.5.

1. *The group* $\mathrm{SU}(\mathcal{A}_F)$ *is, up to an Abelian finite group,*

$$\mathrm{SU}(\mathcal{A}_F) \sim \mathrm{U}(1) \times \mathrm{SU}(2) \times \mathrm{SU}(3).$$

2. *The adjoint action* $u \to \mathrm{Ad}(u)$ *(cf. (21)) of* $\mathrm{SU}(\mathcal{A}_F)$ *in* $\mathcal{H}_F$ *coincides with the standard model action on elementary quarks and leptons.*

One shows [7] that the sum of the irreducible odd bimodules is of the form

$$\mathcal{M}_F = (\pi_L \oplus \pi_R \oplus \pi_R^3 \oplus \pi_L^3) \oplus (\pi_L \oplus \pi_R \oplus \pi_R^3 \oplus \pi_L^3)^0. \qquad (32)$$

This $\mathcal{A}_{LR}$-bimodule $\mathcal{M}_F$ is of dimension $2 \cdot (2 + 2 + 2 \times 3 + 2 \times 3) = 32$ and the adjoint action gives the gauge action of the standard model for one generation, with the following labels for the basis elements of $\mathcal{M}_F$,

$$\begin{pmatrix} \nu_L & \nu_R \\ e_L & e_R \end{pmatrix}$$

for the term $\pi_L \oplus \pi_R$,

$$\begin{pmatrix} u_L^j & u_R^j \\ d_L^j & d_R^j \end{pmatrix}$$

for the term $\pi_R^3 \oplus \pi_L^3$ (with color indices j) and the transformation $q \to \bar{q}$ to pass to the contragredient bimodules. With these labels one checks that the adjoint action of the $\mathrm{U}(1)$ factor is given by multiplication of the basis vectors f by the following powers of $\lambda \in \mathrm{U}(1)$:

	e	ν	u	d
f_L	-1	-1	$\frac{1}{3}$	$\frac{1}{3}$
f_R	-2	0	$\frac{4}{3}$	$-\frac{2}{3}$

6.3. The classification of Dirac operators

To be precise we adopt the following.

Definition 6.6. *A Dirac operator is a self-adjoint operator D in $\mathcal{H}_F$ commuting with J_F, $\mathbb{C}_F = \{(\lambda, \lambda, 0)\} \in \mathcal{A}_F$, anticommuting with γ_F and fulfilling the order one condition $[[D, a], b^0] = 0$ for any $a, b \in \mathcal{A}_F$.*

The physics meaning of the condition of commutation with $\mathbb{C}_F$ is to ensure that one gauge vector boson (the photon) remains massless.

In order to state the classification of Dirac operators we introduce the following notation, let M_e, M_ν, M_d, M_u and M_R be three by three matrices, we then let $D(M)$ be the operator in $\mathcal{H}_F$ given by

$$D(M) = \begin{bmatrix} S & T^* \\ T & \bar{S} \end{bmatrix} \tag{33}$$

where

$$S = S_\ell \oplus (S_q \otimes 1_3) \tag{34}$$

and in the basis (ν_R, e_R, ν_L, e_L) and (u_R, d_R, u_L, d_L),

$$S_\ell = \begin{bmatrix} 0 & 0 & M_\nu^* & 0 \\ 0 & 0 & 0 & M_e^* \\ M_\nu & 0 & 0 & 0 \\ 0 & M_e & 0 & 0 \end{bmatrix} \qquad S_q = \begin{bmatrix} 0 & 0 & M_u^* & 0 \\ 0 & 0 & 0 & M_d^* \\ M_u & 0 & 0 & 0 \\ 0 & M_d & 0 & 0 \end{bmatrix} \tag{35}$$

while the operator T is 0 except on the subspace $\mathcal{H}_{\nu_R} \subset \mathcal{H}_F$ with basis the ν_R which it maps, using the matrix M_R, to the subspace $\mathcal{H}_{\bar{\nu}_R} \subset \mathcal{H}_F$ with basis the $\bar{\nu}_R$.

Theorem 6.7.

1. *Let D be a Dirac operator. There exist 3×3 matrices M_e, M_ν, M_d, M_u and M_R, with M_R symmetric, such that $D = D(M)$.*
2. *All operators $D(M)$ (with M_R symmetric) are Dirac operators.*
3. *The operators $D(M)$ and $D(M')$ are conjugate by a unitary operator commuting with $\mathcal{A}_F$, γ_F and J_F iff there exists unitary matrices V_j and W_j such that*

$$M_e' = V_1 M_e V_3^*, \qquad M_\nu' = V_2 M_\nu V_3^*, \qquad M_d' = W_1 M_d W_3^*,$$
$$M_u' = W_2 M_u W_3^*, \qquad M_R' = V_2 M_R \bar{V}_2^*.$$

In particular Theorem 6.7 shows that the Dirac operators give all the required features, such as

- mixing matrices for quarks and leptons,
- unbroken color,
- see-saw mechanism for right handed neutrinos.

7. The spectral action for $M \times F$ and the standard model

We now consider a four-dimensional smooth compact Riemannian manifold M with a fixed spin structure and recall that it is fully encoded by its Dirac spectral geometry $(\mathcal{A}_1, \mathcal{H}_1, D_1) = (C^\infty(M), L^2(M, S), \partial_M)$. We then consider its product with the above finite geometry $(\mathcal{A}_2, \mathcal{H}_2, D_2) = (\mathcal{A}_F, \mathcal{H}_F, D_F)$. With $(\mathcal{A}_j, \mathcal{H}_j, D_j)$ of KO-dimensions 4 for $j = 1$ and 6 for $j = 2$, the product geometry is given by the rules

$$\mathcal{A} = \mathcal{A}_1 \otimes \mathcal{A}_2, \quad \mathcal{H} = \mathcal{H}_1 \otimes \mathcal{H}_2, \quad D = D_1 \otimes 1 + \gamma_1 \otimes D_2, \quad \gamma = \gamma_1 \otimes \gamma_2, \quad J = J_1 \otimes J_2.$$

Note that it matters that J_1 commutes with γ_1 to check that J commutes with D. The KO-dimension of the finite space F is $6 \in \mathbb{Z}/8$ and thus the KO-dimension of the product geometry $M \times F$ is now $2 \in \mathbb{Z}/8$. In other words according to (12) the commutation rules are

$$J^2 = -1, \quad JD = DJ, \quad \text{and} \quad J\gamma = -\gamma J. \tag{36}$$

Let us now explain how these rules allow to define a natural antisymmetric bilinear form on the even part $\mathcal{H}^+ = \{\xi \in \mathcal{H} \,,\, \gamma\xi = \xi\}$ of $\mathcal{H}$.

Proposition 7.1. *On a real spectral geometry of KO-dimension $2 \in \mathbb{Z}/8$, the following equality defines an antisymmetric bilinear form on $\mathcal{H}^+ = \{\xi \in \mathcal{H} \,,\, \gamma\xi = \xi\}$,*

$$A_D(\xi', \xi) = \langle J\xi', D\xi \rangle, \quad \forall \xi, \xi' \in \mathcal{H}^+. \tag{37}$$

The above trilinear pairing between D, ξ and ξ' is gauge invariant under the adjoint action (cf. (21)) of the unitary group of $\mathcal{A}$,

$$A_D(\xi', \xi) = A_{D_u}(\mathrm{Ad}(u)\xi', \mathrm{Ad}(u)\xi), \quad D_u = \mathrm{Ad}(u)\, D\, \mathrm{Ad}(u^*). \tag{38}$$

Now the Pfaffian of an antisymmetric bilinear form is best expressed in terms of the functional integral involving anticommuting "classical fermions" which at the formal level means that

$$\mathrm{Pf}(A) = \int e^{-\frac{1}{2} A(\xi)} \, D[\xi].$$

It is the use of the Pfaffian as a square root of the determinant that allows to solve the Fermion doubling puzzle which was pointed out in [21]. The solution obtained by a better choice of the KO-dimension of the space F and hence of $M \times F$ is not unrelated to the point made in [17].

Theorem 7.2. *Let M be a Riemannian spin 4-manifold and F the finite non-commutative geometry of KO-dimension 6 described above. Let $M \times F$ be endowed with the product metric.*

1. *The unimodular subgroup of the unitary group acting by the adjoint representation $\mathrm{Ad}(u)$ in $\mathcal{H}$ is the group of gauge transformations of SM.*
2. *The unimodular inner fluctuations A of the metric[2] are parameterized exactly by the gauge bosons of SM (including the Higgs doublet).*

---

[2]The *unimodular* inner fluctuations are obtained by restricting to those A which are traceless, *i.e.*, fulfill the condition $\mathrm{Tr}(A) = 0$.

3. *The full standard model (see the explicit formula in §9) minimally coupled with Einstein gravity is given in Euclidean form by the action functional*

$$S = \mathrm{Tr}(f(D_A/\Lambda)) + \frac{1}{2}\langle J\xi, D_A\xi\rangle, \quad \xi \in \mathcal{H}^+$$

applied to unimodular inner fluctuations $D_A = D + A + JAJ^{-1}$ of the metric.

We take f even and positive with $f^{(n)}(0) = 0$ for $n \geq 1$ for definiteness. Note also that the components of ξ anticommute so the antisymmetric form does not vanish. The proof is given in [7] which is a variant of [4] (*cf.* [18] for a detailed version). After turning off gravity to simplify and working in flat space (after Wick rotation back to Lorentzian signature) one gets the Lagrangian of §9 whose agreement with that of §1 can hardly be fortuitous. It is obtained in Euclidean form and all the signs are the physical ones, provided the test function f is positive which is the natural condition to get a sensible exponent in the functional integral. The positivity of the test function f ensures the positivity of the action functional before taking the asymptotic expansion. In general, this does not suffice to control the sign of the terms in the asymptotic expansion. In our case, however, this determines the positivity of the momenta f_0, f_2, and f_4. The explicit calculation then shows that this implies that the signs of all the terms are the expected physical ones.

We obtain the usual Einstein–Hilbert action with a cosmological term, and in addition the square of the Weyl curvature and a pairing of the scalar curvature with the square of the Higgs field. The Weyl curvature term does not affect gravity at low energies, as explained in §10.6 below.

The fermion doubling problem is resolved by the use of the Pfaffian, we checked that part for the Dirac mass terms, and trust that the same holds for the Majorana mass terms. There is one subtle point which is the use of the following chiral transformation:

$$U = e^{i\frac{\pi}{4}\gamma_5}$$

to transform the Fermionic part of the action to the traditional one, *i.e.*, the Euclidean action for Fermi fields (*cf.* [8]). While this transformation is innocent at the classical level, it is nontrivial at the quantum level and introduces some kind of Maslov index in the transition from our form of the Euclidean action to the more traditional one. We shall now give more details on the bosonic part of the action.

8. Detailed form of the bosonic action

We shall now give the precise form of the bosonic action, the calculation [7] is entirely similar to [4] with new terms appearing from the presence of M_R.

One lets $f_k = \int_0^\infty f(u)\, u^{k-1} du$ for $k > 0$ and $f_0 = f(0)$. Also

$$a = \text{Tr}(M_\nu^* M_\nu + M_e^* M_e + 3(M_u^* M_u + M_d^* M_d)) \tag{39}$$

$$b = \text{Tr}((M_\nu^* M_\nu)^2 + (M_e^* M_e)^2 + 3(M_u^* M_u)^2 + 3(M_d^* M_d)^2)$$

$$c = \text{Tr}(M_R^* M_R)$$

$$d = \text{Tr}((M_R^* M_R)^2)$$

$$e = \text{Tr}(M_R^* M_R M_\nu^* M_\nu).$$

The spectral action is given by a computation entirely similar to [4] which yields:

$$S = \frac{1}{\pi^2}\left(48\, f_4\, \Lambda^4 - f_2\, \Lambda^2\, c + \frac{f_0}{4}\, d\right) \int \sqrt{g}\, d^4x \tag{40}$$

$$+ \frac{96\, f_2\, \Lambda^2 - f_0\, c}{24\pi^2} \int R\, \sqrt{g}\, d^4x$$

$$+ \frac{f_0}{10\,\pi^2} \int \left(\frac{11}{6} R^* R^* - 3\, C_{\mu\nu\rho\sigma}\, C^{\mu\nu\rho\sigma}\right) \sqrt{g}\, d^4x$$

$$+ \frac{(-2\, a\, f_2\, \Lambda^2 + e\, f_0)}{\pi^2} \int |\varphi|^2\, \sqrt{g}\, d^4x$$

$$+ \frac{f_0}{2\,\pi^2} \int a\, |D_\mu\varphi|^2\, \sqrt{g}\, d^4x$$

$$- \frac{f_0}{12\,\pi^2} \int a\, R\, |\varphi|^2\, \sqrt{g}\, d^4x$$

$$+ \frac{f_0}{2\,\pi^2} \int \left(g_3^2\, G_{\mu\nu}^i\, G^{\mu\nu i} + g_2^2\, F_{\mu\nu}^\alpha\, F^{\mu\nu\alpha} + \frac{5}{3}\, g_1^2\, B_{\mu\nu}\, B^{\mu\nu}\right) \sqrt{g}\, d^4x$$

$$+ \frac{f_0}{2\,\pi^2} \int b\, |\varphi|^4\, \sqrt{g}\, d^4x$$

where (a, b, c, d, e) are defined above and $D_\mu\varphi$ is the minimal coupling. A simple change of variables as in [4], namely

$$\mathbf{H} = \frac{\sqrt{a\, f_0}}{\pi}\, \varphi\,, \tag{41}$$

so that the kinetic term becomes[3]

$$\int \frac{1}{2} |D_\mu\mathbf{H}|^2\, \sqrt{g}\, d^4x$$

and

$$\frac{g_3^2\, f_0}{2\pi^2} = \frac{1}{4}\,, \quad g_3^2 = g_2^2 = \frac{5}{3}\, g_1^2\,. \tag{42}$$

---

[3]Here we differ slightly from [4] by a factor of $\sqrt{2}$ to match the conventions of Veltman [26].

transforms the bosonic action into the form:

$$S = \int d^4x \, \sqrt{g} \left[\frac{1}{2\kappa_0^2} R + \alpha_0 \, C_{\mu\nu\rho\sigma} \, C^{\mu\nu\rho\sigma} \right. \tag{43}$$
$$+ \quad \gamma_0 + \tau_0 \, {}^*R^*R + \delta_0 \, R_{;\mu}{}^{\mu}$$
$$+ \frac{1}{4} \, G^i_{\mu\nu} \, G^{\mu\nu i} + \frac{1}{4} \, F^\alpha_{\mu\nu} \, F^{\mu\nu\alpha} + \frac{1}{4} \, B_{\mu\nu} \, B^{\mu\nu}$$
$$\left. + \frac{1}{2} |D_\mu \mathbf{H}|^2 - \mu_0^2 |\mathbf{H}|^2 - \frac{1}{12} \, R \, |\mathbf{H}|^2 + \lambda_0 |\mathbf{H}|^4 \right]$$

where

$$\frac{1}{\kappa_0^2} = \frac{96 \, f_2 \, \Lambda^2 - f_0 \, c}{12 \, \pi^2} \tag{44}$$

$$\mu_0^2 = 2 \, \frac{f_2 \, \Lambda^2}{f_0} - \frac{e}{a} \tag{45}$$

$$\alpha_0 = -\frac{3 \, f_0}{10 \, \pi^2} \tag{46}$$

$$\tau_0 = \frac{11 \, f_0}{60 \, \pi^2} \tag{47}$$

$$\delta_0 = -\frac{2}{3} \, \alpha_0 \tag{48}$$

$$\gamma_0 = \frac{1}{\pi^2} \left(48 \, f_4 \, \Lambda^4 - f_2 \, \Lambda^2 \, c + \frac{f_0}{4} \, d \right) \tag{49}$$

$$\lambda_0 = \frac{\pi^2}{2 \, f_0} \frac{b}{a^2} = \frac{b \, g^2}{a^2} \tag{50}$$

9. Detailed form of the spectral action without gravity

To make the comparison easier we Wick rotate back to Minkowski space and after turning off gravity by working in flat space (and addition of gauge fixing terms[4]) the spectral action, after the change of variables summarized in table 1, is given by the following formula:

$$\mathcal{L}_{SM} = -\tfrac{1}{2} \partial_\nu g^a_\mu \partial_\nu g^a_\mu - g_s f^{abc} \partial_\mu g^a_\nu g^b_\mu g^c_\nu - \tfrac{1}{4} g_s^2 f^{abc} f^{ade} g^b_\mu g^c_\nu g^d_\mu g^e_\nu - \partial_\nu W^+_\mu \partial_\nu W^-_\mu -$$
$$M^2 W^+_\mu W^-_\mu - \tfrac{1}{2} \partial_\nu Z^0_\mu \partial_\nu Z^0_\mu - \tfrac{1}{2c_w^2} M^2 Z^0_\mu Z^0_\mu - \tfrac{1}{2} \partial_\mu A_\nu \partial_\mu A_\nu - i g c_w (\partial_\nu Z^0_\mu (W^+_\mu W^-_\nu -$$
$$W^+_\nu W^-_\mu) - Z^0_\nu (W^+_\mu \partial_\nu W^-_\mu - W^-_\mu \partial_\nu W^+_\mu) + Z^0_\mu (W^+_\nu \partial_\nu W^-_\mu - W^-_\nu \partial_\nu W^+_\mu)) -$$
$$i g s_w (\partial_\nu A_\mu (W^+_\mu W^-_\nu - W^+_\nu W^-_\mu) - A_\nu (W^+_\mu \partial_\nu W^-_\mu - W^-_\mu \partial_\nu W^+_\mu) + A_\mu (W^+_\nu \partial_\nu W^-_\mu -$$
$$W^-_\nu \partial_\nu W^+_\mu)) - \tfrac{1}{2} g^2 W^+_\mu W^-_\mu W^+_\nu W^-_\nu + \tfrac{1}{2} g^2 W^+_\mu W^-_\nu W^+_\mu W^-_\nu + g^2 c_w^2 (Z^0_\mu W^+_\mu Z^0_\nu W^-_\nu -$$
$$Z^0_\mu Z^0_\mu W^+_\nu W^-_\nu) + g^2 s_w^2 (A_\mu W^+_\mu A_\nu W^-_\nu - A_\mu A_\mu W^+_\nu W^-_\nu) +$$
$$g^2 s_w c_w (A_\mu Z^0_\nu (W^+_\mu W^-_\nu - W^+_\nu W^-_\mu) - 2 A_\mu Z^0_\mu W^+_\nu W^-_\nu) - \tfrac{1}{2} \partial_\mu H \partial_\mu H - 2 M^2 \alpha_h H^2 -$$
$$\partial_\mu \phi^+ \partial_\mu \phi^- - \tfrac{1}{2} \partial_\mu \phi^0 \partial_\mu \phi^0 - \beta_h \left(\frac{2M^2}{g^2} + \frac{2M}{g} H + \tfrac{1}{2} (H^2 + \phi^0 \phi^0 + 2\phi^+ \phi^-) \right) +$$

---

[4]We add the Feynman gauge fixing terms just to simplify the form of the gauge kinetic terms.

$$\frac{2M^4}{g^2}\alpha_h - g\alpha_h M\left(H^3 + H\phi^0\phi^0 + 2H\phi^+\phi^-\right) -$$
$$\tfrac{1}{8}g^2\alpha_h\left(H^4 + (\phi^0)^4 + 4(\phi^+\phi^-)^2 + 4(\phi^0)^2\phi^+\phi^- + 4H^2\phi^+\phi^- + 2(\phi^0)^2 H^2\right) -$$
$$gMW_\mu^+ W_\mu^- H - \tfrac{1}{2}g\frac{M}{c_w^2}Z_\mu^0 Z_\mu^0 H -$$
$$\tfrac{1}{2}ig\left(W_\mu^+(\phi^0\partial_\mu\phi^- - \phi^-\partial_\mu\phi^0) - W_\mu^-(\phi^0\partial_\mu\phi^+ - \phi^+\partial_\mu\phi^0)\right) +$$
$$\tfrac{1}{2}g\left(W_\mu^+(H\partial_\mu\phi^- - \phi^-\partial_\mu H) + W_\mu^-(H\partial_\mu\phi^+ - \phi^+\partial_\mu H)\right) + \tfrac{1}{2}g\frac{1}{c_w}Z_\mu^0(H\partial_\mu\phi^0 -$$
$$\phi^0\partial_\mu H) + M(\tfrac{1}{c_w}Z_\mu^0\partial_\mu\phi^0 + W_\mu^+\partial_\mu\phi^- + W_\mu^-\partial_\mu\phi^+) - ig\frac{s_w^2}{c_w}MZ_\mu^0(W_\mu^+\phi^- -$$
$$W_\mu^-\phi^+) + igs_w MA_\mu(W_\mu^+\phi^- - W_\mu^-\phi^+) - ig\frac{1-2c_w^2}{2c_w}Z_\mu^0(\phi^+\partial_\mu\phi^- - \phi^-\partial_\mu\phi^+) +$$
$$igs_w A_\mu(\phi^+\partial_\mu\phi^- - \phi^-\partial_\mu\phi^+) - \tfrac{1}{4}g^2 W_\mu^+ W_\mu^-\left(H^2 + (\phi^0)^2 + 2\phi^+\phi^-\right) -$$
$$\tfrac{1}{8}g^2\frac{1}{c_w^2}Z_\mu^0 Z_\mu^0\left(H^2 + (\phi^0)^2 + 2(2s_w^2 - 1)^2\phi^+\phi^-\right) - \tfrac{1}{2}g^2\frac{s_w^2}{c_w}Z_\mu^0\phi^0(W_\mu^+\phi^- + W_\mu^-\phi^+) -$$
$$\tfrac{1}{2}ig^2\frac{s_w^2}{c_w}Z_\mu^0 H(W_\mu^+\phi^- - W_\mu^-\phi^+) + \tfrac{1}{2}g^2 s_w A_\mu\phi^0(W_\mu^+\phi^- + W_\mu^-\phi^+) +$$
$$\tfrac{1}{2}ig^2 s_w A_\mu H(W_\mu^+\phi^- - W_\mu^-\phi^+) - g^2\frac{s_w}{c_w}(2c_w^2 - 1)Z_\mu^0 A_\mu\phi^+\phi^- - g^2 s_w^2 A_\mu A_\mu\phi^+\phi^- +$$
$$\tfrac{1}{2}ig_s\lambda_{ij}^a(\bar{q}_i^\sigma\gamma^\mu q_j^\sigma)g_\mu^a - \bar{e}^\lambda(\gamma\partial + m_e^\lambda)e^\lambda - \bar{\nu}^\lambda(\gamma\partial + m_\nu^\lambda)\nu^\lambda - \bar{u}_j^\lambda(\gamma\partial + m_u^\lambda)u_j^\lambda - \bar{d}_j^\lambda(\gamma\partial +$$
$$m_d^\lambda)d_j^\lambda + igs_w A_\mu\left(-(\bar{e}^\lambda\gamma^\mu e^\lambda) + \tfrac{2}{3}(\bar{u}_j^\lambda\gamma^\mu u_j^\lambda) - \tfrac{1}{3}(\bar{d}_j^\lambda\gamma^\mu d_j^\lambda)\right) + \frac{ig}{4c_w}Z_\mu^0\{(\bar{\nu}^\lambda\gamma^\mu(1 +$$
$$\gamma^5)\nu^\lambda) + (\bar{e}^\lambda\gamma^\mu(4s_w^2 - 1 - \gamma^5)e^\lambda) + (\bar{d}_j^\lambda\gamma^\mu(\tfrac{4}{3}s_w^2 - 1 - \gamma^5)d_j^\lambda) + (\bar{u}_j^\lambda\gamma^\mu(1 - \tfrac{8}{3}s_w^2 +$$
$$\gamma^5)u_j^\lambda)\} + \frac{ig}{2\sqrt{2}}W_\mu^+\left((\bar{\nu}^\lambda\gamma^\mu(1 + \gamma^5)U^{lep}{}_{\lambda\kappa}e^\kappa) + (\bar{u}_j^\lambda\gamma^\mu(1 + \gamma^5)C_{\lambda\kappa}d_j^\kappa)\right) +$$
$$\frac{ig}{2\sqrt{2}}W_\mu^-\left((\bar{e}^\kappa U^{lep\dagger}{}_{\kappa\lambda}\gamma^\mu(1 + \gamma^5)\nu^\lambda) + (\bar{d}_j^\kappa C_{\kappa\lambda}^\dagger\gamma^\mu(1 + \gamma^5)u_j^\lambda)\right) +$$
$$\frac{ig}{2M\sqrt{2}}\phi^+\left(-m_e^\kappa(\bar{\nu}^\lambda U^{lep}{}_{\lambda\kappa}(1 - \gamma^5)e^\kappa) + m_\nu^\lambda(\bar{\nu}^\lambda U^{lep}{}_{\lambda\kappa}(1 + \gamma^5)e^\kappa)\right) +$$
$$\frac{ig}{2M\sqrt{2}}\phi^-\left(m_e^\lambda(\bar{e}^\lambda U^{lep\dagger}{}_{\lambda\kappa}(1 + \gamma^5)\nu^\kappa) - m_\nu^\kappa(\bar{e}^\lambda U^{lep\dagger}{}_{\lambda\kappa}(1 - \gamma^5)\nu^\kappa)\right) - \tfrac{g}{2}\frac{m_\nu^\lambda}{M}H(\bar{\nu}^\lambda\nu^\lambda) -$$
$$\tfrac{g}{2}\frac{m_e^\lambda}{M}H(\bar{e}^\lambda e^\lambda) + \frac{ig}{2}\frac{m_\nu^\lambda}{M}\phi^0(\bar{\nu}^\lambda\gamma^5\nu^\lambda) - \frac{ig}{2}\frac{m_e^\lambda}{M}\phi^0(\bar{e}^\lambda\gamma^5 e^\lambda) - \tfrac{1}{4}\bar{\nu}_\lambda M_{\lambda\kappa}^R(1 - \gamma_5)\hat{\nu}_\kappa -$$
$$\tfrac{1}{4}\bar{\nu}_\lambda M_{\lambda\kappa}^R(1 - \gamma_5)\hat{\nu}_\kappa + \frac{ig}{2M\sqrt{2}}\phi^+\left(-m_d^\lambda(\bar{u}_j^\lambda C_{\lambda\kappa}(1 - \gamma^5)d_j^\kappa) + m_u^\lambda(\bar{u}_j^\lambda C_{\lambda\kappa}(1 +$$
$$\gamma^5)d_j^\kappa)\right) + \frac{ig}{2M\sqrt{2}}\phi^-\left(m_d^\lambda(\bar{d}_j^\lambda C_{\lambda\kappa}^\dagger(1 + \gamma^5)u_j^\kappa) - m_u^\kappa(\bar{d}_j^\lambda C_{\lambda\kappa}^\dagger(1 - \gamma^5)u_j^\kappa)\right) -$$
$$\tfrac{g}{2}\frac{m_u^\lambda}{M}H(\bar{u}_j^\lambda u_j^\lambda) - \tfrac{g}{2}\frac{m_d^\lambda}{M}H(\bar{d}_j^\lambda d_j^\lambda) + \frac{ig}{2}\frac{m_u^\lambda}{M}\phi^0(\bar{u}_j^\lambda\gamma^5 u_j^\lambda) - \frac{ig}{2}\frac{m_d^\lambda}{M}\phi^0(\bar{d}_j^\lambda\gamma^5 d_j^\lambda).$$

This formula compares nicely with [26], *i.e.*, the Lagrangian of §1. Besides the addition of the neutrino mass terms, and absence of the ghost terms there is only one difference: in the spectral action Lagrangian one gets the term

$$M\left(\frac{1}{c_w}Z_\mu^0\partial_\mu\phi^0 + W_\mu^+\partial_\mu\phi^- + W_\mu^-\partial_\mu\phi^+\right) \tag{51}$$

while in the Veltman's formula [26] one gets instead the following

$$-M^2\phi^+\phi^- - \frac{1}{2c_w^2}M^2\phi^0\phi^0. \tag{52}$$

This difference comes from the gauge fixing term

$$\mathcal{L}_{fix} = -\frac{1}{2}\mathcal{C}^2, \quad \mathcal{C}_a = -\partial_\mu W_a^\mu + M_a\phi_a \tag{53}$$

given by the Feynman-t'Hooft gauge in Veltman's formula [26], indeed one has

$$\mathcal{L}_{fix} = -\frac{1}{2}(\partial_\mu W_a^\mu)^2 - \frac{1}{2c_w^2}M^2\phi^0\phi^0 - M^2\phi^+\phi^-$$
$$+ M\left(\frac{1}{c_w}\phi^0\partial_\mu Z_\mu^0 + \phi^-\partial_\mu W_\mu^+ + \phi^+\partial_\mu W_\mu^-\right). \tag{54}$$

10. Predictions

The conversion table 1 shows that all the mass parameters of the standard model now acquire geometric meaning as components of the non-commutative metric as displayed in the right column. We shall work under the hypothesis that the geometric theory is valid at a preferred scale Λ of the order of the unification scale (*cf.* §10.1) and that the standard model coupled with gravity is just its manifestation when one integrates the high energy modes á la Wilson. Then following [4] one can use the renormalization group equations to run down the various coupling constants. Besides the gauge couplings this will be done for the value of the Higgs quartic self-coupling which gives a rough estimate (around 170 GeV) for the Higgs mass under the "big desert" hypothesis. It is satisfactory that another prediction at unification, namely the mass relation of §10.3 also gives a sensible answer, while similar results hold for the couplings of the gravitational part of the action. But it is of course very likely that instead of the big desert one will meet gradual refinements of the non-commutative geometry $M \times F$ when climbing in energy to the unification scale so that our knowledge of the finite geometry F is still too primitive to make accurate predictions. The "naturalness" problem will be discussed in §10.5. At first sight one might easily confuse the obtained predictions with those of a GUT, but there is a substantial difference since for instance no proton decay is implied in our theory.

10.1. Unification of couplings

The numerical values are similar to those of [4] and in particular one gets the same value of gauge couplings as in grand unified theories SU(5) or SO(10). The three gauge couplings fulfill (42). This means that in the above formula the values of g, g_s and s_w, c_w are fixed exactly as in [4] at

$$g_s = g, \quad \mathrm{tg}(w)^2 = \frac{3}{5}. \tag{55}$$

10.2. See-saw mechanism for neutrino masses

Let us briefly explain the analogue of the see-saw mechanism in our context. We use the notation of §6.3. The restriction of $D(M)$ to the subspace of $\mathcal{H}_F$ with basis

Standard Model	notation	notation	Spectral Action
Higgs Boson	$\varphi = (\frac{2M}{g} + H$ $-i\phi^0, -i\sqrt{2}\phi^+)$	$\mathbf{H} = \frac{1}{\sqrt{2}} \frac{\sqrt{a}}{g}(1 + \psi)$	Inner metric$^{(0,1)}$
Gauge bosons	$A_\mu, Z^0_\mu, W^\pm_\mu, g^a_\mu$	(B, W, V)	Inner metric$^{(1,0)}$
Fermion masses u, ν	m_u, m_ν	$M_u = \delta_u, M_\nu = \delta_\nu$	Dirac$^{(0,1)}$ in u, ν
CKM matrix Masses down	C^κ_λ, m_d	$M_d = C\,\delta_d\,C^\dagger$	Dirac$^{(0,1)}$ in d
Lepton mixing Masses leptons e	$U^{lep}{}_{\lambda\kappa}, m_e$	$M_e = U^{lep}\delta_e U^{lep\dagger}$	Dirac$^{(0,1)}$ in e
Majorana mass matrix	M^R	M_R	Dirac$^{(0,1)}$ in $\nu_R, \bar{\nu}_R$
Gauge couplings	$g_1 = g\,\mathrm{tg}(w),$ $g_2 = g, g_3 = g_s$	$g_3^2 = g_2^2 = \frac{5}{3}g_1^2$	Fixed at unification
Higgs scattering parameter	$\frac{1}{8}g^2\alpha_h, \alpha_h = \frac{m_h^2}{4M^2}$	$\lambda_0 = g^2\frac{b}{a^2}$	Fixed at unification
Tadpole constant	$\beta_h, (-\alpha_h M^2$ $+\frac{\beta_h}{2})\,|\varphi|^2$	$\mu_0^2 = 2\frac{f_2\Lambda^2}{f_0} - \frac{e}{a}$	$-\mu_0^2\,|\mathbf{H}|^2$
Graviton	$g_{\mu\nu}$	$\partial\!\!\!/_M$	Dirac$^{(1,0)}$

TABLE 1. Conversion from Spectral Action to Standard Model

the $(\nu_R, \nu_L, \bar{\nu}_R, \bar{\nu}_L)$ is given by the matrix

$$\begin{bmatrix} 0 & M_\nu^* & M_R^* & 0 \\ M_\nu & 0 & 0 & 0 \\ M_R & 0 & 0 & \bar{M}_\nu^* \\ 0 & 0 & \bar{M}_\nu & 0 \end{bmatrix}. \tag{56}$$

Let us simplify to one generation and let $M_R \sim M$ be a very large mass term – the largest eigenvalue of M_R will be set to the order of the unification scale by the equations of motion of the spectral action – while $M_\nu \sim m$ is much smaller[5]. The eigenvalues of the matrix (56) are then given by

$$\frac{1}{2}\left(\pm M \pm \sqrt{M^2 + 4m^2}\right)$$

---

[5]It is a Dirac mass term, fixed by the Higgs vev.

which gives two eigenvalues very close to $\pm M$ and two others very close to $\pm \frac{m^2}{M}$ as can be checked directly from the determinant of the matrix (56), which is equal to $|M_\nu|^4 \sim m^4$ (for one generation).

10.3. Mass relation $Y_2(S) = 4\,g^2$

Note that the matrices M_u, M_d, M_ν and M_e are only relevant up to an overall scale. Indeed they only enter in the coupling of the Higgs with fermions and because of the rescaling (41) only by the terms

$$k_x = \frac{\pi}{\sqrt{a}\,f_0}\,M_x\,, \quad x \in \{u, d, \nu, e\} \tag{57}$$

which are dimensionless matrices by construction. The conversion for the mass matrices is

$$(k_u)_{\lambda\kappa} = \frac{g}{2M}\,m_u^\lambda\,\delta_\lambda^\kappa \tag{58}$$

$$(k_d)_{\lambda\kappa} = \frac{g}{2M}\,m_d^\mu\,C_{\lambda\mu}\delta_\mu^\rho C_{\rho\kappa}^\dagger$$

$$(k_\nu)_{\lambda\kappa} = \frac{g}{2M}\,m_\nu^\lambda\,\delta_\lambda^\kappa$$

$$(k_e)_{\lambda\kappa} = \frac{g}{2M}\,m_e^\mu\,U^{lep}{}_{\lambda\mu}\delta_\mu^\rho U^{lep\,\dagger}{}_{\rho\kappa}.$$

It might seem at first sight that one can simply use (58) to define the matrices k_x but this overlooks the fact that (57) implies one constraint:

$$\mathrm{Tr}(k_\nu^* k_\nu + k_e^* k_e + 3(k_u^* k_u + k_d^* k_d)) = 2\,g^2\,, \tag{59}$$

using (42) to replace $\frac{\pi^2}{f_0}$ by $2\,g^2$. When expressed in the right-hand side, *i.e.*, the standard model parameters this gives

$$\sum_\lambda (m_\nu^\lambda)^2 + (m_e^\lambda)^2 + 3\,(m_u^\lambda)^2 + 3\,(m_d^\lambda)^2 = 8\,M^2 \tag{60}$$

where M is the mass of the W boson. Thus with the standard notation ([19]) for the Yukawa couplings, so that the fermion masses are $m_f = \frac{1}{\sqrt{2}}y_f\,v$, $v = \frac{2M}{g}$ the relation reads

$$\sum_\lambda (y_\nu^\lambda)^2 + (y_e^\lambda)^2 + 3\,(y_u^\lambda)^2 + 3\,(y_d^\lambda)^2 = 4\,g^2. \tag{61}$$

Neglecting the other Yukawa coupling except for the top quark, and imposing the relation (61) at unification scale, then running it downwards using the renormalization group one gets the boundary value $\frac{2}{\sqrt{3}}g \sim 0.597$ for y_t at unification scale which gives a Fermi scale value of the order of $y_0 =\sim 1.102$ and a top quark mass of the order of $\frac{1}{\sqrt{2}}y_0\,v \sim 173\,y_0$ GeV. This is fine since a large neglected tau neutrino Yukawa coupling (allowed by the see-saw mechanism) similar to that of the top quark, lowers the value at unification by a factor of $\sqrt{\frac{3}{4}}$ which has the effect of lowering the value of y_0 to $y_0 \sim 1.04$. This yields an acceptable value for the top quark mass (whose Yukawa coupling is $y_0 \sim 1$), given that we still neglected all other smaller Yukawa couplings.

10.4. The Higgs scattering parameter

The change of notation for the Higgs fields using the conversion Table 1 gives

$$\mathbf{H} = \frac{1}{\sqrt{2}}\frac{\sqrt{a}}{g}(1+\psi) = (\frac{2M}{g} + H - i\phi^0, -i\sqrt{2}\phi^+)\,. \tag{62}$$

One gets a specific value of the Higgs scattering parameter α_h, as in [4] (which agrees with [19]),

$$\alpha_h = \frac{8\,b}{a^2} \tag{63}$$

(with the notations (39)) which is of the order of $\frac{8}{3}$ if there is a dominating top mass. The numerical solution to the RG equations with the boundary value $\lambda_0 = 0.356$ at $\Lambda = 10^{17}$ GeV gives $\lambda(M_Z) \sim 0.241$ and a Higgs mass of the order of 170 GeV.

10.5. Naturalness

The hypothesis that what we see is the low energy average of a purely geometric theory valid at unification scale Λ needs to be confronted with the "naturalness" problem coming from the quadratically divergent graphs of the theory. This problem already arises in a ϕ^4 theory (in dimension 4) with classical potential $V_0(\phi) = \frac{1}{2}m^2\phi^2 + \frac{\lambda}{4}\phi^4$. Recall that the *effective potential* which is the first term in the expansion of the effective action in powers of the derivatives of the classical field ϕ around the constant field $\phi = \phi_c$

$$S_{eff}(\phi) = \int \left[-V(\phi) + \frac{1}{2}(\partial_\mu\phi)^2 Z(\phi) + \cdots\right] d^D x \tag{64}$$

can be expressed as the following sum over 1PI diagrams with zero external momenta:

$$V(\phi_c) = V_0(\phi_c) - \sum_{\Gamma \in \,1PI} \hbar^L \frac{U(\Gamma(p_1 = 0, \ldots, p_N = 0))}{\sigma(\Gamma)} \frac{\phi_c^N}{N!} \tag{65}$$

where ϕ_c is viewed as a real variable, and $V_0(\phi_c)$ is the classical potential. By construction the quantum corrections are organized in increasing powers of $\hbar$ and these correspond to the loop number of the 1PI graphs. At the one-loop level and for a polynomial interaction, one finds that the unrenormalized value gives ([25], equation 2.64)

$$V(\phi_c) = V_0(\phi_c) + \frac{\hbar}{2} \int \log(1 + \frac{V_0^{''}(\phi_c)}{k^2}) \frac{d^D k}{(2\pi)^D} + O(\hbar^2). \tag{66}$$

In dimension $D = 4$ the integral diverges in the ultraviolet due to the two terms

$$\frac{V_0^{''}(\phi_c)}{k^2} - \frac{V_0^{''\,2}(\phi_c)}{2\,k^4} \tag{67}$$

in the expansion of the logarithm at large momentum k. If the classical potential V_0 is at most quartic the divergences can be compensated by adding suitable counterterms in the classical potential. Thus, in particular, if one uses a ultraviolet cutoff Λ and considers the ϕ^4 theory with classical potential $V_0(\phi) = \frac{1}{2}m^2\phi^2 + \frac{\lambda}{4}\phi^4$, one gets a quadratic divergence of the form

$$\frac{\Lambda^2}{32\pi^2}(3\lambda\phi_c^2 + m^2) - \frac{\log\Lambda}{32\pi^2}(V_0''(\phi_c))^2, \tag{68}$$

whose elimination requires adjusting the classical potential as a function of the cutoff Λ as

$$(V_0 + \delta V_0)(\phi) = V_0(\phi) - \frac{\Lambda^2}{32\pi^2}(3\lambda\phi^2) + \frac{\log\Lambda}{32\pi^2}(6m^2\lambda\phi^2 + 9\lambda^2\phi^4), \tag{69}$$

where we ignored an irrelevant (but Λ-dependent) additive constant.

This shows very clearly that, in order to obtain a Λ-independent effective potential, one needs the bare action to depend upon Λ with a large negative quadratic term of the form $-\frac{\Lambda^2}{32\pi^2}(3\lambda\phi^2)$ at the one-loop level. This is precisely the type of term present in the spectral action in the case of the standard model. The presence of the other quadratic divergences coming from the Yukawa coupling of the scalar field with fermions alters the overall sign of the quadratic divergence only at small enough Λ. However, as shown in [7] §5.7, it comes back to the above sign when Λ gets above 10^{10} GeV and in particular when it is close to the unification scale. This leaves open the possibility of using the quadratically divergent mass term of the spectral action to account for the naturalness problem (but an accurate account would require at least some fine tuning of the unification scale, and a better understanding of the running of the Newton constant).

10.6. Gravitational terms

We now discuss the behavior of the gravitational terms in the spectral action, namely

$$\int \left(\frac{1}{2\kappa_0^2} R + \alpha_0 C_{\mu\nu\rho\sigma} C^{\mu\nu\rho\sigma} + \gamma_0 + \tau_0 R^* R^* - \xi_0 R |\mathbf{H}|^2\right)\sqrt{g}\, d^4x. \tag{70}$$

The traditional form of the Euclidean higher derivative terms that are quadratic in curvature is

$$\int \left(\frac{1}{2\eta} C_{\mu\nu\rho\sigma} C^{\mu\nu\rho\sigma} - \frac{\omega}{3\eta} R^2 + \frac{\theta}{\eta} E\right)\sqrt{g}\, d^4x, \tag{71}$$

with $E = R^* R^*$ the topological term which is the integrand in the Euler characteristic

$$\chi(M) = \frac{1}{32\pi^2}\int E\sqrt{g}\, d^4x = \frac{1}{32\pi^2}\int R^* R^* \sqrt{g}\, d^4x. \tag{72}$$

The running of the coefficients of the Euclidean higher derivative terms in (71), determined by the renormalization group equation, is gauge independent and known and we computed in [7] their value at low scale starting from the initial value prescribed by the spectral action at unification scale. We found that the infrared behavior of these terms approaches the fixed point $\eta = 0$, $\omega = -0.0228$, $\theta = 0.327$. The coefficient η goes to zero in the infrared limit, sufficiently slowly, so that, up to scales of the order of the size of the universe, its inverse remains $O(1)$. On the other hand, $\eta(t)$, $\omega(t)$ and $\theta(t)$ have a common singularity at an energy scale of the order of 10^{23} GeV, which is above the Planck scale. Moreover, within the energy scales that are of interest to our model $\eta(t)$ is neither too small nor too large (it does not vary by more than a single order of magnitude between the Planck scale and infrared energies). This implies in particular that the extra terms (besides the Einstein-Hilbert term) do not have any observable consequence at low energy.

The discussion of the Newton constant is much more tricky since its running is scheme dependent. Under the very conservative hypothesis that it does not run much from our scale to the unification scale one finds (*cf.* [7]) that for a unification scale of 10^{17} GeV an order one tuning ($f_2 \sim 5f_0$) of the moments of the test function f suffices to get an acceptable value for the Newton constant.

11. Final remarks

The above approach to physics can be summarized as a strategy to interpret the complicated input of the phenomenological Lagrangian of gravity coupled with matter as coming from a fine structure (of the form $M \times F$) in the geometry of space-time. Extrapolating this to unification scale (*i.e.*, assuming the big desert) gives predictions which can be compared with experiment. Of course we do not believe that the big desert is there and a key test when "new physics" will be observed is to decide wether it will be possible to interpret the new terms of the Lagrangian in the same manner from non-commutative spaces and the spectral action. This type of test already occurred with the new neutrino physics coming from the Kamiokande experiment and for quite some time I believed that the new terms would simply not fit with the spectral action principle. It is only thanks to the simple idea of decoupling the KO-dimension from the metric dimension that the problem was resolved (this was also done independently by John Barrett [1] with a similar solution).

At a more fundamental level the fact that the action functional can be obtained from spectral data suggests that instead of just looking at the inner fluctuations of a product metric on $M \times F$, one should view that as a special case of a fully unified theory at the operator theoretic level, *i.e.*, a kind of spectral random matrix theory where the operator D varies in the symplectic ensemble (corresponding to the commutation with $i = \sqrt{-1}$ and J that generate the quaternions). The first basic step is to understand how to extend the computations of the spontaneous symmetry breaking of the electroweak sector of SM [25] to the full gravitational sector. The natural symmetry group of the spectral action is the unitary symplectic group, corresponding as above to the commutation with $i = \sqrt{-1}$ and J. In the forthcoming book with M. Marcolli [13] we develop an analogy between the spontaneous symmetry breaking which is the key of our work in number theory (the theory of $\mathbb{Q}$-lattices) and the missing SSB for gravity. One of the simple ideas that emerge from the mere existence of the analogy is that the geometry of space-time is a notion which probably stops making sense when the energy scale (*i.e.*, the temperature) is above the critical value of the main phase transition. In particular it seems an ill fated goal to try and quantize gravity as a fundamental theory in a fixed background, the idea being that the very notion of space-time ceases to make sense at high enough energy scales. We refer to the last section of [13] for a more detailed discussion of this point.

References

[1] J. Barrett, *A Lorentzian version of the non-commutative geometry of the standard model of particle physics* hep-th/0608221.

[2] C. Bordé, *Base units of the SI, fundamental constants and modern quantum physics*, Phil. Trans. R. Soc. A 363 (2005), 2177–2201.

[3] A. Chamseddine, A. Connes, *Universal Formula for Non-commutative Geometry Actions: Unification of Gravity and the Standard Model*, Phys. Rev. Lett. **77** (1996), 4868–4871.

[4] A. Chamseddine, A. Connes, *The Spectral Action Principle*, Comm. Math. Phys. **186** (1997), 731–750.

[5] A. Chamseddine, A. Connes, *Scale Invariance in the Spectral Action*, hep-th/0512169 to appear in Jour. Math. Phys.

[6] A. Chamseddine, A. Connes, *Inner fluctuations of the spectral action*, hep-th/0605011.

[7] A. Chamseddine, A. Connes, M. Marcolli, *Gravity and the standard model with neutrino mixing*, hep-th/0610241.

[8] S. Coleman, *Aspects of symmetry*, Selected Erice Lectures, Cambridge University Press, 1985.

[9] A. Connes, *Non-commutative geometry*, Academic Press (1994).

[10] A. Connes, *Non-commutative geometry and reality*, Journal of Math. Physics **36** no. 11 (1995).

[11] A. Connes, *Gravity coupled with matter and the foundation of non-commutative geometry*, Comm. Math. Phys. (1995)

[12] A. Connes, *Non-commutative Geometry and the standard model with neutrino mixing*, hep-th/0608226.

[13] A. Connes, M. Marcolli *Non-commutative Geometry, Quantum fields and Motives*, Book in preparation.

[14] A. Connes, H. Moscovici, *The local index formula in non-commutative geometry*, GAFA, Vol. **5** (1995), 174–243.

[15] L. Dąbrowski, A. Sitarz, *Dirac operator on the standard Podleś quantum sphere*, Non-commutative Geometry and Quantum Groups, Banach Centre Publications 61, Hajac, P.M. and Pusz, W. (eds.), Warszawa: IMPAN, 2003, pp. 49–58.

[16] S. Giddings, D. Marolf, J. Hartle, *Observables in effective gravity*, hep-th/0512200.

[17] J. Gracia-Bondia, B. Iochum, T. Schucker, *The standard model in non-commutative geometry and fermion doubling*. Phys. Lett. **B 416** no. 1-2 (1998), 123–128.

[18] D. Kastler, *Non-commutative geometry and fundamental physical interactions: The Lagrangian level*, Journal. Math. Phys. 41 (2000), 3867-3891.

[19] M. Knecht, T. Schucker *Spectral action and big desert* hep-ph/065166

[20] O. Lauscher, M. Reuter, *Asymptotic Safety in Quantum Einstein Gravity: nonperturbative renormalizability and fractal spacetime structure*, hep-th/0511260.

[21] F. Lizzi, G. Mangano, G. Miele, G. Sparano, *Fermion Hilbert space and Fermion Doubling in the Non-commutative Geometry Approach to Gauge Theories* hep-th/9610035.

[22] J. Mather, *Commutators of diffeomorphisms. II*, Comment. Math. Helv. 50 (1975), 33–40.

[23] R.N. Mohapatra, P.B. Pal, *Massive neutrinos in physics and astrophysics*, World Scientific, 2004.

[24] M. Rieffel, *Morita equivalence for C^*-algebras and W^*-algebras*, J. Pure Appl. Algebra, 5 (1974), 51–96.

[25] M. Sher, *Electroweak Higgs potential and vacuum stability*, Phys. Rep. Vol.179 (1989) N.5-6, 273–418.

[26] M. Veltman, *Diagrammatica: the path to Feynman diagrams*, Cambridge Univ. Press, 1994.

Alain Connes
IHÉS
35, route de Chartres
F-91440 Bures-sur-Yvette
France

and

Collège de France
3, rue d'Ulm
F-75005 Paris
France
e-mail: alain@connes.org